The Theory of Matrices
in Numerical Analysis

Alston S. Householder

Professor of Mathematics, *University of Tennessee*

The Theory of Matrices
in Numerical Analysis

Dover Publications, Inc. | New York

Published in Canada by General Publishing Company, Ltd., 30 Lesmill Road, Don Mills, Toronto,
Ontario.
Published in the United Kingdom by Constable
and Company, Ltd., 10 Orange Street, London WC 2.

This Dover edition, first published in 1975, is an
unabridged, corrected republication of the work
originally published by the Blaisdell Publishing
Company, a division of Ginn and Company, New
York, in 1964.

International Standard Book Number: 0-486-61781-5
Library of Congress Catalog Card Number: 74-83763

Manufactured in the United States of America
Dover Publications, Inc.
180 Varick Street
New York, N. Y. 10014

TO MY WIFE

Preface

This book represents an effort to select and present certain aspects of the theory of matrices that are most useful in developing and appraising computational methods for solving systems of linear equations (including the inversion of matrices) and for finding characteristic roots. The solution of linear inequalities and the problems of linear programing are not explicitly considered since there are special difficulties inherent in these problems that are largely combinatorial in character and require a quite different approach.

The list of titles at the end of the book should provide convincing evidence that the problems that are treated here are of considerable interest to numerical analysts, and also to mathematicians. This list is culled from perhaps twice as many titles contained in the author's files: these files are certainly not complete, and the number of publications grows at an accelerating rate. The reason is clear. A finite digital computer can be applied to the solution of functional equations and infinite systems only when finite approximations have been made; and usually the first step toward solving a nonlinear system is to linearize. Thus finite linear systems stand at the heart of all mathematical computation. Moreover as science and technology develop, and computers become more powerful, systems to be handled become larger and require techniques that are more refined and efficient.

The purpose here is not to develop specific computational layouts or flowcharts, nor is much done explicitly in the way of operational counts or error analysis. These can be found elsewhere in the literature and some specific references will be made in the appropriate places, but particular mention can be made here of papers and forthcoming books by J. H. Wilkinson. In this book the first chapter develops a variety of notions, many classical but not often emphasized in the literature, which will be applied in the subsequent chapters. Chapter 2 develops the theory of norms which plays an increasingly important role in all error analysis and elsewhere. Chapter 3 makes immediate application to localization theorems, important in providing bounds for errors in characteristic roots, and develops some other useful results. The last four chapters survey the known methods, attempting to show the mathematical principles that underlie them, and the mathematical relations among them.

It has been assumed that the reader is familiar with the general principles of matrix algebra: addition, subtraction, multiplication, and inversion; the Cayley-Hamilton theorem, characteristic roots and vectors, and normal forms; and the related notions of a vector space, linear dependence, rank, and the like. An outline of this theory can be found in Householder (1953); fuller development can be found in Birkhoff and MacLane (1953), in MacDuffee (1943), and in many other standard text books. However, the Lanczos algorithm as developed in Chapter 1 provides a proof of the Cayley-Hamilton theorem; and the contents of Chapter 6 provide constructive derivations of some of the normal forms.

The author is indebted to many people who have contributed in many ways to this book. At the risk of doing an injustice to others who are not named, he would like to mention in particular R. C. F. Bartels, F. L. Bauer, Ky Fan, George Forsythe, Hans Schneider, R. S. Varga, and J. H. Wilkinson. Their comments, criticisms, suggestions, and encouragement, whether by correspondence, by conversation, or both, have been most helpful and stimulating. Grateful acknowledgment is made also to Mae Gill and Barbara Luttrell for the painstaking job of typing.

Table of Contents

4. The Solution of Linear Systems: Methods of Successive Approximation

5. Direct Methods of Inversion

6. Proper Values and Vectors: Normalization and Reduction of the Matrix

7. Proper Values and Vectors: Successive Approximation

The Theory of Matrices
in Numerical Analysis

CHAPTER 1

Some Basic Identities and Inequalities

1.0. Objectives; Notation. In this chapter and the next will be developed some of the basic tools and relations that will be utilized repeatedly in subsequent chapters. At the outset a matrix of quite simple form will be introduced. It is the result of subtracting a matrix of rank 1 at most from the identity matrix, and any matrix of this form will be called here an "elementary matrix." Such matrices will play a fundamental role in inversion, and also in the reduction of a matrix by similarity transformations to facilitate the evaluation of its characteristic roots. These will be applied immediately to obtain certain factorization theorems, which are of basic importance both for inversion and for equation solving.

The method of least squares can be interpreted geometrically as the projection of an arbitrary given vector upon a certain subspace; and each step in certain iterative methods can likewise be interpreted as a projection. Hence a general form of a projection operator will be obtained, and the formation of this operator makes some use of the factorization theorems just, obtained. A further application of the factorization theorems will be made in the following section, where certain classical determinantal identities and inequalities will be derived. These will be required in the analysis and derivation of certain iterative methods, and also for more general theoretical considerations.

In the last two sections of this chapter will be introduced certain polynomials associated with an arbitrary matrix and vector. One class is defined by an algorithm due to Lanczos who introduced it originally as a first step toward finding the characteristic values and vectors of a matrix; indeed, the algorithm is much used for this purpose. But the same algorithm occurs in the method of conjugate gradients for solving a linear system of equations [Chapter 5], and in the least-squares fitting of polynomials to experimental data [Exercise 54]. For this reason, and in order not to break the continuity later on, the Lanczos algorithm is developed here, instead of in Chapter 6 where it might seem more appropriate. Another advantage of having the Lanczos algorithm here is the following: Although the reader is assumed to be acquainted with the Jordan normal form of a matrix, its derivation from the tridiagonal form to which the Lanczos algorithm leads is quite direct. In fact, it is necessary only to verify that a certain matrix that will be exhibited in Chapter 6, does, in fact, transform a tridiagonal matrix to the Jordan form. Hence the reader who wishes to do so may follow this path at the outset to formulate a constructive derivation of the Jordan normal form of an arbitrary matrix. Even the Cayley-Hamilton theorem is a simple consequence, and is not presupposed.

The polynomials to which the Lanczos algorithm leads, and which may be called the Lanczos polynomials, are closely related to another set which will be called orthogonal, for reasons that will be brought out at the end of the chapter. The two sets can coincide when the matrix is Hermitian. The orthogonal polynomials are also defined implicitly by a certain algorithm to be discussed more fully in Chapter 6, but they will find application in Chapter 3 in connection with the localization of the characteristic roots of a matrix.

Certain notational conventions will be observed throughout this book, and for the sake of continuity these will be listed here.

Except for dimensions and indices, or when otherwise indicated, lower case Greek letters represent scalars; lower case Latin letters column vectors; capital letters, Greek or Latin, matrices. However, the notation $C_{n,i}$ will be used for the number of combinations of n things i at a time. The superscript H (for Hermitian) is for conjugate transpose, T for transpose, C for conjugate, and A for adjoint. In Section 1.3 a general reciprocal will be defined to be denoted by the superscript I. Since the general reciprocal of a nonsingular matrix is the true reciprocal, the same superscript will sometimes be used to denote the inverse. In general a_i is the ith column of A, α_{ij} the element in position (i, j), but e_i is the ith column of the identity I, and

$$e = \sum e_i.$$

The elements of x are ξ_i; the elements of y are η_i.

The matrix $|\, A\, |$ has elements $|\, \alpha_{ij}\, |$, and the vector $|\, x\, |$ has elements $|\, \xi_i\, |$, that is, the absolute values of the elements of A and x itself;

$$x \leq y \leftrightarrow \xi_i \leq \eta_i \quad \text{for every } i;$$
$$A \leq B \leftrightarrow \alpha_{ij} \leq \beta_{ij} \quad \text{for every } i \text{ and } j.$$

Particular matrices are I,

$$J = (e_2, e_3, \cdots, e_n, 0),$$
$$K = J + J^{\mathrm{T}}.$$

The determinant of A is $\delta(A)$, and the trace of A is $\tau(A)$. $A \begin{pmatrix} i, j, \cdots \\ k, l, \cdots \end{pmatrix}$ is the determinant whose elements are taken from rows $i, j, \cdots$ and from columns $k, l, \cdots$. The *proper values* of A (these will often be called the *roots* of A) are $\lambda_i(A)$, often ordered

$$|\, \lambda_1(A)\, | \geq \cdots \geq |\, \lambda_n(A)\, |;$$

the statement $\lambda = \lambda(A)$ signifies that λ is one of the proper values; the *singular values* of A are $\sigma_i(A) \geq 0$ where

$$\sigma_i^2(A) = \lambda_i(A^H A);$$

the *spectral radius* of A is

$$\rho(A) = \max_i |\lambda_i(A)|.$$

If $\lambda = \lambda(A)$, then x *belongs* to λ in case $x \neq 0$ and

$$Ax = \lambda x.$$

In that case x is a *proper vector*. But $y \neq 0$ is a *princpial vector* belonging to λ if for some ν,

$$(A - \lambda I)^\nu y = 0$$

and ν is its degree. A proper vector is therefore a principal vector of degree 1.

1.1. Elementary Matrices. A matrix of the form

(1) $$E(u, v; \sigma) = I - \sigma u v^H$$

will be called *elementary*. Since

(2) $$E(u, v; \sigma)E(u, v; \tau) = E(u, v; \sigma + \tau - \sigma\tau v^H u)$$

it follows that for

(3) $$\sigma^{-1} + \tau^{-1} = v^H u,$$
$$E(u, v; \sigma) = E^{-1}(u, v; \tau).$$

By an element-wise expansion it is easy to verify that the determinant has the value

(4) $$\delta[E(u, v; \sigma)] = 1 - \sigma v^H u.$$

Let (P, Q) be a nonsingular matrix of order n such that $P^H Q = 0$. Thus the columns of P and Q are bases for mutually orthogonal and complementary subspaces. Then for any vector a and any vector v such that $v^H a \neq 0$, it is possible to choose σ and u so that for any vector b such that

(5) $$Ea = b,$$

it is true that

(6) $$P^H b = P^H a, \qquad Q^H b = 0.$$

This is to say that a is transformed by E into a vector b which is orthogonal to the space of Q, and whose projection on the space of P is the same as that of a. To see this, one has only to multiply through (5) by P^H and Q^H, apply (6), and observe that u satisfies the equations

$$P^H u = 0$$

$$Q^H u = Q^H a/(\sigma v^H a)$$

whose matrix $(P, Q)^H$ is nonsingular, by hypothesis.

A matrix, or the transpose of a matrix

(7) $L_i(l_i) = E(l_i, e_i; 1),   e_j{}^T l_i = 0,   j \leq i,$

will be called an *elementary triangular matrix*, and

(8) $L_i^{-1}(l_i) = L_i(-l_i),   \delta(L_i) = 1.$

A matrix

(9) $I_{ij} = E(e_i - e_j, e_i - e_j; 1)$

will be called an *elementary permutation matrix*. Evidently

(10) $I_{ii} = I,   \delta(I_{ij}) = -1,   i \neq j.$

The effect of multiplying a matrix on the left by I_{ij}, $i \neq j$, is to interchange ith and jth rows; that of multiplying on the right is to interchange columns.

An *elementary Hermitian* (or *unitary*) *matrix* is of the form

(11) $H(w) = E(w, w; 2),   w^H w = 1.$

Such a matrix is Hermitian, unitary, and also involutory

(12) $H = H^H = H^{-1},   \delta(H) = -1.$

It reflects the space in the hyperplane orthogonal to w. *If a and b are vectors of equal length, $a^H a = b^H b$, and such that $a^H b = b^H a$ (i.e., is real), then it is always possible to choose w so that*

$$Ha = b.$$

In fact, if $a^H b$ is real, set

$$2\mu w = a - b,$$

and obtain the real scalar 2μ by normalizing $a - b$. It is verified that $\mu = w^H a = -w^H b$. This means that the reflection is made in the hyperplane bisecting the angle from a to b. Analogously, any components of a can be held fixed and the reflection made in the complementary subspace if b lies in that subspace.

1.2. Some Factorizations. By applying the orthogonality relations (1.1.7) it is verified directly that

(1) $L = L_1(l_1)L_2(l_2)L_3(l_3) \cdots = I - l_1 e_1{}^T - l_2 e_2{}^T - l_3 e_3{}^T - \cdots.$

Since any *unit lower triangular matrix* (i.e., one in which every diagonal element $= 1$) can be written in the form indicated on the right, it follows that *any unit lower triangular matrix can be factored as in* (1). Since the transpose of a lower triangular matrix is upper triangular the factorization of a unit upper triangular matrix follows immediately. A unit upper triangular matrix can also be written

$$R = I - r_2 e_2{}^T - r_3 e_3{}^T - \cdots, \qquad e_i{}^T r_j = 0, \qquad i \geq j;$$

an alternative factorization is

$$R = \cdots (I - r_3 e_3{}^T)(I - r_2 e_2{}^T),$$

the transpose yielding an alternative factorization of a unit lower triangular matrix. Any nonsingular triangular matrix can be written as the product of a diagonal matrix by a unit triangular matrix. Hence *any nonsingular lower triangular matrix can be factored as a product of elementary lower triangular matrices and a diagonal matrix.*

Let

(2) $$A = A^{(0)} = (a_1^{(0)}, a_2^{(0)}, \cdots, a_m^{(0)}) \neq 0$$

be any matrix of n rows and m columns. Certain algorithms will be developed for expressing A in the form

(3) $$A = MP,$$

where, if p is the rank of A, then M is a matrix of p linearly independent columns and P a matrix of p linearly independent rows. Let j_1 be the smallest index for which $a_{j_1}^{(0)} \neq 0$, and let $e_{i_1}^T a_{j_1}^{(0)} \neq 0$. That is, the element in position i_1 is nonnull. Then in $I_{1i_1} A^{(0)}$, the element in position $(1, j_1)$ is nonnull, and for some L_1^{-1}, possibly the identity, the matrix

$$A^{(1)} = L_1^{-1} I_{1i_1} A^{(0)}$$

is null in the first $j_1 - 1$ columns, and null below the first element in the next column.

It may be that every row below the first in $A^{(1)}$ is null, in which case the algorithm is complete. If not, and if $n = 2$, let $A^{(2)} = A^{(1)}$ and the algorithm is again complete. Otherwise, pick out the first column containing a nonnull element below the first. If this is $a_{j_2}^{(1)}$, then for some i_2,

$$e_2{}^T I_{2i_2} a_{j_2}^{(1)} \neq 0.$$

Hence for some L_2^{-1} the matrix

$$A^{(2)} = L_2^{-1} I_{2i_2} A^{(1)}$$

is null in the first $j_1 - 1$ columns, null below the first element in the next $j_2 - j_1$ columns, and null below the second element in the next column. Proceed thus until reaching $A^{(p)}$ where either $p = n$ or else all rows below the pth are null. The result is that A is expressed in the form

$$A = M' A^{(p)},$$

where

$$M' = I_{1i_1} L_1 I_{2i_2} L_2 \cdots$$

and is nonsingular, and $A^{(p)}$ is a matrix whose first p rows are linearly independent and all other rows, if any, are null. In the latter event, the product remains the same if the $n - p$ null rows of $A^{(p)}$ are dropped, along with the last $n - p$ columns of M. Call the resulting matrices P and M, respectively, and the required form (3) is obtained. Clearly A itself is of rank p.

If A is square and nonsingular, it may be that one can take

$$i_1 = 1, i_2 = 2, i_3 = 3, \cdots,$$

which is to say that no permutation is required at any stage. In that event, M is a unit lower triangular matrix, P upper triangular. *When a non-singular matrix A can be expressed as a product of a unit lower triangular matrix by an upper triangular matrix, the factorization in that form is unique.* In fact, unique determinantal expressions will be obtained in Section 1.4 for the elements of the two matrices. But suppose A is nonsingular and

$$A = M_1 P_1 = M_2 P_2,$$

where M_1 and M_2 are unit lower triangular matrices, and P_1 and P_2 are upper triangular matrices. Then

$$M_2^{-1} M_1 = P_2 P_1^{-1}.$$

But the product on the left is unit lower triangular, and that on the right is upper triangular. Hence both are diagonal, and, furthermore, in the product on the left each diagonal element $= 1$. Hence, each product can only be I, and therefore $M_1 = M_2$, $P_1 = P_2$.

As a method for evaluating $\delta(A)$ when A is square, the method of triangularization is the method of Chió. As a step toward the inversion of A (to be completed by inverting the triangles) it is Gaussian elimination. However, as a practical algorithm it is not to be recommended for either purpose in this simple form unless the exact arithmetic operations are performed throughout (without rounding or truncation). The computational precautions will be discussed later.

The elementary Hermitians can be used for a different factorization. Again consider any matrix

$$A = A^{(0)} = (a_1^{(0)}, a_2^{(0)}, \cdots, a_m^{(0)}) \neq 0$$

of n rows and m columns. Let $a_{j_1}^{(0)}$ be the first nonnull column, and form $H_1 = H(w_1)$ so that $H_1 a_{j_1}^{(0)}$ is some multiple of e_1. Hence let

$$A^{(1)} = H_1 A^{(0)}.$$

Observe that no permutation is required. If any row below the first in $A^{(1)}$ is nonnull, let $a_{j_2}^{(1)}$ be the first column having a nonvanishing component orthogonal to e_1. Form $H_2 = H(w_2)$ so that the component of $a_{j_2}^{(1)}$ orthogonal to e_1 is reflected into a multiple of e_2. If $n = 2$, then $H_2 = I$. By continuing thus there is formed eventually a matrix $A^{(p)}$ such that

$$A = HA^{(p)},$$
$$H = H_1 H_2 H_3 \cdots,$$

and either $p = n$ or else every row below the pth in $A^{(p)}$ is null. The matrix H is, of course, not Hermitian, but it is unitary. If $p < n$, drop the null rows in $A^{(p)}$ and the corresponding columns of H, calling the resulting matrices R and W, respectively. Then

(4) $A = WR, \quad W^H W = I,$

where R has p linearly independent rows and W has p linearly independent columns.

In case A is nonsingular, W is unitary and R is an upper triangle. Suppose, in this event, one obtains in any way two factorizations

$$A = W_1 R_1 = W_2 R_2,$$

where R_1 and R_2 are both upper triangles, and W_1 and W_2 are both unitary. Then

$$W_2{}^H W_1 = R_2 R_1{}^{-1}.$$

Hence $R_2 R_1{}^{-1}$ is both unitary and upper triangular. But it is easy to verify that such a matrix is diagonal (cf. Exercise 4),

$$W_2{}^H W_1 = R_2 R_1{}^{-1} = D,$$

and, being unitary and diagonal,

$$|D| = I,$$

which is to say that the diagonal elements are points on the unit circle. Hence

$$W = W_1 = W_2 D, \qquad R_2 = DR_1 = DR,$$
$$A = WR = (WD^{-1})(DR).$$

Thus when A is nonsingular and factored as in (4), each column of W and each row of R is uniquely determined up to a scalar factor which is a point on the unit circle.

If A is itself unitary, R will be diagonal. Hence, in particular, *every unitary matrix can be expressed as a product of elementary Hermitians and of a diagonal matrix with elements on the unit circle.*

The factorization (4) can be achieved in another way. Suppose the columns of A are linearly independent. The first column of W is a normalized first column of A (note that the normalization is unaffected by the introduction of a point on the unit circle as a scalar factor). The second column of W is a unit vector chosen so that a_2, the second column of A, is a linear combination of the two columns of W. The third column of W is a unit vector so chosen that a_3 is a linear combination of these three columns of W. It is easily verified that on completing the algorithm the desired factorization will have been achieved. This method is known as *Schmidt orthogonalization*.

1.3. Projections, and the General Reciprocal.

Let the columns of the rectangular matrix F be linearly independent. The vector Fx is a vector in the space spanned by the columns of F, and Fx is the orthogonal projection of a vector a on this space in case $a - Fx$ is orthogonal to F:

$$F^H(a - Fx) = 0.$$

Hence Fx is the orthogonal projection of a on the space of F in case

$$(1) \qquad x = (F^H F)^{-1} F^H a,$$

and the inverse indeed exists because of the linear independence of the columns of F (this follows directly from the theorem of corresponding matrices, Section 1.4). Hence

$$(2) \qquad P_F = F(F^H F)^{-1} F^H$$

is called *the projector for the space of F* since, for any a, $P_F a$ is the orthogonal projection of a upon the space of F. It is *the* projector, and not just *a* projector, since if M is nonsingular, and F is replaced by FM, one obtains the same P_F. This can be verified directly. Note that P_F is Hermitian, and furthermore that

$$(3) \qquad P_F{}^2 = P_F,$$

that is, P_F is idempotent. Moreover,

$$(I - P_F)^2 = I - P_F, \qquad (I - P_F)P_F = 0,$$

and $I - P_F$ is the projector for the subspace complementary to that of F.

Any Hermitian idempotent matrix P is a projector. To prove this, let P have rank p, and express P in the form

$$P = FR^H,$$

where each of F and R has p linearly independent columns. Since P is Hermitian and idempotent,

$$P = FR^H = RF^H = FR^H RF^H.$$

Hence

$$F^H FR^H = F^H FR^H RF^H,$$

and since $F^H F$ is nonsingular, therefore

$$R^H = R^H RF^H.$$

Hence

$$R^H R = R^H RF^H FR^H R.$$

Again, $R^H R$ is nonsingular, hence

$$F^H F R^H R = I;$$

hence

$$P = F R^H R F^H = F(F^H F)^{-1} F^H = P_F,$$

and P is a projector.

Now consider an arbitrary matrix A of rank r, and let

$$A = F R^H$$

where each of F and R has r linearly independent columns. It is legitimate to denote

$$P_A = P_F = F(F^H F)^{-1} F^H.$$

Next consider the matrix defined by

(4)
$$A^I = R(R^H R)^{-1} (F^H F)^{-1} F^H.$$

By direct verification it is found that

$$A A^I = F(F^H F)^{-1} F^H,$$

and also that

$$A^I A = R(R^H R)^{-1} R^H$$

which is similarly the projector $P_A{}^H$ for A^H. Moreover

(5)
$$A A^I A = A, \qquad A^I A A^I = A^I.$$

Because of this last property A^I is known as the *general reciprocal* (E. H. Moore). It can be shown to be unique. Formally it solves the least squares problem, given a vector a, to find the vector x such that if

$$d = a - Ax,$$

then $d^H d$ is a minimum, since the required vector Ax is the orthogonal projection of a upon the space of A. Hence the required solution is

$$x = A^I a,$$

and it satisfies

$$A^H A x = A^H a.$$

If G is positive definite Hermitian, the product $x^H G y$ is a natural generalization of the ordinary scalar product $x^H y = x^H I y$. It will be shown in Section 1.4 that any positive definite matrix G can be factored in the form

(6)
$$G = B^H B,$$

with B a square, nonsingular matrix. Hence

$$x^H G y = (Bx)^H (By),$$

so that the generalized scalar product of x and y is the ordinary scalar product of Bx and By, and x and y can be considered as representing in an oblique coordinate system the vectors whose representations are Bx and By in an orthonormal system. The vectors x and y will be considered orthogonal with respect to G in case

$$x^H G y = 0.$$

If the columns of F are linearly independent, then Fx is the orthogonal projection with respect to G of the vector a in case $a - Fx$ is orthogonal to F with respect to G:

$$F^H G(a - Fx) = 0,$$

(7) $$x = (F^H G F)^{-1} F^H G a.$$

Hence the generalized projector, with respect to G, is

(8) $$P_{F;G} = F(F^H G F)^{-1} F^H G.$$

It is idempotent, but not Hermitian. It is unchanged when F is replaced by FM, where M is nonsingular. Evidently

$$(I - P_{F;G})P_{F;G} = 0,$$
$$(I - P_{F;G})^2 = I - P_{F;G}.$$

1.4. Some Determinantal Identities. Suppose A is nonsingular and the factorization

$$A = LP$$

is possible with L unit lower triangular and P upper triangular. A diagonal factor D can be removed from P so that

(1) $$A = LDR$$

where R is unit upper triangular. Determinantal expressions for the elements of L, D, and R will now be obtained.

Let the matrices be partitioned

(2) $$\begin{pmatrix} A_{11} & A_{12} \\ A_{21} & A_{22} \end{pmatrix} = \begin{pmatrix} L_{11} & 0 \\ L_{21} & L_{22} \end{pmatrix} \begin{pmatrix} D_{11} & 0 \\ 0 & D_{22} \end{pmatrix} \begin{pmatrix} R_{11} & R_{12} \\ 0 & R_{22} \end{pmatrix}$$

with A_{11}, L_{11}, D_{11}, and R_{11} square matrices of order p. Since L_{11} and R_{11} are unit triangles,

$$\delta(L_{11}) = \delta(R_{11}) = 1.$$

Moreover, by direct multiplication it follows that

$$A_{11} = L_{11} D_{11} R_{11}.$$

Hence

$$\delta(A_{11}) = \delta(D_{11}).$$

By letting p range from 1 to n it follows that if

$$D = \mathrm{diag}\,(\delta_1, \delta_2, \cdots, \delta_n),$$

then

(3) $$\delta_p = A\begin{pmatrix} 1, 2, \cdots, p \\ 1, 2, \cdots, p \end{pmatrix} \Big/ A\begin{pmatrix} 1, 2, \cdots, p-1 \\ 1, 2, \cdots, p-1 \end{pmatrix}.$$

Next, observe that

$$(A_{11}, A_{12}) = L_{11}(D_{11}R_{11}, D_{11}R_{12}).$$

Again, since $\delta(L_{11}) = 1$, any pth-order determinant formed from the columns of the matrix on the left is equal to the determinant formed from the corresponding columns of the rectangular matrix on the right. In particular, let the determinants be formed of the first $p - 1$ columns along with any other. The determinant on the right is again triangular and equal to $\delta_1 \delta_2 \cdots \delta_{p-1}$ multiplied by the last element of the adjoined column. Let this be $\delta_p \rho_{pi}$, $i \ge p$. It follows, on applying (3) that

(4) $$\rho_{pi} = A\begin{pmatrix} 1, 2, \cdots, p-1, p \\ 1, 2, \cdots, p-1, i \end{pmatrix} \Big/ A\begin{pmatrix} 1, 2, \cdots, p \\ 1, 2, \cdots, p \end{pmatrix}, \qquad i \ge p.$$

Analogously one finds

(5) $$\lambda_{ip} = A\begin{pmatrix} 1, 2, \cdots, p-1, i \\ 1, 2, \cdots, p-1, p \end{pmatrix} \Big/ A\begin{pmatrix} 1, 2, \cdots, p \\ 1, 2, \cdots, p \end{pmatrix}, \qquad p \le i.$$

This also demonstrates the uniqueness of the factorization, when it can be made, since the elements of the factor matrices are given by explicit formulas.

Now consider the factorization (still with A nonsingular)

$$A = WP,$$

where W is unitary and P upper triangular. This can be written

(6) $$A = WDR,$$

where D is diagonal and R unit upper triangular. Then

(7) $$H = A^H A = R^H D^H D R.$$

Hence formulas (3) and (4) can be applied to give the diagonal elements of $D^H D$ and the elements of R:

(8) $$\bar{\delta}_p \delta_p = H\begin{pmatrix} 1, 2, \cdots, p \\ 1, 2, \cdots, p \end{pmatrix} \Big/ H\begin{pmatrix} 1, 2, \cdots, p-1 \\ 1, 2, \cdots, p-1 \end{pmatrix},$$

(9) $$\rho_{pi} = H\begin{pmatrix} 1, 2, \cdots, p-1, p \\ 1, 2, \cdots, p-1, i \end{pmatrix} \Big/ H\begin{pmatrix} 1, 2, \cdots, p \\ 1, 2, \cdots, p \end{pmatrix}, \qquad i \ge p.$$

Now write (6) in the partitioned form

$$\begin{pmatrix} A_{11} & A_{12} \\ A_{21} & A_{22} \end{pmatrix} = \begin{pmatrix} W_{11} & W_{12} \\ W_{21} & W_{22} \end{pmatrix} \begin{pmatrix} D_{11}R_{11} & D_{11}R_{12} \\ 0 & D_{22}R_{22} \end{pmatrix},$$

and consider, in particular,

$$(10) \qquad \begin{pmatrix} A_{11} \\ A_{21} \end{pmatrix} = \begin{pmatrix} W_{11} \\ W_{21} \end{pmatrix} D_{11}R_{11}.$$

There is no restriction in assuming D real and nonnegative, since if Δ is diagonal with points on the unit circle, then $W\Delta$ is unitary, and Δ can be chosen so that $\Delta^{-1}D$ is real and nonnegative. Hence each δ_i is the square root of the right member of (8). Hence *any determinant formed from complete rows of*

$$\begin{pmatrix} W_{11} \\ W_{21} \end{pmatrix}$$

is equal to the determinant formed from the same rows of

$$\begin{pmatrix} A_{11} \\ A_{21} \end{pmatrix}$$

divided by

$$H^{\frac{1}{2}} \begin{pmatrix} 1, 2, \cdots, p \\ 1, 2, \cdots, p \end{pmatrix}.$$

From (3) one obtains *a necessary and sufficient condition for the unique factorization* (1), which *is that every principal minor in the sequence*

$$A \begin{pmatrix} 1 \\ 1 \end{pmatrix}, A \begin{pmatrix} 12 \\ 12 \end{pmatrix}, A \begin{pmatrix} 123 \\ 123 \end{pmatrix}, \cdots, A \begin{pmatrix} 1, 2, \cdots, n-1 \\ 1, 2, \cdots, n-1 \end{pmatrix}$$

shall be different from zero. In case A is singular the conditions are sufficient to assure factorization, but not uniqueness. The factorization (6) is possible for any A with D and A having the same rank p, as was shown in the derivation of (1.2.4).

A Hermitian matrix A is said to be *positive definite* in case

$$x \neq 0 \rightarrow x^{\mathrm{H}}Ax > 0;$$

it is *nonnegative semidefinite* in case $x^{\mathrm{H}}Ax \geq 0$ for every x. For *non-Hermitian matrices* the concept is *not defined*. If A is positive definite (nonnegative semidefinite) then $-A$ is *negative definite* (*nonpositive semidefinite*) and conversely. There will be no occasion to speak of negative definite or nonpositive semidefinite matrices here, hence when a matrix is said to be definite (semidefinite) it will be understood to be positive (nonnegative).

Any principal submatrix of a semidefinite (definite) matrix is itself semi-definite (definite), for consider

$$(\bar{\eta}, y^H)\begin{pmatrix} \alpha & a^H \\ a & A \end{pmatrix}\begin{pmatrix} \eta \\ y \end{pmatrix} = \alpha\eta\bar{\eta} + \bar{\eta}a^Hy + \eta y^Ha + y^HAy.$$

With $\eta = 0$, $y \neq 0$, it is apparent that the entire matrix cannot be semidefinite (definite) unless A is. In particular, a definite matrix can have no vanishing diagonal element, and *a diagonal element of a semidefinite matrix can vanish only if all other elements of the same row and column also vanish*. For if $\alpha = 0$, take $y = a$. Thus by making $\eta = \bar{\eta}$ real and negative and sufficiently large, $2\eta a^Ha + a^HAa$ can be made negative unless $a = 0$.

Any semidefinite matrix can be expressed in the form

$$A = B^HB;$$

moreover, B *can be taken upper triangular*. In fact, the factorization

$$A = R^HDR$$

is possible with R unit upper triangular, $D \geq 0$, real and diagonal, and *if A is definite the factorization in this form is unique.*

Suppose the theorem verified for all matrices of order n or less, and consider a bordered matrix of order $n + 1$. The diagonal element may be assumed nonnull or there is nothing to prove. The identity

$$\begin{pmatrix} \alpha^2 & \alpha a^H \\ \alpha a & A \end{pmatrix} = \begin{pmatrix} \alpha & 0 \\ a & I \end{pmatrix}\begin{pmatrix} 1 & 0 \\ 0 & \Delta \end{pmatrix}\begin{pmatrix} \alpha & a^H \\ 0 & I \end{pmatrix},$$

$$\Delta = A - aa^H$$

is easily verified. Moreover, Δ is at least semidefinite if the matrix on the left is, for

$$(-\alpha^{-1}y^Ha, \, y^H)\begin{pmatrix} \alpha^2 & \alpha a^H \\ \alpha a & A \end{pmatrix}\begin{pmatrix} -\alpha^{-1}a^Hy \\ y \end{pmatrix} = y^H\Delta y.$$

Since Δ is semidefinite of order n, it has a triangular factorization; hence so has the matrix of order $n + 1$. The uniqueness for a definite matrix follows from (8) and (9).

Let A be nonsingular and let $C = A^{-1}$. Let A and C be similarly partitioned,

$$A = \begin{pmatrix} A_{11} & A_{12} \\ A_{21} & A_{22} \end{pmatrix}, \quad C = \begin{pmatrix} C_{11} & C_{12} \\ C_{21} & C_{22} \end{pmatrix}.$$

Hence, A_{11} and C_{11} being square matrices of order $p \geq 1$,

$$A_{11}C_{11} + A_{12}C_{21} = I, \quad A_{11}C_{12} + A_{12}C_{22} = 0,$$
$$A_{21}C_{11} + A_{22}C_{21} = 0, \quad A_{21}C_{12} + A_{22}C_{22} = I.$$

Now consider the product

$$\begin{pmatrix} A_{11} & A_{12} \\ A_{21} & A_{22} \end{pmatrix} \begin{pmatrix} C_{11} & 0 \\ C_{21} & I_{22} \end{pmatrix} = \begin{pmatrix} I & A_{12} \\ 0 & A_{22} \end{pmatrix}.$$

It follows that

(11) $\delta(A)\delta(C_{11}) = \delta(A_{22}).$

Analogously

$$\delta(A)\delta(C_{22}) = \delta(A_{11}),$$

and, in fact, by interchanging rows or columns or both in A and repeating with the matrix so formed and its inverse, a theorem of Jacobi is obtained: *The ratio of any minor of $\delta(A^T)$ to the complementary minor of $\delta(A^{-1})$ is equal to $\delta(A)$.* By complementary minor is meant that one remaining after the deletion of the rows and columns containing the minor itself.

Next, consider

$$\begin{pmatrix} A_{11} & A_{12} \\ A_{21} & A_{22} \end{pmatrix} \begin{pmatrix} X_{11} & C_{12} \\ X_{21} & C_{22} \end{pmatrix} = \begin{pmatrix} A_{11}X_{11} + A_{12}X_{21} & 0 \\ A_{21}X_{11} + A_{22}X_{21} & I \end{pmatrix},$$

where X_{11} and X_{21} are arbitrary matrices whose dimensions are those of A_{11} and A_{21}, respectively. Now the Laplace expansion of the determinant of the second matrix on the left will include the particular term

$$\delta(C_{22})\delta(X_{11}).$$

But when this is multiplied by $\delta(A)$ and the Jacobi theorem applied, it becomes

$$\delta(A_{11})\delta(X_{11}).$$

In general, the Laplace expansion of the determinant of the second matrix on the left when multiplied by $\delta(A)$ will consist of the sum of all possible products of determinants of order p formed from rows of

$$\begin{pmatrix} X_{11} \\ X_{21} \end{pmatrix}$$

by the corresponding determinants formed from columns of

$$(A_{11} \quad A_{12}).$$

On the other hand this is equal to the determinant of

$$A_{11}X_{11} + A_{12}X_{21} = (A_{11} \quad A_{12})\begin{pmatrix} X_{11} \\ X_{21} \end{pmatrix}.$$

This is called the *theorem of corresponding minors* (Binet-Cauchy). It can be stated in the following form: *If X and Y are matrices of p columns and* *n rows each, $p \leq n$, then the determinant*

$$\delta(Y^T X)$$

is equal to the sum of the $C_{n,p}$ products of pairs of pth-order determinants that can be formed by selecting p rows from Y and the same p rows of X. The theorem justifies calling $\delta(A_{11}X_{11} + A_{12}X_{21})$ the inner product of the two rectangular matrices.

An immediate corollary is the following: *If A and B are nonnegative semidefinite Hermitian matrices of the same order, then*

(12) $$\delta(A + B) \geq \delta(A) + \delta(B).$$

First, suppose A and B are positive definite and factored as in (7)

$$A = P^H P, \qquad B = R^H R,$$

with P and R upper triangles. Then

$$A + B = P^H P + R^H R = (P^H, R^H)\begin{pmatrix} P \\ R \end{pmatrix}.$$

But the determinant $\delta(P^H P + R^H R)$ is the sum of the squares of the moduli of the determinants formed from columns of (P^H, R^H) and the result follows with the strict inequality. By continuity, the conclusion still holds in the case of semidefiniteness but with possible equality.

Another corollary is a generalization of an inequality of Hadamard: *Let A be nonnegative semidefinite, and let it be partitioned in the form*

$$A = \begin{pmatrix} A_{11} & A_{12} \\ A_{21} & A_{22} \end{pmatrix},$$

where A_{11} and A_{22} are square. Then

(13) $$\delta(A) \leq \delta(A_{11})\delta(A_{22}).$$

As a corollary to this is the original inequality of Hadamard:

(14) $$\delta(A) \leq \prod_1^n \alpha_{ii},$$

where α_{ii} are the diagonal elements of A.

To prove the more general assertion, let A be factored into triangular factors, $A = LL^H$, and let L be partitioned as was A above:

$$L = \begin{pmatrix} L_{11} & 0 \\ L_{21} & L_{22} \end{pmatrix}.$$

Then

$$\delta(L) = \delta(L_{11})\delta(L_{22})$$

and

$$\delta(A) = \delta(L_{11})\delta(L_{22})\delta(L_{11}^H)\delta(L_{22}^H) = \delta(L_{11}L_{11}^H)\delta(L_{22}L_{22}^H).$$

But

$$A_{11} = L_{11}L_{11}{}^{\mathrm{H}}, \qquad A_{22} = L_{21}L_{21}{}^{\mathrm{H}} + L_{22}L_{22}{}^{\mathrm{H}},$$

and since

$$\delta(L_{22}L_{22}{}^{\mathrm{H}}) \leq \delta(L_{21}L_{21}{}^{\mathrm{H}} + L_{22}L_{22}{}^{\mathrm{H}}) = \delta(A_{22}),$$

the theorem follows.

In a given square matrix A of order n, consider any set of p rows and form all $C_{n,p}$ determinants by selecting p columns at a time. Arrange the values of these determinants in some fixed order in a row. Do likewise for every other set of p distinct rows, the ordering of the rows being the same as the ordering of the columns. The result is a matrix of order $C_{n,p}$ called the pth compound of A. This will be designated $A^{(p)}$. If B is also a matrix of order n, let

$$AB = C.$$

Then it follows from the theorem of corresponding minors that

$$A^{(p)}B^{(p)} = C^{(p)},$$

provided the same ordering is used in forming all compounds. In particular, therefore,

$$(A^{-1})^{(p)} = (A^{(p)})^{-1};$$

hence if U is unitary (orthogonal) then $U^{(p)}$ is also unitary (orthogonal). Also if T is (unit) triangular and $T^{(p)}$ given a lexicographic ordering, then $T^{(p)}$ is (unit) triangular.

For any A, along with $A^{(p)}$ consider also $A^{(n-p)}$, ordering in such a way that in forming the product $A^{(p)}A^{(n-p)}$ each element of $A^{(p)}$, considered as a minor determinant from $\delta(A)$, is multiplied by its complementary minor. Then

$$A^{(p)}A^{(n-p)} = \delta(A)I.$$

Hence

$$\delta(A^{(p)})\delta(A^{(n-p)}) = \delta^{C_{n,p}}(A).$$

But $\delta(A)$, considered as a polynomial in the elements of A, is irreducible. Hence $\delta(A^{(p)})$ and $\delta(A^{(n-p)})$ are both powers of $\delta(A)$. By considering the special case $A = \alpha I$ it appears that (Cauchy-Sylvester)

$$(15) \qquad \qquad \delta(A^{(p)}) = \delta^{C_{n-1,p-1}}(A).$$

The *adjoint* of A is the compound of order $n - 1$, the elements being ordered in an obvious fashion:

$$A^{\mathrm{A}} = A^{(n-1)}.$$

The preceding relation gives

$$\delta(A^{\mathrm{A}}) = \delta^{n-1}(A),$$

and, in fact,

$$AA^{\mathrm{A}} = \delta(A)I,$$

$$A^{-1} = \delta^{-1}(A)A^{\mathrm{A}}.$$

One further identity will be derived for later use. Let the square matrix A be partitioned

$$A = \begin{pmatrix} A_{11} & A_{12} \\ A_{21} & A_{22} \end{pmatrix},$$

and suppose A_{11} is nonsingular. Then a factorization can be made:

$$A = \begin{pmatrix} I_{11} & 0 \\ L_{21} & I_{22} \end{pmatrix} \begin{pmatrix} A_{11} & A_{12} \\ 0 & P_{22} \end{pmatrix},$$

$$L_{21} = A_{21}A_{11}^{-1}, \qquad P_{22} = A_{22} - A_{21}A_{22}^{-1}A_{12}.$$

This matrix factorization implies the determinantal factorization

(16) $$\delta(A) = \delta(A_{11})\delta(A_{22} - A_{21}A_{11}^{-1}A_{12}),$$

or

(16') $$\delta(A)/\delta(A_{11}) = \delta(A_{22} - A_{21}A_{11}^{-1}A_{12})$$

which expresses the quotient of a determinant by one of its minors. Naturally rows or columns or both could be permuted in any way to bring any other minor into the denominator.

As a special case of (16), taking A_{22} a scalar, $A_{21} = a'$, a row vector, and $A_{12} = a$, a column vector,

(17) $$\delta \begin{pmatrix} A & a \\ a' & \alpha \end{pmatrix} - \alpha\delta(A) - a'A^{\lambda}a,$$

which remains valid even when A is singular.

1.5. The Lanczos Algorithm for Tridiagonalization. This algorithm produces, for any matrix A, matrices B and C such that

(1) $$C^H B = P$$

is diagonal,

(2) $$C^H A B = \begin{pmatrix} L_1 & 0 & 0 & \cdots \\ 0 & L_2 & 0 & \cdots \\ 0 & 0 & L_3 & \cdots \\ \cdots & \cdots & \cdots & \cdots \end{pmatrix} = PL,$$

and each L_i is of the form

(3) $$L_i = \begin{pmatrix} * & * & 0 & 0 & \cdots \\ * & * & * & 0 & \cdots \\ 0 & * & * & * & \cdots \\ 0 & 0 & * & * & \cdots \\ \cdots & \cdots & \cdots & \cdots & \cdots \end{pmatrix},$$

the $*$'s designating possibly nonnull elements. Since P will turn out to be nonsingular, it will follow that B is also nonsingular and

$$C^{\mathrm{II}} = PB^{-1}.$$

Consequently (2) can be written

$$B^{-1}AB = L,$$

and the form of L is the same as that of PL. This will show that *any matrix is similar to a tridiagonal matrix.*

First let $v_1 = v \neq 0$ be any nonnull vector, and form the *Krylov sequence* of vectors defined by

$$(4) \qquad\qquad v_{i+1} = Av_i = A^i v_1.$$

In this sequence, there will be a first vector that is expressible as a linear combination of the preceding ones. Let this be $v_{\nu+1}$. Hence for some polynomial

$$\psi(\lambda) = \psi_1 + \psi_2\lambda + \cdots + \psi_\nu\lambda^{\nu-1} + \lambda^\nu$$

of minimal degree $\nu \leq n$,

$$\psi(A)v = 0.$$

This polynomial will be said to *annihilate* v and to be *minimal* for v. If $\omega(\lambda)$ also annihilates v,

$$\omega(A)v = 0,$$

then $\psi(\lambda)$ divides $\omega(\lambda)$, since if

$$\omega(\lambda) = \sigma(\lambda)\psi(\lambda) + \rho(\lambda)$$

where $\rho(\lambda)$ is the remainder after dividing ω by ψ, hence of degree $< \nu$, it follows that

$$\rho(A)v = 0.$$

But $\psi(\lambda)$ is minimal for v, hence $\rho(\lambda) = 0$.

Now of all vectors v there is at least one for which the degree ν is maximal, since for any vector $\nu \leq n$. Let v be such a vector. It is to be shown that $\psi(A)x = 0$ for any vector x, and hence

$$\psi(A) = 0.$$

Thus $\psi(\lambda)$ is the *minimal polynomial* for the matrix A.

To prove it, consider any vector u such that u and v are linearly independent. Let its minimal polynomial be $\phi(\lambda)$. If $\omega(\lambda)$ is the lowest common multiple of ϕ and ψ, then ω annihilates every vector in the plane of u and v, since

$$\omega(A)(\lambda u + \mu v) = \lambda\omega(A)u + \mu\omega(A)v = 0.$$

Hence ω contains as a divisor the minimal polynomial of every vector in the plane. But ω is of degree $2n$ at most, hence has only finitely many divisors. Hence some pair of independent vectors, a and b, in the plane, possess the same minimal polynomial, and by the same argument, this polynomial annihilates every vector in the plane of a and b. But this can only be ψ which is minimal for v and hence of all minimal polynomials has maximum degree. Hence ψ annihilates every vector in the plane of u and v, and since u was any vector whatever, other than v, ψ annihilates every vector in space.

Since ψ annihilates every vector, it annihilates in particular every e_i, hence

$$\psi(A) = \psi(A)I = 0.$$

Thus the matrix A satisfies the equation

$$\psi(\lambda) = 0,$$

and does not satisfy any equation of lower degree. Hence ψ is the minimal polynomial for A. Likewise $\bar{\psi}$, the polynomial obtained when every coefficient of ψ is replaced by its conjugate, is the minimal polynomial for A^H. Thus $\psi(\lambda)$ annihilates also every row vector, and there is no polynomial of lower degree which has this property.

Some further consequences can be drawn from the argument. For practical applications of the Lanczos algorithm, it is important to know that ψ *is the minimal polynomial for "almost every" vector*. In fact, the minimal polynomial for any vector is a divisor of ψ, and the vectors annihilated by any divisor $\neq \psi$ form a linear proper subspace. Since there are only finitely many proper divisors of ψ, there can be only finitely many proper subspaces of vectors annihilated by polynomials of degree $< \nu$. This proves the assertion.

To return to the Krylov sequence of vectors (4), suppose v_1 has ψ as its minimal polynomial. The Lanczos algorithm does not form the v_i explicitly, but forms instead certain linear combinations b_i, each b_i being a linear combination of $v_1, v_2, \cdots, v_i$ only. In particular, $b_1 = v$. This can be expressed by saying that

(5)
$$Ab_1 = b_1\tau_{11} + b_2,$$
$$Ab_2 = b_1\tau_{12} + b_2\tau_{22} + b_3,$$
$$Ab_3 = b_1\tau_{13} + b_2\tau_{23} + b_3\tau_{33} + b_4,$$

$$\cdots\cdots\cdots\cdots$$

where for $i < \nu$, the last term in the expression for Ab_i is b_{i+1}, but

(5')
$$Ab_\nu = b_1\tau_{1\nu} + \cdots + b_\nu\tau_{\nu\nu}.$$

But this means that a matrix

$$B = (b_1, b_2, \cdots, b_\nu)$$

has been formed such that

$$AB = B(J + T)$$

where T is upper triangular. In like manner, if $c_1 = w$, one can form

$$C = (c_1, c_2, \cdots, c_\nu)$$

such that

$$A^H C = C(J + S)$$

where S is also upper triangular. In this process, the last column of T and the last column of S are determined, but the first $\nu - 1$ columns are arbitrary.

It will now be required that τ_{11} and σ_{11} be selected so that $c_1^H b_2 = 0$ and $b_1^H c_2 = 0$; that $\tau_{12}, \tau_{22}, \sigma_{12},$ and σ_{22}, be selected so that $c_1^H b_3 = c_2^H b_3 = 0$ and $b_1^H c_3 = b_2^H c_3 = 0$, and, in general, it can be verified that the elements τ_{ij} and σ_{ij} can be selected in sequence so that

$$C^H B = P$$

where P is diagonal, provided only $c_i^H b_i \neq 0$ for $i = 1, 2, \cdots, \nu$. On the other hand, it is not possible to obtain a $c_{\nu+1}$ and a $b_{\nu+1}$ such that $c_{\nu+1}^H b_{\nu+1} \neq 0$, since then we would have formed implicitly more than ν linearly independent vectors in the Krylov sequence.

It will be shown now that b_1 and c_1 *can be chosen so that* $c_i^H b_i \neq 0$ *for* $i \leq \nu$.

For this purpose and for later use, let

$$\eta_i = c_1^H A^i b_1,$$

$$(6) \qquad H_{i+1} = \begin{pmatrix} \eta_0 & \eta_1 & \cdots & \eta_i \\ \eta_1 & \eta_2 & \cdots & \eta_{i+1} \\ \cdots\cdots\cdots\cdots\cdots \\ \eta_i & \eta_{i+1} & \cdots & \eta_{2i} \end{pmatrix}.$$

It will be shown that the determinant of this Hankel matrix H_{i+1} has the value

$$(7) \qquad \delta(H_{i+1}) = c_1^H b_1 \, c_2^H b_2 \cdots c_{i+1}^H b_{i+1} = \rho_1 \, \rho_2 \cdots \rho_{i+1}.$$

For suppose $i \leq \nu$. It follows from equations (5) that b_{j+1} is obtained by adding to $A^j b_1$ a linear combination of the vectors $b_1, Ab_1, \cdots, A^{j-1}b_1$. Hence, for $j \leq i$, if this linear combination of the first j columns of $\delta(H_{i+1})$ is added to the next column, the value of the determinant is unchanged, and the elements of the columns can be written

$$c_1^H b_{j+1}, c_1^H A b_{j+1}, c_1^H A^2 b_{j+1}, \cdots.$$

Let this be done for every column but the first. Likewise, if a suitable linear combination of the first j rows is added to the next, that row becomes

$$c_{j+1}^H b_1, \; c_{j+1}^H b_2, \; c_{j+1}^H b_3, \; \cdots$$

and every element vanishes but the one on the diagonal. The result follows from applying this to each row but the first. For $i \geq \nu$ the matrix is singular, and since $b_{\nu+1} = 0$, (7) holds trivially.

Now note that, for example

$$H_3 = \begin{pmatrix} c_1^H \\ c_1^H A \\ c_1^H A^2 \end{pmatrix} (b_1, \, A b_1, \, A^2 b_1),$$

and if $\nu \geq 3$, both matrices on the right are of rank 3, hence this product cannot be identically singular. In fact, the determinant of the product is a rational function of the elements of c_1^H and b_1 which does not vanish identically. Hence the determinant will be, in general, nonvanishing. A similar factorization in general establishes the assertion.

Suppose, therefore, that the vectors have been chosen as required. It follows that

$$C^H A B = P(J + T),$$

$$B^H A^H C = P^H(J + S).$$

Hence

$$P(J + T) = (J^T + S^H)P.$$

But the matrix on the left is null below the subdiagonal, and that on the right is null above the superdiagonal. Hence T and S are nonnull only along and just above the diagonal. Let

$$T = \Delta + \Gamma, \qquad S = \Delta' + \Gamma',$$

where Δ and Δ' are diagonal, Γ and Γ' are nilpotent. Then

$$P(J + \Delta + \Gamma) = (J^T + \Delta'^H + \Gamma'^H)P.$$

Hence

$$P\Gamma = J^T P, \qquad \Gamma = P^{-1} J^T P,$$
$$P\Delta = \Delta'^H P, \qquad \Delta = \Delta'^H$$
$$PJ = \Gamma'^H P \qquad \Gamma'^H = P J P^{-1}.$$

Hence

$$T = \Delta + P^{-1} J^T P, \qquad S^H = \Delta + P J P^{-1},$$

where Δ is diagonal, and the result is

$$(8) \quad AB = B(J + \Delta + P^{-1} J^T P), \qquad C^H A = (J^T + \Delta + P J P^{-1})C^H.$$

In case $\nu = n$, the required reduction is now complete and

$$L = J + \Delta + P^{-1} J^T P.$$

If not, then among all vectors b and c satisfying

$$C^H b = 0, \qquad c^H B = 0,$$

there is a b_1' and a c_1' whose minimal polynomial is of highest degree $\nu' \leq \nu$. Moreover, that minimal polynomial is a divisor of $\psi(\lambda)$. In particular, it is possible to choose b_1' and c_1' so that the algorithm applied to them continues until there are formed matrices B' and C' of rank ν' and such that

$$C'^H B' = P'$$

is nonsingular. Hence

$$AB' = B'(J + \Delta' + P'^{-1}J^T P') = B'L'.$$

Moreover, it can be shown sequentially that every vector b_i' is orthogonal to all columns of C, hence $C^H B' = 0$, and likewise that $B^H C' = 0$. Hence

$$\begin{pmatrix} C^H \\ C'^H \end{pmatrix} (B, B') = \begin{pmatrix} P & 0 \\ 0 & P' \end{pmatrix}$$

and both matrices on the left are of rank $\nu + \nu'$. Now

$$A(B, B') = (B, B') \begin{pmatrix} L & 0 \\ 0 & L' \end{pmatrix}.$$

If $\nu + \nu' = n$, the reduction is complete. Otherwise select b_1'' and c_1'' orthogonal to (C, C') and (B, B') respectively and proceed. Eventually nonsingular matrices

$$(B, B', B'', \cdots), \qquad (C, C', C'', \cdots)$$

are obtained and with an obvious change of notation the form (2) is established.

Equations (5) show that b_{i+1} is expressible in the form

(9) $$b_{i+1} = \chi_i(A)b_1,$$

where $\chi_i(\lambda)$ is a monic polynomial of degree i (i.e., one whose leading coefficient is unity):

$$\chi_i(\lambda) = \lambda^i + \chi_{i,i-1}\lambda^{i-1} + \cdots.$$

Also c_{i+1} is expressible as the result of operating upon c_1 by a monic polynomial of degree i. Furthermore, given b_1 and c_1, the vector b_{i+1} is characterized by the property of being orthogonal to

$$c_1, c_2, \cdots, c_i,$$

hence to the vectors

$$c_1, A^H c_1, \cdots, (A^H)^{i-1} c_1.$$

Then if the Hankel matrix H_i defined in (6) is nonsingular, it will be shown that

$$
(10) \qquad \chi_i(\lambda) = \delta(H_i^{-1})\delta
\begin{pmatrix}
\eta_0 & \eta_1 & \cdots & \eta_i \\
\eta_1 & \eta_2 & \cdots & \eta_{i+1} \\
\cdots\cdots\cdots\cdots\cdots \\
\eta_{i-1} & \eta_i & \cdots & \eta_{2i-1} \\
1 & \lambda & \cdots & \lambda^i
\end{pmatrix}.
$$

In fact, this polynomial is certainly monic, and the scalar

$$
c_1^{\mathrm{H}} A^j \chi_i(A) b_1
$$

can be formed by replacing the row of powers of λ by

$$
\eta_j, \eta_{j+1}, \cdots, \eta_{j+i},
$$

which, for $j < i$, coincides with one of the rows above it. In like manner it can be shown that

$$
(11) \qquad c_{i+1}^{\mathrm{H}} = c_1^{\mathrm{H}} \chi_i(A),
$$

for the same polynomial χ_i. From (5') it follows that

$$
\chi_\nu(A) b_1 = 0,
$$

hence $\chi_\nu(\lambda)$ *is the minimal polynomial for A.*

The polynomials $\chi_i(\lambda)$ will be called the *Lanczos polynomials for A determined by* b_1 *and* c_1. They can be formed recursively by the relations

$$
(12) \qquad
\begin{aligned}
&\chi_0(\lambda) = 1, \\
&\chi_1(\lambda) = (\lambda - \delta_1)\chi_0, \\
&\chi_2(\lambda) = (\lambda - \delta_2)\chi_1 - (\rho_2/\rho_1)\chi_0, \\
&\chi_3(\lambda) = (\lambda - \delta_3)\chi_2 - (\rho_3/\rho_2)\chi_1, \\
&\cdots\cdots\cdots\cdots\cdots\cdots\cdots\cdots \\
&\chi_\nu(\lambda) = (\lambda - \delta_\nu)\chi_{\nu-1} - (\rho_\nu/\rho_{\nu-1})\chi_{\nu-2},
\end{aligned}
$$

where the δ_i and ρ_i are diagonal elements of Δ and of P, since if λ is replaced by A throughout, and each $\chi_i(A)$ multiplied by b_1, these become the relations defining the b_i. Moreover, the polynomial $\chi_j(\lambda)$, $j \leq \nu$, is the characteristic polynomial for the principal submatrix formed from the first j rows and columns of L, as can be verified inductively. Let

$$
B_j = (b_1, b_2, \cdots, b_j), \qquad C_j = (c_1, c_2, \cdots, c_j)
$$

and let L_j and P_j be the corresponding submatrices of L and P. Then

$$
\begin{aligned}
&C_j^{\mathrm{H}}(\lambda I - A)B = (P_j, 0)(\lambda I - L), \\
&C_j^{\mathrm{H}}(\lambda I - A)B_j = P_j(\lambda I - L_j), \\
&P_j^{-1}C_j^{\mathrm{H}}(\lambda I - A)B_j = \lambda I - L_j.
\end{aligned}
$$

Hence

(13) $$\chi_j(\lambda) = \delta[P_j^{-1}C_j^{\mathrm{H}}(\lambda I - A)B_j].$$

1.6. Orthogonal Polynomials.   Let $v \neq 0$ and form again the Krylov sequence

(1) $$v_1 = v, v_2 = Av_1, v_3 = Av_2, \cdots, v_{i+1} = Av_i = A^iv_1, \cdots.$$

The scalar products

(2) $$\mu_{i,j} = v_{i+1}^{\mathrm{H}}v_{j+1}$$

will be called *moments*, and the matrix

(3) $$M_{i+1} = \begin{pmatrix} \mu_{00} & \mu_{01} & \cdots & \mu_{0i} \\ \mu_{10} & \mu_{11} & \cdots & \mu_{1i} \\ \cdots\cdots\cdots\cdots\cdots \\ \mu_{i0} & \mu_{i1} & \cdots & \mu_{ii} \end{pmatrix}$$

the *moment matrix* of order $i + 1$. It is Hermitian, and if the matrix A is Hermitian and $b_1 = c_1 = v_1$, it coincides with the Hankel matrix H_{i+1} (1.5.6). Likewise, the monic polynomials

$$\phi_0(\lambda) = 1,$$

(4) $$\phi_i(\lambda) = \delta(M_i^{-1})\delta \begin{pmatrix} \mu_{00} & \mu_{01} & \cdots & \mu_{0i} \\ \mu_{10} & \mu_{11} & \cdots & \mu_{1i} \\ \cdots\cdots\cdots\cdots\cdots\cdots \\ \mu_{i-1,0} & \mu_{i-1,1} & \cdots & \mu_{i-1,i} \\ 1 & \lambda & \cdots & \lambda^i \end{pmatrix}, \qquad i \geq 1$$

coincide under the same conditions with the Lanczos polynomials $\chi_i(\lambda)$ (1.5.10). In general, these polynomials have the property that

(5) $$v_j^{\mathrm{H}}\phi_i(A)v = 0, \qquad j \leq i,$$

since the scalar product is the result of replacing the row of powers of λ on the right of (4) by the row

$$\mu_{j-1,0}, \mu_{j-1,1}, \cdots, \mu_{j-1,i}.$$

Hence if

(6) $$p_1 = v_1, \qquad p_{i+1} = \phi_i(A)v_1,$$

then, according to (5), each p_i is orthogonal to v_j for $j < i$. But p_j is equal to v_j plus a linear combination of $v_1, v_2, \cdots, v_{j-1}$. Hence p_i is orthogonal to each p_j for $j < i$, or, in general,

$$p_i^{\mathrm{H}}p_j = 0, \qquad i \neq j.$$

Evidently $v_j - p_j$ is the orthogonal projection of v_j upon the space of $v_1, v_2, \cdots, v_{j-1}$, and hence if ν is the degree of the minimal polynomial of $v = v_1$, then $p_{\nu+1} = 0$. Hence $\phi_\nu(\lambda)$ is the minimal polynomial for v, and therefore, if $b_1 = v_1$, $\phi_\nu = \chi_\nu$ independently of c_1, provided only c_1 is chosen to yield a maximal Lanczos sequence.

Assuming the minimal polynomial for v to be of degree ν, let

(7) $$V_\rho = (v_1, \cdots, v_\rho), \qquad P_\rho = (p_1, \cdots, p_\rho), \qquad \rho \le \nu.$$

From the construction of the p_i, it is evident that there exists a unit upper triangular matrix T_ρ of order ρ such that

(8) $$V_\rho = P_\rho T_\rho, \qquad P_\rho^H P_\rho = D_\rho^2,$$

where D_ρ^2 is diagonal. In fact, if

(9) $$Q_\rho = P_\rho D_\rho^{-1}, \qquad Q_\rho^H Q_\rho = I,$$

that is, the columns of Q_ρ are the normalized columns of P_ρ, and the factorization

$$V_\rho = Q_\rho(D_\rho T_\rho)$$

of V_ρ is identical with the factorization (1.2.4) of A and can be carried out in the same way. However, a construction analogous to that of Lanczos (1.5.5) is possible except that in the recursion none of the terms can be expected to vanish.

In this case it is required to form the columns of P_ν in sequence, starting with $p_1 = v_1$, in such a way that

(10) $$A P_\nu = P_\nu S_\nu$$

where $S_\nu - J$ is upper triangular (and J is as defined in Section 1.0). Thus

(11)
$$Ap_1 = p_1\sigma_{11} + p_2,$$
$$Ap_2 = p_1\sigma_{12} + p_2\sigma_{22} + p_3,$$
$$\cdots\cdots\cdots\cdots\cdots\cdots\cdots\cdots\cdots$$
$$Ap_\nu = p_1\sigma_{1\nu} + p_2\sigma_{22} + \cdots + p_\nu\sigma_{\nu\nu}.$$

The algorithm proceeds in a straightforward manner, with p_1 given, and

$$p_1^H p_1 = \delta_1^2,$$
$$p_1^H A p_1 = \sigma_{11}\delta_1^2, \qquad p_2^H A p_1 = p_2^H p_2 = \delta_2^2,$$
$$\cdots\cdots\cdots\cdots\cdots\cdots\cdots\cdots\cdots$$

This construction establishes that the p_{i+1} are expressible in the form (6) for some polynomials $\phi_i(\lambda)$; but the $\phi_i(\lambda)$ are unique and, indeed, they have already been given in the form (4); hence the construction shows that the $\phi_i(\lambda)$ satisfy the recursion

$$
\begin{aligned}
\phi_1(\lambda) &= (\lambda - \sigma_{11})\phi_0(\lambda), \\
\phi_2(\lambda) &= (\lambda - \sigma_{22})\phi_1(\lambda) - \sigma_{12}\phi_0(\lambda), \\
\phi_3(\lambda) &= (\lambda - \sigma_{33})\phi_2(\lambda) - \sigma_{23}\phi_1(\lambda) - \sigma_{13}\phi_0(\lambda), \\
&\cdots\cdots\cdots\cdots\cdots\cdots\cdots\cdots\cdots\cdots\cdots\cdots\cdots\cdots\cdots \\
\phi_\nu(\lambda) &= (\lambda - \sigma_{\nu\nu})\phi_{\nu-1}(\lambda) - \cdots - \sigma_{1\nu}\phi_0(\lambda).
\end{aligned}
\tag{12}
$$

The relation (10) is equivalent to

$$(\lambda I - A)P_\nu = P_\nu(\lambda I - S_\nu),$$

or to

$$(\lambda I - A)Q_\nu = Q_\nu D_\nu(\lambda I - S_\nu)D_\nu^{-1},$$

and this implies

$$Q_\nu^H(\lambda I - A)Q_\nu = D_\nu(\lambda I - S_\nu)D_\nu^{-1}.$$

But this in turn implies

$$Q_\rho^H(\lambda I - A)Q_\rho = D_\rho(\lambda I - S_\rho)D_\rho^{-1}, \qquad \rho \le \nu \tag{13}$$

where S_ρ is the principal submatrix of the first ρ rows and columns of S_ν. It can be verified inductively that

$$\phi_\rho(\lambda) = \delta(\lambda I - S_\rho), \tag{14}$$

and since

$$Q_\rho^H(\lambda I - A)Q_\rho = \lambda I - Q_\rho^H A Q_\rho,$$

it follows that $\phi_\rho(\lambda)$ *is the characteristic polynomial for* $Q_\rho^H A Q_\rho$, $\rho \le \nu$.

It follows from (2), (3), and (8) that

$$M_\rho = V_\rho^H V_\rho = T_\rho^H D_\rho^2 T_\rho,$$

hence, since T_ρ is a unit triangle, that

$$\delta(M_\rho) = \delta(D_\rho^2) = \delta_1^2 \delta_2^2 \cdots \delta_\rho^2.$$

Consequently

$$
\begin{aligned}
p_1^H p_1 &= \mu_{00} = \delta(M_1), \\
p_\rho^H p_\rho &= \delta(M_\rho)/\delta(M_{\rho-1}) \\
&= v_\rho^H v_\rho - v_\rho^H V_{\rho-1}^H (V_{\rho-1}^H V_{\rho-1})^{-1} V_{\rho-1} v_\rho, \qquad \rho > 1,
\end{aligned}
\tag{15}
$$

the latter form being a consequence of (1.4.16).

Let ω be any characteristic root of $Q_\rho^H A Q_\rho$; let $u \ne 0$ be a vector belonging to it,

$$Q_\rho^H A Q_\rho u = \omega u, \tag{16}$$

and consider the vector

$$Q_\rho u = P_\rho D_\rho^{-1} u = V_\rho T_\rho^{-1} D_\rho^{-1} u.$$

If

$$(17) \qquad a = T_\rho^{-1} D_\rho^{-1} u,$$

then

$$(18) \qquad Q_\rho u = V_\rho a = \alpha(A)v,$$

where $\alpha(\lambda)$ is some polynomial of degree $\rho - 1$ at most, having the elements of a as coefficients. Now

$$r = (A - \omega I)\alpha(A)v = (A - \omega I)Q_\rho u,$$

and

$$Q_\rho{}^H r = Q_\rho{}^H (A - \omega I) Q_\rho u = 0$$

because of (16); hence r is a linear combination of the vectors $v_1, \cdots, v_{\rho+1}$ and is orthogonal to the vectors $v_1, \cdots, v_\rho$; hence r is a scalar multiple of $p_{\rho+1}$. From this it follows that $(\lambda - \omega)\alpha(\lambda)$ is a constant multiple of $\phi_\rho(\lambda)$, or that $\alpha(\lambda)$, *as defined by* (17) *and* (18) *is equal to* $\phi_\rho(\lambda)/(\lambda - \omega)$, except possibly for a constant factor. In particular, then $\alpha(\lambda)$ is of degree $\rho - 1$ exactly. In case

$$(19) \qquad e_\rho{}^T a = 1,$$

the polynomial $\alpha(\lambda)$ is monic and

$$(20) \qquad \phi_\rho(\lambda) = (\lambda - \omega)\alpha(\lambda).$$

The normalization (19) is equivalent to

$$e_\rho{}^T T_\rho^{-1} D_\rho^{-1} u = 1.$$

But T_ρ, and therefore T_ρ^{-1}, are unit triangles; hence

$$e_\rho{}^T T_\rho^{-1} = e_\rho{}^T;$$

therefore (19) is equivalent to

$$(21) \qquad e_\rho{}^T u = \delta_\rho = (p_\rho{}^T p_\rho)^{\frac{1}{2}}.$$

Hence, *if u satisfies* (16) *and* (21), *then the polynomial* $\alpha(\lambda)$ *defined by* (17) *and* (18) *satisfies* (20).

REFERENCES

MacDuffee (1946) gives an excellent survey of matrix theory up to the early thirties. The most complete modern development is Gantmaher (1956), and the translation Gantmacher (1959), both of which contain extensive bibliographies. Bellman (1960) gives much interesting material and a number of references. Turnbull (1929), and Turnbull and Aitken (1930) can both be recommended highly. Wedderburn (1934) is often referred to. Marcus (1960) gives an excellent outline.

For determinants, the references include only some of the classical papers, but in this area there is no reference superior to Muir's monumental *Theory of Determinants*, and it cannot be recommended too highly. Much of matrix theory originated in the form of determinantal identities, and, conversely, many determinantal relations become more readily apprehended when seen as interpretations of relations among matrices. The determinantal relations here presented are of this class, which is not surprising since they are also the ones that seem to be most useful in the numerical manipulation of matrices. For further relations, derived in much the same spirit, see Price (1947). Reference should also be made to Gantmaher once more and to Turnbull, although the derivations there are more in the classical manner.

The Hadamard inequality, which is in fact due to Kelvin (see Muir's *Theory*, volume 4; also *Contributions*), has given rise to developments in several directions: Szasz (1917), de Bruijn (1956), Fage (1946, who seems to have rediscovered Szasz's extension), Mirsky (1957), and others.

The use of matrices of the form here called "elementary" is implicit in Turnbull and Aitken (1930), Wedderburn (1934), Turing (1948), and doubtless others. Gastinel (1960) has introduced a closely related form. For the use of elementary Hermitian matrices in this context, see Householder (1958d).

Projections are common tools in the literature on functional analysis. The general reciprocal was introduced by E. H. Moore in 1920 (see R. W. Barnard's introduction to Moore's *General Analysis*). However, it is now being rediscovered and revivified (Penrose, 1955; Sheffield, 1958; Greville, 1959; Lanczos, 1958). Further use of projections will be made in Chapter 4.

The Lanczos algorithm is well known in the theory of orthogonal polynomials, but Lanczos (1950) seems to have been the first to apply it to the reduction of matrices. The Krylov sequence is so named here because Krylov (1931) seems to have been the first to use it as a means for finding the characteristic polynomial (the method will be described in Chapter 6), and because several other methods, including that of Lanczos, in fact make use of a Krylov sequence in disguised form. The sequence will also appear in Chapter 3 in the inclusion theorems, along with the orthogonal polynomials $\phi(\lambda)$. The Lanczos polynomials $\chi(\lambda)$, as well as the orthogonal polynomials, will reappear in Chapter 7, and they have numerous other applications: Rutishauser (1953, 1954, 1957), and Henrici (1958).

Orthogonal polynomials themselves have an extensive literature which will not be cited here, but their association with matrices seems to be relatively new. Other references will be given later, but at this point mention may be made of the papers by Bauer and Samelson (1957), Rutishauser (1954), Vorob'ev (1958), and Bauer and Householder (1960a). They are also used by Stiefel (1958), but in a rather different way.

PROBLEMS AND EXERCISES

1. Apply the Lanczos reduction to prove the Cayley-Hamilton theorem, that every square matrix A satisfies its own characteristic equation $\delta(\lambda I - A) = 0$, i.e., that $\phi(A) = 0$ where $\phi(\lambda) = \delta(\lambda I - A)$; and also, that any root of the characteristic equation is a root of the minimal equation.

2. Carry out explicitly the method outlined at the end of Section 2 for factoring a matrix A into the product of a unitary matrix by an upper triangular matrix; also the equivalent factorization into the product of a matrix of mutually orthogonal columns by a unit upper triangular matrix.

3. Show that if $Ax = \lambda x$, $x \neq 0$, there exists an elementary Hermitian matrix H such that

$$HAHe_1 = \lambda e_1,$$

and thence show that for any matrix A there exists a unitary matrix Ω such that

$$\Omega^H A \Omega = T$$

with T upper (lower) triangular. Show, moreover, that the diagonal elements of T are the proper values of A, and that by proper choice of Ω these can be ordered arbitrarily.

4. If A is both normal (i.e., $A^H A = A A^H$) and triangular, then A is diagonal.

5. If A in Exercise 3 is normal, then T is diagonal; in particular, if A is Hermitian, T is also real, or if A is unitary, then $|T| = I$.

6. If T is upper triangular, and $\tau_{ii} \neq \tau_{jj}$, then σ can be chosen so that

for $j > i$ $\qquad e_i{}^T E(e_i, e_j; -\sigma) T E(e_i, e_j; \sigma) e_j = 0.$

Hence for any matrix A there exists a matrix $X = \Omega F$, where Ω is a product of elementary Hermitians, and F a product of elementary matrices of the form $E(e_i, e_j; \sigma)$, such that

$$X^{-1}AX = \operatorname{diag}(R_1, R_2, \cdots),$$

each R_i being upper (lower) triangular and having all diagonal elements equal.

7. If all proper values of A are distinct, it is normalizable (i.e., similar to a normal matrix).

8. A nilpotent matrix can have only 0 as a proper value; any proper value of an idempotent matrix is either 0 or a root of unity.

9. If A is nilpotent of degree ν, $A^\nu = 0 \neq A^{\nu-1}$, there exists a vector $v = v_1$ such that in the Krylov sequence $v_{\nu+1} = 0 \neq v_\nu$. If

$$V = (v_1, v_2, \cdots, v_\nu),$$

then

$$AV = VJ.$$

10. If $\nu < n$ in Exercise 9, and if $b_1 = v_1$ and a suitable c_1 are taken to start the Lanczos reduction, then, as in the general case, $B = VT$, where T is a unit upper triangular matrix, and

$$AB = B(T^{-1}JT).$$

If v_1' is chosen so that $c^H v_1' = 0$, then also

$$AV' = V'J',$$

where J' is of order $\nu' \leq \nu$. Hence show that any nilpotent matrix is similar to a matrix

$$\text{diag}\,(J, J', \cdots).$$

11. Combine the results of Exercise 6 and Exercise 10 to show that any matrix is similar to a matrix of the form

$$\text{diag}\,(\lambda_1 I + J_1, \lambda_2 I + J_2, \cdots)$$

where each J_i is a matrix J (Jordan normal form).

12. If A and B are both normalizable, the following three conditions are equivalent:

(i) $AB = BA$;

(ii) there exists a nonsingular matrix X such that $X^{-1}AX$ and $X^{-1}BX$ are both diagonal;

(iii) there exists a normalizable matrix C and polynomials $\alpha(\lambda)$ and $\beta(\lambda)$ such that $A = \alpha(C)$ and $B = \beta(C)$.

13. Expand

$$\delta \begin{pmatrix} \lambda I & -A \\ -B & \lambda I \end{pmatrix}$$

in two ways to show that the characteristic polynomials of AB and BA are equal up to a power of λ as a factor.

14. By a similar device show that

$$\delta(I - AB) = \delta(I - BA).$$

15. Let

$$AP = PM$$

where M is nonsingular of order ρ and P is of rank ρ. Let

$$V^H P = I.$$

Then if

$$B = A - PMV^H = A(I - PV^H),$$

it follows that

$$\delta(\lambda I - A) = \lambda^{-\rho}\delta(\lambda I - B)\delta(\lambda I - M).$$

16. If A and B are square matrices, there exists an $X \neq 0$ satisfying $AX = XB$ if and only if A and B have in common at least one proper value. Hence the equation $AX - XB = C \neq 0$ has a unique solution if and only if A and B have in common no proper value.

17. If $AB = BA$ and B is nilpotent, then $\delta(A + B) = \delta(A)$ (Frobenius).

18. The trace of A^{ν} vanishes for every integer $\nu > 0$ if and only if A is nilpotent.

19. For any matrix A (not necessarily square) there exist elementary Hermitians $H(u)$ (Section 1.1) and $H(v)$, scalars τ and σ with $|\tau| = |\sigma| = 1$, and a scalar $\alpha \geq 0$ such that

$$AH(v)e_1 = \alpha \bar{\sigma} \tau H(u)e_1,$$
$$A^H H(u)e_1 = \alpha \bar{\tau} \sigma H(v)e_1.$$

Hence there exist unitary matrices U and V such that $U^H A V$ is a nonnegative real diagonal matrix (whose diagonal elements are the singular values of A).

20. Any matrix A is expressible in the ("polar") form $A = H\Omega$, where Ω is unitary and H is Hermitian and semidefinite. If A is nonsingular, H is positive definite and the representation is unique.

21. If A is normalizable, its roots are all real if and only if there exists a positive definite Hermitian matrix H such that HA is Hermitian.

22. If each submatrix is square in

$$A = \begin{pmatrix} A_{11} & A_{12} \\ A_{21} & A_{22} \end{pmatrix}$$

and $A_{11}A_{21} = A_{21}A_{11}$, then

$$\delta(A) = \delta(A_{11}A_{22} - A_{21}A_{12}).$$

Generalize this result.

23. Let

$$A = \begin{pmatrix} P & Q \\ 0 & I \end{pmatrix}.$$

If $P - I$ is nonsingular, then for any integer $\nu \geq 0$,

$$A^{\nu} = \begin{pmatrix} P^{\nu} & Q_{\nu} \\ 0 & I \end{pmatrix}$$

$$Q_{\nu} = (P^{\nu} - I)(P - I)^{-1}Q.$$

If P also is nonsingular the formula is valid also for $\nu < 0$.

24. Let $\phi(\lambda)$ and $\psi(\lambda)$ be mutually prime polynomials, and let $A = \phi(M)$, $B = \psi(M)$. Then if $ABx = 0$, there exist vectors y and z such that $By = Az = 0$ and $x = y + z$.

25. If $A = \lambda I + J$ is of order n, and $\phi(\lambda)$ is any polynomial, then

$$\phi(A) = \phi(\lambda)I + \phi'(\lambda)J + \cdots + \phi^{(n-1)}(\lambda)J^{n-1}/(n - 1)!.$$

26. If $A_\lambda = \lambda I - uu^H$, $u^H u = 1$, then

$$A_\lambda{}^A = - \lambda^{n-2} A_{1-\lambda}.$$

27. If the matrix

$$H = \begin{pmatrix} A & B \\ B^H & C \end{pmatrix}$$

is positive definite, then for any R,

$$\delta(H)/\delta(A) \le \delta(C + R^H B + B^H R + R^H A R),$$

with equality if and only if $AR + B = 0$ (Fan).

28. Let

$$A = \begin{pmatrix} A_{11} & A_{12} & A_{13} & A_{14} \\ A_{21} & A_{22} & A_{23} & A_{24} \\ A_{31} & A_{32} & A_{33} & A_{34} \\ A_{41} & A_{42} & A_{43} & A_{44} \end{pmatrix}$$

be nonnegative semidefinite with the A_{ii} square. Using the notation

$$(i) = \delta(A_{ii}), \ (ij) = \delta\begin{pmatrix} A_{ii} & A_{ij} \\ A_{ji} & A_{jj} \end{pmatrix}, \cdots,$$

then

$$(ij)(jk)(kl) \ge (ijk)(jkl).$$

29. Prove by induction (Szasz, Mirsky, Fage): If A is positive definite, and π_ν is the product of all νth-order principal minor determinants of $\delta(A)$, then

$$\pi_1 \ge \pi_2{}^{C_{n-1,1}^{-1}} \ge \pi_3{}^{C_{n-1,2}^{-1}} \ge \cdots \ge \pi_n.$$

Moreover, the inequalities are all strict unless A is diagonal.

30. If V is unitary, then there is a Hermitian matrix A such that $V = \exp(iA)$.

31. Let $A^T = -A$ and real. Then $\delta(A)$ is a perfect square, and vanishes when A is of odd order.

32. If

$$\delta\begin{pmatrix} A & a \\ a' & \beta \end{pmatrix} = 0,$$

then

$$\delta\begin{pmatrix} A & a \\ a' & \alpha \end{pmatrix} = (\alpha - \beta)\delta(A).$$

33. For A positive definite, apply Exercise 30 to A^A to show (Mirsky)

$$\delta^{-1}(A) < \delta^{-n/(n-1)} \pi_1{}^{C_{n-1,1}^{-1}} < \delta^{-n/(n-2)} \pi_2{}^{C_{n-1,2}^{-1}} < \cdots < \delta^{-n}(A) \pi_{n-1}.$$

34. (Wedderburn) The matrix $A - \tau u v^{\mathrm{H}}$ is of rank less than that of A if and only if $u = Ax$, $v = A^{\mathrm{H}}y$, and

$$\tau^{-1} = y^{\mathrm{H}}Ax \neq 0.$$

35. If in a Krylov sequence $v_{\nu+1} = 0 \neq v_\nu$, then $v_1, v_2, \cdots, v_\nu$ are linearly independent.

36. If $A = A^2$, then rank $(A) +$ rank $(I - A) = n$.

37. If $ABx = 0$ implies $Bx = 0$, then rank $(AB) =$ rank (B) and conversely.

38. If rank $(A) =$ rank (BA), then rank $(AC) =$ rank (BAC).

39. If rank $(A^\nu) =$ rank $(A^{\nu+1})$, then rank $(A^{\nu+1}) =$ rank $(A^{\nu+2}) = \cdots$.

40. The sequence n, rank (A), rank (A^2), $\cdots$, is convex.

41. Obtain a proof of the Cayley-Hamilton theorem by verifying it for a triangular matrix and applying Exercise 3.

42. Any projector P can be written in the form

$$P = VV^{\mathrm{H}}, \qquad V^{\mathrm{H}}V = I.$$

43. Of the three conditions
  (i) H is Hermitian,
  (ii) H is unitary,
  (iii) H is involutory (i.e., $H^2 = I$),
and any two imply the third. Such a matrix will be called a reflector (see below).

44. If H is a reflector, then $P = (I - H)/2$ is a projector, and conversely, if P is a projector, then $H = I - 2P$ is a reflector. For any a, $-Ha$ is the reflection of a in the space of P.

45. A reflector is expressible as a product of mutually commutative elementary Hermitians, and conversely, any such product is a reflector.

46. Using Exercise 3, show that for any square matrix A and any polynomial $\omega(\lambda)$, the proper values of $\omega(A)$ are $\omega(\lambda_i)$.

47. A matrix of the form

$$M = \begin{pmatrix} 0 & M_{12} & 0 \\ 0 & 0 & M_{23} \\ M_{31} & 0 & 0 \end{pmatrix},$$

or one that can be reduced to that form by a permutational transformation, is said to be cyclic of index 3 ("Property A" considered in Chapter 4 is the property of being cyclic of index 2). Show that if the diagonal null blocks are of order n_1, n_2, and n_3, then

$$\delta(\lambda^{\mathrm{T}} - M) = \lambda^{-2n_1+n_2+n_3}\delta(\lambda^3 I - M_{12}M_{23}M_{31})$$
$$= \lambda^{n_1-2n_2+n_3}\delta(\lambda^3 I - M_{23}M_{31}M_{12})$$
$$= \lambda^{n_1+n_2-2n_3}\delta(\lambda^3 I - M_{31}M_{12}M_{23}).$$

Generalize.

48. Show that with suitable nonsingularity assumptions the following factorization is possible,

$$\begin{pmatrix} P_1 & Q_1 & 0 & 0 & \cdots \\ R_1 & P_2 & Q_2 & 0 & \cdots \\ 0 & R_2 & P_3 & Q_3 & \cdots \\ 0 & 0 & R_3 & P_4 & \cdots \\ \multicolumn{5}{c}{\cdots\cdots\cdots\cdots\cdots} \end{pmatrix}$$

$$= \begin{pmatrix} I & 0 & 0 & 0 & \cdots \\ \Gamma_1 & I & 0 & 0 & \cdots \\ 0 & \Gamma_2 & I & 0 & \cdots \\ 0 & 0 & \Gamma_3 & I & \cdots \\ \multicolumn{5}{c}{\cdots\cdots\cdots\cdots\cdots} \end{pmatrix} \begin{pmatrix} B_1 & Q_1 & 0 & 0 & \cdots \\ 0 & B_2 & Q_2 & 0 & \cdots \\ 0 & 0 & B_3 & Q_3 & \cdots \\ 0 & 0 & 0 & B_4 & \cdots \\ \multicolumn{5}{c}{\cdots\cdots\cdots\cdots\cdots} \end{pmatrix},$$

all blocks being square and of the same order, and exhibit the recursion for finding the B_i and Γ_i. Hence show further that, if all matrices P_i, R_i, Q_i are commutative, and there are ν^2 blocks in each matrix, then the determinant of the matrix on the left is $\delta(\Psi_\nu)$ where

$$\Psi_0 = I,$$
$$\Psi_1 = P_1,$$
$$\Psi_i = P_i\Psi_{i-1} - R_{i-1}Q_{i-1}\Psi_{i-2}.$$

49. Define the polynomials $\psi_\nu(\lambda, \rho)$ by the recursion

$$\psi_0 = 1,$$
$$\psi_1 = \lambda,$$
$$\psi_i = \lambda\psi_{i-1} - \rho^2\psi_{i-2},$$

and show that if $\lambda = 2\rho\cos 2\theta$, then

$$\psi_\nu = \rho^\nu \sin 2(\nu + 1)\theta/\sin 2\theta;$$

hence that if $K = J + J^{\mathrm{T}}$, its proper values are

$$\lambda_\nu(K) = 2\cos 2\theta_\nu, \qquad 2\theta_\nu = 2\nu\theta_1 = \nu\pi/(n + 1)$$

(this matrix arises repeatedly in the numerical solution of differential equations).

50. In Exercise 49, let $R_i = Q_i = -I$, $P_i = B$. Show that the determinant is equal to that of $\psi_\nu(B, I)$, where $\psi_\nu(\lambda, \rho)$ is defined above, and that the characteristic equation is

$$\delta[\psi_\nu(\lambda I - B, I)] = 0.$$

Let β_i be any proper value of B. Then $\psi_\nu(\lambda - \beta_i, 1)$ is a divisor of the characteristic polynomial, hence any proper value of the original matrix is of the form

$$\beta_i + 2 \cos 2\theta_k, \qquad 2\theta_k = k\pi/(\nu + 1)$$

(this matrix arises in the solution of the Laplace equation).

51. The direct product of matrices A and B is defined as

$$A \times B = \begin{pmatrix} \alpha_{11}B & \alpha_{12}B & \alpha_{13}B & \cdots \\ \alpha_{21}B & \alpha_{22}B & \alpha_{23}B & \cdots \\ \alpha_{31}B & \alpha_{32}B & \alpha_{33}B & \cdots \end{pmatrix}.$$

$$\cdots\cdots\cdots\cdots\cdots\cdots$$

Show that, if dimensions conform,

$$(A_1 \times B_1)(A_2 \times B_2) = (A_1 A_2) \times (B_1 B_2).$$

Hence show that:

(i) If A and B are of order n and m, respectively, then

$$\delta(A \times B) = \delta^m(A)\delta^n(B) = \delta(B \times A).$$

(ii) If $\alpha_i = \lambda_i(A)$, $\beta_j = \lambda_j(B)$, then

$$\alpha_i \beta_j = \lambda_{ij}(A \times B) = \lambda_{ij}(B \times A).$$

52. Generalize the last result as follows (Williamson). Let $\alpha_{ij}(\lambda)$ be a rational fraction for every $i, j \leq n$, and let

$$A(\lambda) = (\alpha_{ij}(\lambda)).$$

If $\alpha_{ij}(B)$ exists for every i, j, and if $\beta_i = \lambda_i(B)$, then for every i the proper values of $A(\beta_i)$ are proper values of $A(B)$ (such forms occur in the finite difference approximations to partial differential equations where often the matrix K occurs for B, and often a linear function of it for A).

53. Polynomials $\phi_i(\lambda)$ are said to be orthogonal on the set

$$\lambda_0, \lambda_1, \cdots, \lambda_n$$

in case

$$\sum_{k=0}^{n} \phi_i(\lambda_k)\phi_j(\lambda_k) = 0, \qquad i \neq j.$$

Show that if the polynomials are monic with $\phi_0 = 1$, they satisfy a recursion of the form

$$\phi_1 = (\lambda - \alpha_0)\phi_0,$$
$$\phi_2 = (\lambda - \alpha_1)\phi_1 - \beta_0\phi_0,$$
$$\phi_3 = (\lambda - \alpha_2)\phi_2 - \beta_1\phi_1,$$
$$\cdots\cdots\cdots\cdots\cdots\cdots\cdots\cdots$$

and determine the α_i and β_j. These polynomials are useful in the least-squares fitting of experimental data by means of polynomials.

54. Show that the Lanczos reduction can be carried out (Section 1.5) with

$$AB = B(T - J), \qquad C^H A = C^H(S - J),$$

obtain the forms that correspond to (1.5.8), and write down in detail the recursive relations for obtaining the vectors b_i and c_i with the associated scalars.

CHAPTER 2

Norms, Bounds, and Convergence

2.0. The Notion of a Norm. For many purposes it is convenient to be able to associate with any vector or matrix a single nonnegative scalar that in some sense provides a measure of its magnitude. Such a scalar would be analogous to the modulus of a complex number. The Euclidean length of a vector, and the Euclidean "length" of a matrix, when the matrix is regarded as a vector in n^2-space, are possible choices. We could say that a vector or a matrix is small when its Euclidean length is small.

On considering the scalar as a function of the vector or matrix, and writing down the properties it seems natural for such a function to possess, it is seen that many functions are possible. For a vector the Euclidean length is, indeed, one; for a matrix, there are disadvantages, and it will find little use here, although it occurs frequently in the literature. But even for a vector, the Euclidean length is not the most convenient in all situations.

In this chapter a class of functions called *norms* will be defined geometrically and shown to possess the properties to be expected of a reasonable measure of magnitude of a vector. With each norm will be associated a class of functions of matrices, any one of which might be considered a measure of magnitude, but among which there is a unique one, called the *bound*, whose properties are especially convenient and which will therefore be used most frequently here. The Euclidean length of a matrix is a member of that general class that is associated with the Euclidean length of a vector, but it is not a bound as the word is being used here.

Matrices having no negative elements occur frequently in practice, and also in theoretical discussions, and it turns out that they possess properties of particular interest that can be deduced rather readily from a consideration of certain norms and bounds. Hence the first application of the theory of norms will be made to a study of these matrices. The second application will be to the development of criteria of convergence of series and sequences of matrices and vectors. These questions will recur in later chapters. In Chapter 3, another type of application will be made, in this case to the localization of the characteristic roots of a matrix.

Since norms will be defined geometrically by means of certain convex bodies, it is necessary to begin by summarizing some elementary properties of convex bodies.

2.1. Convex Sets and Convex Bodies.   A *convex set* K, in real or complex space, is a set of points such that

$$x, y \in K, \qquad 0 \leq \lambda \leq 1 \Rightarrow \lambda x + (1 - \lambda)y \in K;$$

hence with every two points of K the entire segment joining them is also in K. A *convex body* in n-space is a closed, bounded, convex set with interior points. If H and K are convex, their intersection $H \cap K$ is convex, and, more generally, the intersection of all sets of any collection of convex sets is convex.

The intersection of all convex sets containing an arbitrary set K is the *convex hull* of K and will be denoted K_c. Since the entire space is convex, there is at least one convex set containing K. If K is convex, then $K = K_c$. As an alternative definition, K_c is the intersection of all half-spaces containing K, where by a half-space is meant

(1) $$S(u, v) = [x \mid \mathrm{Re}\ u^H x \leq v],$$

with u any vector and v any real scalar. The half-space $S(u, v)$ is bounded by and contains the hyperplane

(2) $$P(u, v) = [x \mid \mathrm{Re}\ u^H x = v].$$

If K is bounded, then for any u, there is a least v such that $K \subset S(u, v)$; for this v, $P(u, v)$ contains at least one boundary point of K but no interior points, and is called a *support plane* $P(K; u)$. If $K \nsubseteq P(K; u)$, then $P(K; u)$ divides the space into two half-spaces, with P the common boundary, one containing K and the other containing only boundary points of K.

The union $H \cup K$ of two sets H and K is not necessarily convex even when H and K are, but the convex hull $(H \cup K)_c$ will be of frequent use. For simplicity it will be denoted $H \vee K$. If K consists of the single point k, K will be denoted (k). A *polytope* is the convex hull of a finite number of points $\underset{i}{\vee}(k_i)$.

A set K such that

$$k \in K, \qquad \mid \omega \mid \leq 1 \Rightarrow \omega k \in K$$

is said to be *equilibrated*. For any set K the convex hull of all ωk with $\mid \omega \mid \leq 1$ and $k \in K$ will be denoted $[K]$ and called the *equilibration* of K. In particular $[(x)]$ will be denoted simply $[x]$, and this represents a disk in the complex plane with center at the origin and x on the boundary. The convex hull of any finite number of sets $[k_i]$ will be called an *equilibrated polytope*.

For any two sets H and K, not necessarily distinct, the *Minkowski sum* is

$$H + K = [h + k \mid h \in H, k \in K].$$

If H and K are convex, $H + K$ is convex, and if H or K is a convex body, then $H + K$ is a convex body. Any parallelotope can be represented in the form $\Sigma[h_i]$, where the h_i are linearly independent.

For any matrix A and set K, define

$$AK = [Ak \mid k \in K].$$

If K is convex, AK is convex, and if A is nonsingular and K a convex body, then AK is a convex body. By αK will be meant $(\alpha I)K$.

The following relations are easily verified for H and K convex (Exercise 1):

$$
\begin{aligned}
& (A + B)K \subset AK + BK, \qquad A(H + K) = AH + AK, \\
& A(H \cup K) = AH \cup AK, \qquad A(H \cap K) \subset AH \cap AK, \\
& \alpha \geq 0, \qquad \beta \geq 0 \Rightarrow (\alpha + \beta)K = \alpha K + \beta K, \\
& H \subset K \Rightarrow AH \subset AK.
\end{aligned}
$$

(3)

2.2. Norms and Bounds. If K is an equilibrated convex body, there exists a $\nu > 0$ such that $x \in \nu K$, since the origin is interior to K. Then the *norm* of x with respect to K will be

$$\| x \|_K = \mathrm{glb}\,[\nu \mid \nu \geq 0, x \subset \nu K]$$

and the *least upper bound*, or simply *bound*, of A with respect to K will be

$$\mathrm{lub}_K (A) = \mathrm{glb}\,[\alpha \mid \alpha \geq 0, AK \subset \alpha K].$$

Evidently $x \in \| x \|_K K$ and $AK \subset \mathrm{lub}_K (A)K$. The subscript K will be omitted when there is no ambiguity. The following properties can be verified for any norm and bound:

I. $\quad x \neq 0 \Rightarrow \| x \| > 0, \qquad\qquad A \neq 0 \Rightarrow \mathrm{lub}\,(A) > 0;$

II. $\quad \| \alpha x \| = | \alpha | \| x \|, \qquad\qquad\quad \mathrm{lub}\,(\alpha A) = | \alpha | \mathrm{lub}\,(A);$

III. $\quad \| x + y \| \leq \| x \| + \| y \|, \qquad \mathrm{lub}\,(A + B) \leq \mathrm{lub}\,(A) + \mathrm{lub}\,(B);$

IV. $\quad \| Ax \| \leq \| x \| \mathrm{lub}\,(A), \qquad\quad \mathrm{lub}\,(AB) \leq \mathrm{lub}\,(A)\,\mathrm{lub}\,(B).$

Moreover, *for any matrix A there is at least one $x \neq 0$ such that*

$$\| Ax \| = \| x \| \mathrm{lub}\,(A).$$

Evidently

$$\mathrm{lub}\,(I) = 1.$$

Proof of III will be given, since the other parts are sufficiently evident. Let

$$x \in \nu K, \qquad y \in \mu K.$$

Then

$$x + y \in \nu K + \mu K = (\nu + \mu)K,$$

hence $\| x + y \| \leq \nu + \mu$. By letting $\nu = \| x \|$, $\mu = \| y \|$, this establishes III_1.

Next, let
$$AK \subset \alpha K, \qquad BK \subset \beta K.$$
Then
$$(A + B)K \subset AK + BK \subset \alpha K + \beta K = (\alpha + \beta)K.$$
Hence
$$\text{lub } (A + B) \leq \alpha + \beta,$$
and if $\alpha = \text{lub } A$, $\beta = \text{lub } B$, this establishes III_2.

If K_1 and K_2 are both equilibrated convex bodies, and

(1)
$$K_1 \subset \kappa_1 K_2, \qquad K_2 \subset \kappa_2 K_1,$$
then

(2)
$$\begin{aligned} \| x \|_2 \leq \kappa_1 \| x \|_1, \qquad \text{lub}_2 (A) \leq \kappa_1 \kappa_2 \text{lub}_1 (A), \\ \| x \|_1 \leq \kappa_2 \| x \|_2, \qquad \text{lub}_1 (A) \leq \kappa_1 \kappa_2 \text{lub}_2 (A), \end{aligned}$$

where the numerical subscripts refer to the two convex bodies. Thus, let
$$\alpha_1 = \text{lub}_1 (A), \qquad \alpha_2 = \text{lub}_2 (A).$$
Then the first of inclusions (1) implies that
$$AK_1 \subset \kappa_1 AK_2,$$
and since
$$AK_2 \subset \alpha_2 K_2,$$
therefore
$$AK_1 \subset \alpha_2 \kappa_1 K_2 \subset \alpha_2 \kappa_1 \kappa_2 K_1,$$
because of the second inclusion (1). But this implies that
$$\alpha_1 \leq \alpha_2 \kappa_1 \kappa_2,$$
which is the last of (2). An analogous argument leads to the inequalities on the left. As a particular case,
$$K_1 = \kappa K_2, \qquad \kappa > 0 \Rightarrow \| x \|_2 = \kappa \| x \|_1, \qquad \text{lub}_1 (A) = \text{lub}_2 (A).$$

This theorem implies that all norms are equivalent, in the sense that if the sequence
$$\| x_0 - x \|_{K_1}, \qquad \| x_1 - x \|_{K_1}, \qquad \| x_2 - x \|_{K_1}, \cdots$$
vanishes in the limit, then also the sequence
$$\| x_0 - x \|_{K_2}, \qquad \| x_1 - x \|_{K_2}, \qquad \| x_2 - x \|_{K_2}, \cdots$$
vanishes in the limit, and in either event the sequence of vectors x_i can be said to have the limit x. Analogous remarks apply to matrices. Likewise a sequence that is bounded in one norm is bounded in every norm.

For the special case $K_1 = \kappa K_2$, the vector norms differ by the scalar factor κ but the matrix bounds are identical. In applications (e.g., in solving differential equations) it is sometimes necessary to consider spaces of increasingly high order, and to introduce a factor n^{-1} into the vector norms to keep them bounded. This has no effect on the matrix bounds.

If P is nonsingular, then

(3) $\quad H = PK \Rightarrow \| x \|_H = \| P^{-1}x \|_K, \quad \text{lub}_H (A) = \text{lub}_K (P^{-1}AP).$

Thus, $x \in \| x \|_H H$; hence $P^{-1}x \in \| x \|_H K$. Hence $\| P^{-1}x \|_K \leq \| x \|_H$. Analogously $\| x \|_H \leq \| P^{-1}x \|_K$. This theorem is of particular interest when P is a matrix of proper vectors of A.

Given any bounded real-valued function $\| x \|$ of the elements of a vector x satisfying conditions (I_1), (II_1), and (III_1) (i.e., the first of each of the three pairs of conditions) the set of points

$$K = [x \mid \| x \| \leq 1]$$

is an equilibrated convex body and

$$\| x \|_K = \| x \|.$$

Thus, to prove convexity, if $x \in K$ and $y \in K$, then, for $0 \leq \lambda \leq 1$,

$$\| \lambda x + (1 - \lambda)y \| \leq \| \lambda x \| + \| (1 - \lambda)y \| = \lambda \| x \| + (1 - \lambda) \| y \|$$
$$\leq \lambda + (1 - \lambda) = 1.$$

Hence $\| \lambda x + (1 - \lambda)y \| \leq 1$, which is to say that $\lambda x + (1 - \lambda)y \in K$. Clearly K is equilibrated since, if $x \in K$ and $| \omega | \leq 1$, then $\| \omega x \| = | \omega | \| x \| \leq 1$; hence $\omega x \in K$.

Hence properties (I_1), (II_1), and (III_1) completely characterize a vector norm. It is natural therefore to consider the four properties (I_2), (II_2), (III_2), and (IV_2) as characterizing a matrix norm. Thus consider a real valued function $\| A \|$ of the elements of a matrix A satisfying

(I_2) $A \neq 0 \Rightarrow \| A \| > 0;$

(II_2) $\| \alpha A \| = | \alpha | \| A \|;$

(III_2) $\| A + B \| \leq \| A \| + \| B \|;$

(IV_2) $\| AB \| \leq \| A \| \| B \|.$

Such a function will be called a *matrix norm*. A matrix norm, however, is not necessarily a bound. For example, it can be verified that the Euclidean length of a matrix

$$\tau^{\frac{1}{2}}(A^H A)$$

is a matrix norm, but

$$\tau^{\frac{1}{2}}(I^H I) = \tau^{\frac{1}{2}}(I) = n^{\frac{1}{2}},$$

whereas it has been seen that $\text{lub}_K (I) = 1$ for any K.

If any matrix norm $\| A \|$ and any vector norm $\| x \|$ are related in such a way that for any x and A the condition

(IV$_1$) $\| Ax \| \leq \| A \| \, \| x \|$

is satisfied, then the two norms are said to be *consistent*. It can be verified that the Euclidean matrix norm and the Euclidean vector norm are consistent. Also, for any equilibrated convex body K, lub$_K$ (A), considered as a matrix norm, is consistent with the vector norm $\| x \|_K$. But it is also true that, *given any matrix norm, there exists a consistent vector norm*. In fact, for any vector $k \neq 0$, let

$$\| x \| = \| x k^H \|.$$

Then properties $(I_1) - (IV_1)$ are obvious consequences of $(I_2) - (IV_2)$; the first three of these show $\| x \|$ to be a norm, while the last is the condition for consistency.

Consider, now, any equilibrated convex body K, the associated vector norm $\| x \|$ and bound lub (A), and let $\| A \|$ be any matrix norm consistent with the vector norm. Then *for any matrix A*,

$$\text{lub } (A) \leq \| A \|.$$

In fact, for any A, there exists a vector $x \neq 0$ such that

$$\text{lub } (A) = \| Ax \| \, / \, \| x \|,$$

whereas for any vector x, because of (IV_2),

$$\| A \| \geq \| Ax \| \, / \, \| x \|.$$

This establishes the inequality.

Given an equilibrated convex body K, let

$$K' = \bigcap_{u \in K} S(u, 1),$$

where $S(u, v)$ has been defined in (2.1.1). Otherwise defined,

(4) $K' = [x \mid u \in K \Rightarrow \text{Re } x^H u \leq 1].$

Then K' is a convex body which is closed and equilibrated and is called the *polar* of K.

Since K' is an equilibrated convex body, it is possible to define the norm $\| x \|_{K'}$ and the bound lub$_{K'}$ (A) with respect to K', and since Re $x^H y$ is continuous on the closed, bounded set $\| y \|_K \leq 1$, therefore

(5) $$\| x \|_{K'} = \operatorname*{lub}_{\substack{u \neq 0 \\ u \in K}} \operatorname{Re} x^H u / \| u \|_K = \operatorname*{lub}_{\|u\|_K = 1} \operatorname{Re} x^H u.$$

It follows immediately that, for any x and y,

(6) $$\operatorname{Re} x^H y \leq \| x \|_{K'} \| y \|_K.$$

This will be called the *generalized Cauchy inequality*. Moreover, it can be shown that

(7) $$\| y \|_K = \operatorname*{lub}_{\|v\|_{K'} = 1} \operatorname{Re} v^H y;$$

and since $\operatorname{Re} v^H y = \operatorname{Re} y^H v$, for any y and v, this justifies calling the two norms *dual*, and shows that

$$(K')' = K,$$

which is to say that K *is the polar of its own polar* K'.

If y is held fixed in (6) and x allowed to vary, it follows immediately that

$$\| y \|_K \geq \operatorname*{lub}_{\|v\|_{K'} = 1} \operatorname{Re} v^H y.$$

The reverse inequality can be seen to follow from the convexity of K.

Since K and K' are closed, it follows that for any $x \neq 0$ there exists a $y \neq 0$ and for any $y \neq 0$ there exists an $x \neq 0$ such that the equality holds in (6):

$$\operatorname{Re} x^H y = \| x \|_{K'} \| y \|_K.$$

Vectors x and y so related will be called *dual vectors*.

The bound of A with respect to K is the bound of A^H with respect to K'. In fact, for any vectors x and y, and any matrix A,

(8) $$\operatorname{Re} x^H A y \leq \| x \|_{K'} \| A y \|_K \leq \| x \|_{K'} \| y \|_K \operatorname{lub}_K (A).$$

Select $y \neq 0$ such that the second inequality becomes an equality, then $x \neq 0$ so that the first is also an equality. Hence, for this x and y

$$\operatorname{Re} x^H A y = \| x \|_{K'} \| y \|_K \operatorname{lub}_K (A).$$

But

(9) $$\operatorname{Re} x^H A y \leq \| A^H x \|_{K'} \| y \|_K \leq \| x \|_{K'} \| y \|_K \operatorname{lub}_{K'} (A^H).$$

Hence

$$\operatorname{lub}_K (A) \leq \operatorname{lub}_{K'} (A^H).$$

But (9) holds for any x and y; hence if $x \neq 0$ were chosen to make the second inequality in (9) an equality, and $y \neq 0$ to make the first an equality, then (8) would imply that

$$\operatorname{lub}_{K'} (A^H) \leq \operatorname{lub}_K (A).$$

Therefore

(10) $$\operatorname{lub}_K (A) = \operatorname{lub}_{K'} (A^H).$$

Among the convex bodies of particular interest are the unit sphere

$$S = S' = [x \mid x^H x \leq 1],$$

and the parallelotopes

$$E = \sum_1^n [e_i], \qquad E' = \bigvee_1^n [e_i].$$

As the notation signifies, S is self-polar, and E and E' are mutually polar. Alternative definitions of E and E' are

$$E = [x \mid | \, x \, | \leq e], \qquad E' = [x \mid e^T \mid x \mid \, \leq 1].$$

It can be verified immediately that

$$\| \, x \, \|_E = \max_i | \, \xi_i \, |, \qquad \| \, x \, \|_{E'} = e^T \mid x \mid,$$

and that

$$\text{lub}_E (A) = \| \, | \, A \, | \, e \, \|_E.$$

To compute $\text{lub}_S (A)$ it is necessary to maximize $x^H A^H A x$ subject to the condition $\| \, x \, \|_S{}^2 = x^H x = 1$. But $A^H A$ is Hermitian and at least semi-definite, and hence can be written

$$A^H A = V \Sigma^2 \, V^H$$

where V is unitary and Σ^2 diagonal and real. By taking $x = Vy$, then $y^H y = x^H x = 1$. Thus it is required to maximize $y^H \Sigma^2 \, y$ subject to $y^H y = 1$. But if

$$\Sigma = \text{diag} \, (\sigma_1, \sigma_2, \cdots), \qquad \sigma_1 \geq \sigma_2 \geq \cdots \geq 0,$$

the maximum is $\sigma_1{}^2$; hence

$$\text{lub}_S (A) = \sigma_1 = \rho^{\frac{1}{2}}(A^H A),$$

which is the largest singular value of A. The popularity of the Euclidean matrix norm is due in part to the fact that it is consistent with the Euclidean, or spherical, vector norm, and much more readily computed than is the corresponding bound.

By applying (3), many other convex bodies can be obtained from these three. Thus for any nonsingular matrix P, PS is an ellipsoid, and conversely, any ellipsoid with center at the origin is of the form PS. A norm with respect to an ellipsoid will be called an *ellipsoidal norm*.

If

$$G = \text{diag} \, (\gamma_1, \gamma_2, \cdots, \gamma_n) \geq 0, \qquad \gamma_i > 0,$$

then G is nonsingular, and the polytopes GE and its polar,

$$GE = \sum \gamma_i[e_i], \qquad (GE)' = G^{-1}E' = \bigvee \gamma_i^{-1}[e_i],$$

will be of frequent use (cf. especially Section 2.4). Let

$$Ge = g.$$

Then
$$x \in \nu GE$$
in case
$$|x| \le \nu g.$$
Hence
$$\| x \|_{GE} = \inf [\nu \mid |x| \le \nu g].$$
Likewise
$$\text{lub}_{GE} (A) = \inf [\alpha \mid |A| g \le \alpha g].$$
Again
$$\| x \|_{G^{-1}E'} = g^{\mathrm{T}} |x|,$$
and
$$\text{lub}_{G^{-1}E'} (A) = \inf [\alpha \mid g^{\mathrm{T}} |A| \le \alpha g^{\mathrm{T}}].$$

In view of the role played by the vector g, these norms and bounds will sometimes be called g-norms and g-bounds, in the one case, and g'-norms and g'-bounds in the other, and hence will be designated $\| x \|_g$, $\text{lub}_g (A)$, $\| x \|_{g'}$, $\text{lub}_{g'} (A)$.

Ellipsoidal bounds have some theoretical application (e.g., Chapter 4), even though they are not readily computed. From (3) it follows that $\text{lub}_{PS} (A)$ is the largest singular value of $P^{-1}AP$, therefore the square root of the largest root μ^2 of

$$\delta[(P^{-1}AP)(P^{-1}AP)^{\mathrm{H}} - \mu^2 I] = 0,$$

hence of

(11) $$\delta(AQA^{\mathrm{H}} - \mu^2 Q) = 0, \qquad Q = PP^{\mathrm{H}}$$

Conversely, for any positive definite matrix Q, the largest root μ of (11) is a bound $\text{lub}_K (A)$ of A with respect to some ellipsoid K, since any positive definite matrix Q has a factorization in the form $Q = PP^{\mathrm{H}}$ (Section 1.4), and K can be taken to be PS.

2.3. Norms, Bounds, and Spectral Radii.   Let

$$Ax = \lambda x, \qquad x \ne 0,$$

λ being a proper value of A. Then in any norm

$$\| \lambda x \| = |\lambda| \| x \| \le \| x \| \text{lub} (A) \le \| x \| \| A \|,$$

consistency of the norms presupposed. Hence

$$|\lambda| \le \text{lub} (A) \le \| A \|,$$

and therefore

(1) $$\rho(A) \le \text{lub} (A) \le \| A \|.$$

Thus the spectral radius $\rho(A)$ of a matrix cannot exceed the value of any of its norms.   On the other hand, *for any particular matrix A and any $\epsilon > 0$, there exists a convex body K for which*

(2) $$\text{lub}_K (A) \le \epsilon + \rho(A).$$

It is sufficient to take A in Jordan normal form because of (2.2.3), and to consider as an extreme case the matrix

$$\lambda I + J,$$

where J is defined in Section 1.0.   Let

$$W(\epsilon) = \text{diag}\,(1, \epsilon, \epsilon^2, \cdots).$$

Then

$$W(\epsilon)(\lambda I + J)W^{-1}(\epsilon) = \begin{pmatrix} \lambda & 0 & 0 & \cdots \\ \epsilon & \lambda & 0 & \cdots \\ 0 & \epsilon & \lambda & \cdots \\ \cdot & \cdot & \cdot & \cdot \end{pmatrix} = \lambda I + \epsilon J;$$

hence $\text{lub}_E [W(\lambda I + J)W^{-1}] = \text{lub}_{E'} [W(\lambda I + J)W^{-1}] = |\lambda| + \epsilon$.   This shows that K can be taken to be a polytope.   But K can as well be taken to be an ellipsoid.   In fact, by direct evaluation

$$\text{lub}_E [(\lambda I + \epsilon J)(\lambda I + \epsilon J)^H] = |\lambda|^2 + 2|\lambda\epsilon| + |\epsilon|^2,$$

hence

$$\rho[(\lambda I + \epsilon J)(\lambda I + \epsilon J)^H] \le (|\lambda| + |\epsilon|)^2,$$

and consequently no singular value of $\lambda I + \epsilon J$ can exceed $|\lambda| + |\epsilon|$.

A matrix A will be said to be *of class M* in case there exists a convex body K such that

(3) $$\text{lub}_K (A) = \rho(A).$$

For a diagonal matrix D,

(4) $$\text{lub}_E (D) = \text{lub}_{E'} (D) = \text{lub}_S (D) = \rho(D).$$

Hence any diagonal matrix is of class M, and therefore so is any diagonalizable matrix (i.e., a matrix similar to a diagonal matrix).   On the other hand, $\rho(J) = 0$ whereas in any norm whatever, $\|J\| > 0$.   Hence not all matrices are of class M.

Any convex body K for which (3) holds will be called *minimizing for A*, and the norm will be called *minimal for A*.   It is readily seen that the sphere is minimizing for any normal matrix.   Any matrix for which K is minimizing will be said to *belong to K*.   In case A and B belong to K,

$$\rho(AB) \le \rho(A)\rho(B),$$

(5)

$$\rho(\alpha A + \beta B) \le |\alpha|\,\rho(A) + |\beta|\,\rho(B),$$

although it is not in general true that AB or $\alpha A + \beta B$ also belong to K.

Evidently if A belongs to K, then $P^{-1}AP$ belongs to $P^{-1}K$ because of (2.2.3).

From (4) it is seen that any diagonal matrix belongs to S, to E, and to E', and, in fact, to any K of the form GS, GE, $G^{-1}E'$, where G is a nonsingular diagonal matrix. If K is any equilibrated convex body to which every diagonal matrix belongs, the associated norm and bound will be called *monotonic*, or *absolute*. Such norms possess the following properties:

(i) $D = \mathrm{diag}\,(\delta_1, \delta_2, \cdots, \delta_n) \Rightarrow \mathrm{lub}\,(D) = \max_i |\,\delta_i\,|$;

(ii) $\|\,x\,\| = \|\,|\,x\,|\,\|$;

(iii) $|\,x\,| \le |\,y\,| \Rightarrow \|\,x\,\| \le \|\,y\,\|$;

moreover, *these properties are equivalent.*

For any x, there exists a diagonal D with $|\,D\,| = I$ such that $|\,x\,| = Dx$. If (i) holds, then

$$\mathrm{lub}\,(D) = \mathrm{lub}\,(D^{-1}) = 1.$$

Hence $\|\,|\,x\,|\,\| \le \|\,x\,\|$. But $x = D^{-1}\,|\,x\,|$, hence $\|\,x\,\| \le \|\,|\,x\,|\,\|$. Thus (i) implies (ii).

Suppose (ii) holds, and consider any y for which $\|\,y\,\| = 1$. For any diagonal matrix D with $|\,D\,| = I$, evidently $|\,y\,| = |\,Dy\,|$, hence $\|\,y\,\| = \|\,|\,y\,|\,\| = \|\,Dy\,\|$, and therefore Dy lies on the boundary of K. Hence if $|\,y\,| \ge |\,x\,|$, then the point x can only lie within or on the boundary of K. Hence $\|\,x\,\| \le \|\,y\,\|$, and therefore (ii) implies (iii).

To prove that (iii) implies (i) it can be supposed that $D \ne 0$, and it is no restriction to suppose that

$$\rho(D) = \max |\,\delta_i\,| = 1.$$

Then $|\,Dy\,| \le |\,y\,|$ for any y. Hence

$$\|\,Dy\,\| \le \|\,y\,\|$$

for any y, and therefore

$$\mathrm{lub}\,(D) \le 1.$$

But necessarily

$$\mathrm{lub}\,(D) \ge 1 = \rho(D).$$

Hence D belongs to K and (iii) implies (i). This completes the proof of equivalence.

Any matrix that is similar to a diagonal matrix is of class M, but not conversely. However, *the matrix A is of class M if and only if for every $\lambda_i = \lambda_i(A)$ such that $|\,\lambda_i\,| = \rho(A)$, the number of linearly independent proper vectors belonging to λ_i is equal to its multiplicity.* In fact, if the condition holds, A is similar to a matrix of the form $A' = \mathrm{diag}\,(A_1, A_2)$ where A_1 is diagonal, A_2 is square, and

$$\rho(A_1) > \rho(A_2).$$

In the space of A_1, let K_1 define a monotonic norm; in the space of A_2 there exists a K_2 such that

$$\text{lub}_{K_2} (A_2) \leq \rho(A_1).$$

Let $K = K_1 + K_2$. Then K is an equilibrated convex body and

$$\text{lub}_K (A') = \rho(A_1).$$

To prove that the condition is necessary, it is sufficient to consider the matrix $\lambda I + J$. But J is known not to be of class M, hence assume $\lambda \neq 0$, and there is no restriction in supposing $\lambda = 1$. Suppose K is minimizing for $I + J$ and let $\| x \|_K \leq 1$. Then $\| (I + J)^r x \|_K \leq 1$ for any ν; hence $(I + J)^r x \in K$ for any ν and for any $x \in K$. But K is bounded, whereas the sequence of points $(I + J)^r x$ is unbounded for any $x \neq e_n$.

Somewhat analogous to the (least upper) bound of a matrix is the greatest lower bound associated with any equilibrated convex body K:

(6) $$\text{glb}_K (A) = \sup [\alpha \mid \alpha K \subset AK].$$

Evidently for any x,

(7) $$\| Ax \|_K \geq \| x \|_K \text{glb}_K (A);$$

if A is singular, $\text{glb}_K (A) = 0$, and otherwise

(8) $$\text{glb}_K (A) = 1/\text{lub}_K (A^{-1}).$$

Consequently, for any equilibrated convex body K, any matrix A, and any proper value $\lambda(A)$,

(9) $$\text{glb}_K (A) \leq | \lambda(A) |.$$

Thus the circular ring about the origin bounded by the circles whose radii are $\text{glb}_K (A)$ and $\text{lub}_K (A)$ contain all proper values of A within or on the boundary. This principle will be developed in the next chapter to provide other regions known to contain some, all, or none of the roots of a matrix A.

2.4. Nonnegative Matrices.

An important subclass of the matrices of class M is the class of irreducible nonnegative matrices. They are important not only because they occur frequently in practice, but also for theoretical reasons, stemming largely from the fact that, for any matrix A, $| A |$ is nonnegative, and

$$\rho(A) \leq \rho(| A |),$$

as will be shown later.

A square matrix A is *reducible* in case there exists a permutation matrix P such that

$$P^T A P = \begin{pmatrix} A_{11} & A_{12} \\ 0 & A_{22} \end{pmatrix},$$

where A_{11} and A_{22} are square submatrices; otherwise it is *irreducible*. A *nonnegative matrix* B is a real matrix having no negative elements. A well-known *theorem of Frobenius* will now be proved: *If B is nonnegative and irreducible, then $\beta = \rho(B)$ is a proper value which is simple, and belonging to it is a strictly positive proper vector.* Further, it will follow that B *is of class M, and minimizing for it is an equilibrated polytope of the form GE, where $G \geq 0$ is nonsingular and diagonal.* By applying the same argument to B^T it will follow that also *minimizing for B is a polytope of the form $G'E'$, where $G' \geq 0$ is also nonsingular and diagonal.*

First it will be shown that, if there exists a nonnegative proper vector b for B, then $b > 0$, and b belongs to a proper value β which is equal to $\rho(B)$. Indeed, if

$$(1) \qquad\qquad Bb = \beta b,$$

then clearly $\beta > 0$. Suppose, then, that b has null elements. Apply a permutational transformation if necessary so that the first i elements of b are strictly positive and the last $n - i$ null. Let

$$b^T = (b_1{}^T, 0), \qquad b_1 > 0,$$

and partition B to correspond. Then

$$B_{11}b_1 = \beta b_1,$$
$$B_{21}b_1 = 0.$$

But B is irreducible and $b_1 > 0$, hence $B_{21}b_1 \neq 0$. The contradiction shows that $b > 0$, hence

$$\beta = \text{lub}_b\,(B) \geq \rho(B),$$

but

$$\beta = \lambda(B) \leq \rho(B)$$

and therefore

$$\beta = \rho(B).$$

Consider the set Σ of all vectors $g \geq 0$ such that

$$1 \leq \| g \|_e \leq 2.$$

Evidently Σ is bounded and closed. The set of all scalars γ such that

$$Bg \leq \gamma g, \qquad g \in \Sigma$$

achieves its greatest lower bound for some $g \in \Sigma$, and it can be achieved only for some strictly positive g, since as any element of g vanishes, the least admissible γ becomes infinite. Hence the greatest lower bound cannot be less than $\rho(B)$. It will now be shown that, for any $g > 0$, if

$$(2) \qquad\qquad Bg \leq \gamma g, \qquad Bg \neq \gamma g,$$

then there can be found a g' and a $\gamma' < \gamma$ such that

$$Bg' \leq \gamma'g'.$$

If g' is not in Σ, some positive scalar multiple will be.

Suppose, then, that (2) is satisfied. If $Bg < \gamma g$, then $g' = g$ is effective, since there exists a $\gamma' < \gamma$ for which $Bg \leq \gamma' g$. Hence suppose that Bg and γg are equal in at least one element. Apply a permutational transformation if necessary, so that only the last $n - i$ elements are equal. With appropriate partitioning, then, it can be supposed that

(3)
$$B_{11}g_1 + B_{12}g_2 < \gamma g_1,$$
$$B_{21}g_1 + B_{22}g_2 = \gamma g_2.$$

Since B is irreducible, $B_{21} \neq 0$; hence there is at least one element of g_1, which, if diminished, will replace at least one of the $n - i$ equalities by an inequality, and if it is not diminished too much, none of the existing inequalities will become an equality. Thus $g = g^{(0)}$ can be replaced by a vector $g^{(1)} > 0$ such that

$$Bg^{(1)} \leq \gamma g^{(1)},$$

and among the n inequalities at least $i + 1$ are strict. If not all are strict, the process can be repeated, and after at most $n - i$ applications of the process, a vector $g' > 0$ results as required.

It follows that the minimizing vector g can only be equal to a vector b satisfying (1). If there is also a vector $b' > 0$ such that

$$Bb' = \beta' b',$$

then $\beta' = \rho(B) = \beta$. If $b' \neq b$, there is a greatest scalar α such that $b - \alpha b' \geq 0$. But b and b' are both proper vectors belonging to β, hence $b - \alpha b'$ is a proper vector belonging to β, and since $b - \alpha b'$ has at least one vanishing element and all other elements are nonnegative, therefore $b - \alpha b' = 0$. Thus any nonnegative proper vector is a scalar multiple of b. Moreover, since B is of class M, the number of independent proper vectors belonging to $\beta = \rho(B)$ is equal to its multiplicity (Section 2.3). Hence β is a simple root.

It is now easy to see that, if $|A| \leq B$, then $\rho(A) \leq \rho(B)$. In fact, if B is irreducible, $\beta = \rho(B)$, and $Bb = \beta b > 0$, then $|A| b \leq \beta b$; hence $\text{lub}_b (A) \leq \beta$. If B is reducible and $\epsilon > 0$, then $B + \epsilon e e^{\mathrm{T}}$ is irreducible. Moreover $|A + \epsilon e e^{\mathrm{T}}| \leq B + \epsilon e e^{\mathrm{T}}$. Hence

$$\rho(A + \epsilon e e^{\mathrm{T}}) \leq \rho(B + \epsilon e e^{\mathrm{T}})$$

for every $\epsilon > 0$, hence the inequality holds also for $\epsilon = 0$.

However, a much stronger statement can be made in case B is irreducible. In fact, *if* $|A| \leq B$ *and* B *is irreducible, then* $\rho(A) = \rho(B)$ *if and only if* A *is of the form*

(4)
$$A = \omega D B D^{-1}, \qquad |\omega| = 1, \qquad |D| = I,$$

and $\omega\beta$ *is a proper value of* A. This is Wielandt's lemma.

To prove it, suppose $\rho(A) = \rho(B)$. Then A has a proper value λ such

that $|\lambda| = \beta = \rho(B)$. Let
$$a^H A = \lambda a^H.$$
Then
$$|\lambda a^H| = \beta |a^H| \le |a^H| |A| \le |a^H| B,$$
hence for any $g > 0$,
$$\beta |a^H| g \le |a^H| Bg.$$
But in particular if $Bg = \beta g$, then
$$|a^H| Bg = \beta |a^H| g,$$
and equalities must hold throughout. But then
$$|a^H| B = \beta |a^H|;$$
hence $|a^H| > 0$, and since
$$|a^H| (B - |A|) = 0,$$
therefore
$$B = |A|.$$
If the diagonal D is chosen so that $a^H D = |a^H|$, and if $\lambda = \omega\beta$, then
$$|a^H| D^{-1} A D = \lambda |a^H| = \omega\beta |a^H|,$$
and therefore
$$|a^H| \omega^{-1} D^{-1} A D = \beta |a^H| > 0,$$
and, since $|D^{-1}AD| - |A| - B$, it follows that
$$\big||a^H| D^{-1} A D\big| = |a^H| |D^{-1}AD|.$$
Hence $\omega^{-1} D^{-1} A D$ can only be real and nonnegative, and hence equal to B.

It follows as a corollary that *if A and B are both nonnegative, $A \le B$, $B irreducible, and $\rho(A) = \rho(B)$, then $A = B$.* In particular *if A is any submatrix of B,* not necessarily a principal submatrix, obtained from B by making zero certain rows and columns, *then the spectral radius of A is strictly less than that of B when B itself is irreducible.*

A *stochastic matrix S* satisfies
$$S \ge 0, \qquad Se = e.$$
Any irreducible nonnegative matrix is similar to a scalar multiple of a stochastic matrix. For suppose $B \ge 0$ and is irreducible, and let
$$Bg = \beta g > 0.$$
If the diagonal matrix G is defined by $Ge = g$, then
$$\beta^{-1} G^{-1} B G$$
is stochastic.

If S is stochastic, let σ be any proper value and x the proper vector belonging to it:
$$Sx = \sigma x.$$

Let Q be the convex hull of the elements ξ_i considered as points in the complex plane. Then each element of Sx is a point of Q, and hence $\sigma Q \subset Q$. Conversely if, for any scalar σ, there exists a polygon Q of n vertices such that $\sigma Q \subset Q$, then σ is a proper value of some stochastic matrix of order n. To see this let x be a vector whose elements are the vertices of Q. Since each element of σx is in Q by hypothesis, each element of σx is expressible as an arithmetic mean of the elements of x. But this is to say that there exists a matrix $S \geq 0$ such that the two relations

$$Se = e, \qquad Sx = \sigma x$$

are satisfied. This theorem is due to Dmitriev and Dynkin (1946).

If S is irreducible and stochastic, $\rho(S) = 1$, and $\lambda = 1$ is a simple proper value. Suppose, however, there is a proper value $\sigma \neq 1$ and $|\sigma| = 1$. Let

$$Sx = \sigma x.$$

Then if Q is the convex hull of the ξ_i, $\sigma Q \subset Q$, and hence $\sigma^\nu Q \subset Q$ for any integer $\nu \geq 0$. But σ must be a rotation, hence a root of unity of degree n or less. Let

$$\sigma = \exp 2\pi i/p, \qquad p \leq n.$$

Since $\sigma^{-1}S$ belongs to E and has 1 as a proper value, by Wielandt's lemma

$$(5) \qquad\qquad S = \sigma DSD^{-1}, \qquad |D| = I.$$

Hence

$$\sigma^{-1}D^{-1}SDe = e,$$
$$SDe = \sigma De$$

and therefore De is a proper vector belonging to σ. But also

$$SD^2e = \sigma^2 D^2 e,$$

hence σ^2 is a proper value and D^2e is a proper vector belonging to it. A continuation of the argument shows that $\sigma, \sigma^2, \cdots, \sigma^p = 1$ are all proper values of S. Moreover the vertices of Q are the diagonal elements of D, Q is a regular polygon of νp vertices where $\nu p \leq n$, and one of these vertices is at 1.

It follows from this that after a possible permutational transformation of S, the matrix

$$D = \begin{pmatrix} I & 0 & 0 & 0 & \cdots \\ 0 & \delta I & 0 & 0 & \cdots \\ 0 & 0 & \delta^2 I & 0 & \cdots \\ \cdot & \cdot & \cdot & \cdot & \cdot \end{pmatrix}$$

where, however, the I's are not necessarily of the same order. Here δ is a primitive (νp)th root of unity, and every power of δ appears up to and including the power $\nu p - 1$. If S is partitioned correspondingly,

$$S = \begin{pmatrix} S_{11} & S_{12} & S_{13} & \cdots \\ S_{21} & S_{22} & S_{23} & \cdots \\ \cdot & \cdot & \cdot & \cdot \end{pmatrix}$$

and (5) applied, it develops that only $S_{i,i+\nu} \neq 0$. But if (1.4.16) is applied repeatedly to reduce $\delta(\lambda I - S)$, it can be seen that a matrix of this form has δ as a proper value (Exercise 1.48). Hence $\nu = 1$, and

(6)
$$S = \begin{pmatrix} 0 & S_{12} & 0 & 0 & \cdots \\ 0 & 0 & S_{23} & 0 & \cdots \\ \cdot & \cdot & \cdot & \cdot & \cdot \\ S_{p1} & 0 & 0 & 0 & \cdots \end{pmatrix}.$$

Such a matrix is said to be *cyclic of index p*.

More generally, *if $B \geq 0$ is irreducible with $\beta = \rho(B)$, and has $p - 1$ proper values $\sigma\beta \neq \beta$ with $|\sigma| = 1$, then σ is a pth root of unity and after a possible permutational transformation B can be given the form*

(7)
$$B = \begin{pmatrix} 0 & B_{12} & 0 & 0 & \cdots \\ 0 & 0 & B_{23} & 0 & \cdots \\ \cdot & \cdot & \cdot & \cdot & \cdot \\ B_{p1} & 0 & 0 & 0 & \cdots \end{pmatrix}.$$

2.5. Convergence; Functions of Matrices.
From (2.1.1) it follows that any norm is a continuous function of the elements of a vector or matrix. In particular, if the sequence of vectors x_ν is such that in any norm $\| x_\nu \| \to 0$, then this is true in any other norm, including $\| x_\nu \|_e$; hence in the limit all elements of the x_ν vanish, and conversely. The same applies to a sequence of matrices.

If the matrices A_ν of a sequence are such that the vectors $x_\nu = A_\nu x$ vanish in the limit for whatever x, then the matrices A_ν vanish in the limit, since, in particular, the sequence must vanish when x is any column of I. The case $A_\nu = A^\nu$, the sequence being of powers of a given matrix, will be of particular interest, as will power series, and sequences of matrices of the form

(1)
$$(A^H)^\nu C A^\nu,$$

where C is a constant matrix.

If, with any norm, $\| A \| < 1$, then by property (IV$_1$) it is clear that $A^\nu \to 0$. But this implies that $\rho(A) < 1$. Moreover, if $\rho(A) < 1$, then there exists some norm for which $\| A \| < 1$ and therefore $A^\nu \to 0$. Conversely, let $\rho(A) \geq 1$. Since $\rho(A^\nu) = \rho^\nu(A)$, it follows that $\rho(A^\nu) \geq 1$ for every $\nu \geq 0$. Hence A^ν cannot vanish in the limit.

Next, consider sequences of products of the form (1). It is sufficient to show that such a sequence vanishes in the limit if $\rho(A) < 1$. There exists, then, a convex body K such that

$$\operatorname{lub}_K (A) < 1.$$

Let K' be the polar of K, and let

$$K \subset \kappa K', \qquad K' \subset \kappa' K.$$

Then

$$\operatorname{lub}_K (A^H) = \operatorname{lub}_{K'} (A) \le \kappa \kappa' \operatorname{lub}_K (A)$$

and the same applies to any power. Hence

$$\operatorname{lub}_K [(A^H)^\nu C A^\nu] \le \kappa \kappa' \operatorname{lub}_K (C) \operatorname{lub}_K^{2\nu} (A),$$

and the assertion is proved.

Now take the sequence of partial sums

$$S_\nu = I + A + A^2 \cdots + A^\nu.$$

It will be shown that the partial sums S_ν have a limit S if and only if $\rho(A) < 1$ and in this event $I - A$ is nonsingular and

$$S = (I - A)^{-1}.$$

It is clear that if $\rho(A) < 1$, $I - A$ is nonsingular, since then $1 - \lambda(A) = \lambda(I - A) \ne 0$ for every proper value $\lambda(A)$. Suppose $\rho(A) < 1$. Then

$$(I - A)^{-1} - S_\nu = (I - A)^{-1} A^{\nu+1},$$

$$\| (I - A)^{-1} - S_\nu \| \le \| (I - A)^{-1} \| \, \| A \|^{\nu+1}$$

for any norm, and for some norm $\| A \| < 1$. Hence S_ν has the limit $(I - A)^{-1}$. Conversely, if $\rho(A) \ge 1$, then $\rho(A^\nu) = \rho^\nu(A) \ge 1$ and the series cannot converge.

In general, if the series

$$\psi(\lambda) = \alpha_0 + \alpha_1 \lambda + \alpha_2 \lambda^2 + \cdots$$

converges in any circle about the origin of radius exceeding $\rho(A)$, then the series

$$\alpha_0 + \alpha_1 A + \alpha_2 A^2 + \cdots$$

converges and its limit may be taken to define $\psi(A)$. In particular, $\exp A$ is defined for every A, and $\log (I - A)$ is defined in case $\rho(A) < 1$.

To return to the geometric series, let $\alpha = \| A \| < 1$. Then

(2) $$\| (I - A)^{-1} \| \le 1 + \alpha + \alpha^2 + \cdots = (1 - \alpha)^{-1},$$

and

(3) $$\| (I - A)^{-1} - S_\nu \| \le \alpha^{\nu+1}/(1 - \alpha).$$

Now consider a nonsingular matrix A, and let C represent any supposed approximation to A^{-1}. Let

(4) $$H = I - AC, \qquad \| H \| < 1.$$

Then it is readily verified that

$$\| A^{-1} \| \leq \| C \|/(1 - \| H \|),$$
(5) $$\| A^{-1} - C \| \leq \| CH \|/(1 - \| H \|),$$
$$\| A - C^{-1} \| \leq \| HA \|/(1 - \| H \|).$$

In fact, the proof is made by observing that

$$A^{-1} = C(I - H)^{-1} = C(I + H + H^2 + \cdots),$$
$$C^{-1} = (I - H)^{-1}A = (I + H + H^2 + \cdots)A,$$

the series being convergent. These relations give rigorous bounds for the errors in an approximate inverse.

REFERENCES

The notion of norms is of fundamental importance in functional analysis, but until recently does not often appear in the literature of numerical analysis or of the theory of matrices. Nevertheless, Kantorovich (Kantorovitch 1939, Kantorovič 1948) and Lonseth (1947) use norms effectively, spectral norms are used by von Neumann and Goldstine (1947) but with a different name, and Faddeeva (1950) uses them in convergence proofs. With any norm can be associated a function defining a distance between two points by taking

$$\delta(x, y) = \delta(y, x) = \| x - y \|; \qquad \| x \| = \delta(0, x).$$

Weissinger (1951) introduces norms through distances.

Matrix norms are defined in many different ways in the older literature, but the favorite was the Euclidean norm of the matrix considered as a vector in n^2-space. Wedderburn (1934) calls this the absolute value of the matrix and traces the idea back to Peano in 1887. Other definitions are given by von Neumann (1937), Bowker (1947), Wong (1954), and Ostrowski (1955).

Norms which are not necessarily real numbers, but are required only to be partially ordered, are considered by Kantorovitch (1939) and Schröder (1956). The association of norms with convex bodies was made for abstract spaces by Kolmogoroff (1934), and a further development in the spirit of the present chapter was given by Householder (1958). In de la Garza (1953), the suggestion to adapt the norm to the purpose is implicit, although only polyhedral norms occur. More general and systematic treatments are made by Gautschi (1954) and Ostrowski (1955).

It is apparent that polyhedral norms are intimately associated with the theory of nonnegative matrices, for which the basic results are due to Frobenius (1908, 1909, 1912). For these, and especially for stochastic matrices, there is extensive literature, and probably the best single source is Gantmaher (1954). Special mention should be made, however, of Wielandt (1950).

The unitarily invariant norms (cf. Exercises) were discussed by von Neumann (1937), and applied by Fan and Hoffman (1955). The notion of absolute or monotonic norm is due to Bauer, Stoer, and Witzgall (1961). Fiedler and Pták (1960) introduced a subclass of monotonic norms for developing exclusion theorems. For minimal norms (with respect to a given matrix) see Householder (1959).

For the closely related theory of convex bodies see Bonnesen and Fenchel (1934), Eggleston (1958), Lyusternik (1956), and Yaglom and Boltyanskii (1951).

The algebraic inequality $x^T y \leq \| x \|_s \| y \|_s$ for real vectors is due to Cauchy (1821). In the literature the inequality is often referred to as the Schwarz inequality, or, in Russia, the Bunyakovski inequality, since Schwarz and Bunyakovski extended it to more general operators. Since only the original algebraic form enters here, it seems appropriate to attach Cauchy's name.

The notion of dual norm was developed by von Neumann (1937), but the notion is essentially equivalent to that of polar sets, for which see Bonnesen and Fenchel (1934).

Little use is made here of functions of matrices, but they play an important role in the solution of systems of ordinary differential equations. See Gantmaher (1954), Bellman (1960), and Rinehart (1955, 1956).

Norms have their most obvious application to questions of convergence, which are introduced here but will recur repeatedly in connection with iterative methods. But intimately associated is a classical problem of nonsingularity of certain matrices, or the nonvanishing of certain determinants, since if the series

$$I + B + B^2 + B^3 + \cdots$$

converges, then it converges to the inverse of $I - B$, which is therefore nonsingular. More generally, therefore, $D - B$ is assumed to be nonsingular if $\rho(D^{-1}B) < 1$, hence if $\| D^{-1}B \| < 1$: Taussky (1949). This, in turn, leads to the determination of inclusion and of exclusion theorems for proper values, as will be developed in the next chapter. Closely related to questions of singularity of matrices is the problem of bounding a determinant. In this connection see Price (1951), and a number of papers by Ostrowski, of which only some are listed.

PROBLEMS AND EXERCISES

1. Prove in detail the relations (2.1.3), and show by a counterexample that the equalities

$$(A + B)K = AK + BK, \qquad A(H \cap K) = AH \cap AK$$

do not hold in general. Show also that $\alpha K + \beta K \subset (\alpha + \beta)K$ fails for K not convex.

2. Show in detail that the Euclidean length of a matrix,

$$\| A \| = \tau^{\frac{1}{2}}(A^H A)$$

is consistent with the Euclidean vector norm and is a matrix norm. Show also that if the singular values of A are α_i, then

$$\| A \|^2 = \Sigma \, \alpha_i^2.$$

3. Let K be a bounded, closed, convex body, which contains the origin in its interior, but which is not necessarily equilibrated; define $\| x \|_K$ and $\mathrm{lub}_K (A)$ as in Section 2.2. Show that then (I), (III), and (IV) are satisfied, but that (II) is replaced by the weaker condition

(II′) $\alpha \geq 0 \Rightarrow \| \alpha x \|_K = \alpha \| x \|_K, \qquad \mathrm{lub}_K (\alpha A) = \alpha \, \mathrm{lub}_K (A).$

Moreover, for any matrix A there exists a vector $x \neq 0$ such that

$$\| Ax \|_K = \| x \|_K \, \mathrm{lub}_K (A).$$

4. Let the function $\| x \|$ satisfy (I_1), $(II_1′)$, and (III_1). Show that

$$K = [x \mid \| x \| \leq 1]$$

is a bounded, closed, convex body which contains the origin as an interior point.

5. With K as in Exercise 3, show that for any A,

$$\rho(A) \leq \mathrm{lub}_K (A).$$

6. For any matrix A, if there exists a bounded, closed, convex body K which contains the origin as an interior point, and is such that

$$\mathrm{lub}_K (A) = \rho(A),$$

then there exists an equilibrated, bounded, closed convex body H such that

$$\mathrm{lub}_H (A) = \rho(A).$$

7. Let $\Sigma_1, \Sigma_2, \Sigma_3, \cdots$ be Euclidean spaces of arbitrary dimensionality. Let $K_i \subset \Sigma_i$ be a bounded, closed, convex body which contains the origin in its interior. Let A_{ij} be a matrix such that if $x_j \in \Sigma_j$, then $A_{ij} x_j \in \Sigma_i$. Define

$$\mathrm{lub}_{ij} (A_{ij}) = \inf [\alpha \mid \alpha \geq 0, \, A_{ij} K_j \subset \alpha K_i].$$

Show that:

(I) $A_{ij} \neq 0 \Rightarrow \text{lub}_{ij}(A_{ij}) > 0;$

(II') $\alpha \geq 0 \Rightarrow \text{lub}_{ij}(\alpha A_{ij}) = \alpha \, \text{lub}_{ij}(A_{ij});$

(III) $\text{lub}_{ij}(A_{ij} + B_{ij}) \leq \text{lub}_{ij}(A_{ij}) + \text{lub}_{ij}(B_{ij});$

(IV) $\| A_{ij}x_j \|_{K_i} \leq \| x_j \|_{K_j} \text{lub}_{ij}(A_{ij}); \, \text{lub}_{ik}(A_{ij}A_{jk}) \leq \text{lub}_{ij}(A_{ij}) \text{lub}_{jk}(A_{jk}).$

8. Prove that $\rho(A) < 1$ if and only if there exists a positive definite matrix Q such that $Q - AQA^{\text{H}}$ is also positive definite (Stein).

9. If A is nonsingular, there exists a K such that $\text{glb}_K(A)$ is equal to the least of the moduli of the proper values of A if and only if the number of linearly independent proper vectors belonging to each proper value of minimal modulus is equal to its multiplicity.

10. If A is irreducible and nonnegative and $x \geq 0$ is a proper vector, then x belongs to $\lambda = \rho(A)$.

11. Let $A \geq 0$,
$$A_i = (I - e_i e_i^{\text{T}})A(I - e_i e_i^{\text{T}}),$$
$$\rho = \rho(A), \qquad \rho_i = \rho(A_i).$$
Then
$$\mu > \rho_i, \qquad \delta(\mu I - A) > 0 \Rightarrow \mu > \rho.$$

12. If $A \geq 0$ and $\rho = \rho(A)$, then the following conditions are equivalent:

(i) $\mu > \rho;$

(ii) $\delta(\mu I - A) \neq 0$ and $(\mu I - A)^{-1} \geq 0;$

(iii) there exists a $g > 0$ such that $\mu g > Ag;$

(iv)   every principal submatrix of $\mu I - A$ has a positive determinant;

(v)   every leading principal submatrix of $\mu I - A$ has a positive determinant [a *leading principal submatrix* of order i is that one formed from the first i rows and the first i columns].

13. If $A \geq 0$ is irreducible, then $(\mu I - A)^{-1} > 0$ if and only if $\mu > \rho(A)$.

14. Let $D \geq 0$ be diagonal and $C \geq 0$. Then the following conditions are equivalent (a matrix $D - C$ which satisfies them is called an M-matrix, not to be confused, however, with a matrix of class M):

(i)   Re $\lambda_i(D - C) > 0$ for every $i;$

(ii) $D - C$ is nonsingular and $(D - C)^{-1} \geq 0;$

(iii) there exists a vector $g > 0$ such that $Dg > Cg;$

(iv)   every principal submatrix of $D - C$ has a positive determinant;

(v)   every leading principal submatrix of $D - C$ has a positive determinant.

15. If D and C satisfy the conditions of 14, $| F | \geq D$, and $| K | \leq C$, then $F - K$ is nonsingular, $| (F - K)^{-1} | \leq (D - C)^{-1}$, and $| \delta(F - K) | \geq \delta(D - C)$.

16. If D and C satisfy the conditions in Exercise 14,
$$D = \begin{pmatrix} D_2 & 0 \\ 0 & D_2 \end{pmatrix}, \qquad C = \begin{pmatrix} C_{11} & C_{12} \\ C_{21} & C_{22} \end{pmatrix},$$
then $\delta(D - C) \leq \delta(D_1 - C_{11})\delta(D_2 - C_{22})$.

17. A matrix $A \geq 0$ is *primitive* in case it is irreducible and $|\lambda_i| = \rho(A) \Rightarrow \lambda_i = \rho(A)$. If A is primitive, then A^σ is primitive for every $\sigma > 0$.

18. If $A \geq 0$ is irreducible, then $(\epsilon I + A)^{n-1} > 0$ for every $\epsilon > 0$.

19. If $A \geq 0$ is irreducible and $D \geq 0$ is nonsingular diagonal, then $(D + A)^{n-1} > 0$.

20. If A is irreducible, then for any $x \geq 0$, $y \geq 0$, with $x \neq 0$ and $y \neq 0$, there exists a $\sigma > 0$ such that $x^\mathrm{T} A^\sigma y > 0$. It follows from this in particular that if $\rho(A) < 1$, then $(I - A)^{-1} > 0$.

21. If $A \geq 0$ is primitive, there exists a $\nu > 0$ such that for every $\sigma \geq 0$, $A^{\nu+\sigma} > 0$, and conversely, if such a ν exists, then A is primitive.

22. For any n scalars λ_i, define the scalars μ_i by

$$\prod_1^n (\lambda + \lambda_i) = \sum_0^n C_{n,\nu} \mu_\nu{}^\nu \lambda^{n-\nu}.$$

It is known that if $\lambda_i \geq 0$, then

$$\mu_i \geq \mu_{i+1}$$

(see, e.g., Polya and Szego, or Hardy, Littlewood, and Polya). Hence give Mirsky's proof of Richter's theorem that the Euclidean norm satisfies

$$\| A^\mathrm{A} \| \leq \| A \|^{n-1} n^{(n-2)/2}.$$

23. If $\lambda \neq 0$, there exist scalars $\kappa' \geq 0$ and $\kappa'' \geq 0$ such that, for all positive integers,

$$\kappa' \leq |\lambda|^{-\sigma} \sigma^{-n+1} \operatorname{lub}_e [(\lambda I - J)^\sigma] \leq \kappa''.$$

24. If P is any projector (Section 1.3), then for any admissible K

$$\operatorname{lub}_K (P) \geq 1;$$

moreover, for any projector $P \neq I$ there exist admissible K's for which the inequality holds.

25. Let K be admissible, let K' be its polar, and let A be any matrix. Then

$$\operatorname{lub}_S{}^2 (A) \leq \operatorname{lub}_K (A) \operatorname{lub}_{K'} (A).$$

That is to say, the spherical bound of a matrix cannot exceed the geometric mean of the bounds with respect to mutually polar convex bodies.

26. If A is Hermitian, then

$$\operatorname{lub}_S (A) \leq \operatorname{lub}_e (A) = \operatorname{lub}_{e'} (A) \leq n^{\frac{1}{2}} \operatorname{lub}_S (A).$$

27. For any matrix A let U, V, and M satisfy (Exercise 1.19)

$$A = U M V^\mathrm{H}, \qquad M = \operatorname{diag}(\mu_1, \mu_2, \cdots, \mu_n)$$

$$\mu_1 \geq \mu_2 \geq \cdots \geq \mu_n \geq 0, \qquad U U^\mathrm{H} = I, \qquad V V^\mathrm{H} = I.$$

For any $\nu \leq n$ let

$$A_\nu = U M_\nu V^\mathrm{H}, \qquad M_\nu = \operatorname{diag}(\mu_1, \cdots, \mu_\nu, 0, \cdots, 0).$$

Show that A_ν solves the following problem: Among all matrices of rank ν, find that matrix A_ν which minimizes the Euclidean norm of $A - A_\nu$ (the principal component, or Hotelling method of factor analysis).

28. The series

$$\exp A = I + A + A^2/2! + A^3/3! + \cdots$$

converges for every A, and for any norm

$$\| \exp A \| \leq \exp \| A \|.$$

Moreover:

(i) $\exp (A + B) = \exp A \exp B$ if $AB = BA$;

(ii) $\delta(\exp A) = \exp [\tau(A)]$;

(iii) $\exp A \exp (-A) = I$;

(iv) $Y = \exp (\sigma A)$ satisfies the differential equation

$$dY/d\sigma = AY$$

with initial conditions $Y(0) = I$, provided the matrix A is independent of σ.

29. Let $\mathscr{A}$ be any subalgebra of the algebra of matrices. Then $\nu(A) = \| A \|$ for every $A \in \mathscr{A}$ if and only if $\| A^2 \| = \| A \|^2$ for every $A \in \mathscr{A}$.

30. Let $A \geq 0$ be irreducible and $r > 0$ any positive vector. Let $\sigma > 0$ be such that

$$Ar \geq \sigma r.$$

Then $\rho(A) \geq \sigma$, and the inequality is strict unless $Ar = \sigma r$.

31. If $A \geq 0$ is irreducible and $g = g_0 > 0$, form the sequence

$$g_i = A^\nu g,$$

and let $\alpha_{(\nu)}$ be the greatest scalar and $\alpha^{(\nu)}$ the least scalar with

$$\alpha_{(\nu)} g_\nu \leq A g_\nu \leq \alpha^{(\nu)} g_\nu.$$

Then

$$\alpha_{(0)} \leq \alpha_{(1)} \leq \alpha_{(2)} \leq \cdots \leq \rho(A) \leq \cdots \leq \alpha^{(1)} \leq \alpha^{(0)}.$$

32. If V is unitary, show that $\mathrm{glb}_s (V) = \mathrm{lub}_s (V) = 1$, hence all its proper values lie on the unit circle.

33. For any two matrices R and S, if there exists a vector $x \neq 0$ such that $Rx = Sx$, then for any K,

$$\mathrm{glb}_K (R) \leq \mathrm{lub}_K (S).$$

34. For any matrix norm let $\| e_i e_j^{\mathrm{T}} \| = \epsilon_{ij}$. Then

$$\| A \| \leq \Sigma \Sigma \, | \alpha_{ij} | \, \epsilon_{ij}.$$

Hence obtain Ostrowski's proof that a norm is a continuous function of the elements (property IV is not required).

35. Let $\mu(A) = \max_{ij} |\, \alpha_{ij}\,|$. Then

$$\| A \| = n\mu(A)$$

is a norm.

36. Let $\nu_1(A)$ and $\nu_2(A)$ satisfy I_2, II_2 (or $\mathrm{II}_2{}'$), and III_2, but not necessarily IV_2 (in Ostrowski's terms these are "generalized norms"). There exist scalars σ_1 and σ_2, independent of A, such that

$$\sigma_1 \leq \nu_1(A)/\nu_2(A) \leq \sigma_2.$$

37. Let $\nu(A)$ be a generalized norm (see above). There exists a scalar σ such that

$$\| A \| = \sigma\nu(A)$$

is a norm (i.e., satisfies $\mathrm{I}_2 - \mathrm{IV}_2$) (Gastinel).

38. Show that if $\nu(A)$ satisfies $\mathrm{II}_2{}'$, III_2, IV_2, and $\mathrm{I}_2{}'$: $\nu(A) \geq 0$ and there exists an A_0 such that $\nu(A_0) > 0$, then $\nu(A)$ satisfies also I_2 and is a norm (the theorem is due essentially to Ostrowski, who, however, assumes the somewhat stronger condition II_2).

39. Exercise 23 implies that, if $\nu(A)$ is any generalized norm (Exercise 36) and $\lambda \neq 0$, then there exist scalars $\kappa' \geq 0$ and $\kappa'' \geq 0$ (not the same as in Exercise 23) such that

$$\kappa' \leq |\, \lambda \,|^{-\sigma}\, \sigma^{-n+1}\nu[(\lambda I - J)^\sigma] \leq \kappa''.$$

This in turn implies that for any matrix A,

$$\lim \nu^{1/\sigma}(A^\sigma) = \rho(A).$$

40. Let $\| A \|$ be the Euclidean norm of A. Then

$$\inf_S \| S^{-1}AS \| = \Sigma \,|\, \lambda_i(A) \,|^2,$$

and the bound is attained for some S if and only if A is normalizable (Mirsky).

41. Let $S = (I - J^{\mathrm{T}})^{-1} - (I - J)^{-1}$. Then

$$2\delta(\lambda I - S) = (\lambda + 1)^n + (\lambda - 1)^n,$$

$$\rho(S) = \cot \pi(2n)^{-1}.$$

42. Let S be as in Exercise 41, P be skew-symmetric, and $|\, P \,| \leq |\, S \,|$. Then $\rho(P) \leq \rho(S) \leq [n(n - 1)/2]^{\frac{1}{2}}$.

43. Let the real matrix $A = R + Q$ where R is symmetric and Q skew-symmetric. If

$$|\, Q \,| \leq \beta \,|\, S \,|$$

then (Pick, Bendixson)

$$|\, \mathrm{Im}\, \lambda_i(A) \,| \leq \beta \cot \pi(2n)^{-1} \leq \beta\sqrt{n(n - 1)/2}.$$

44. If $A = R + iQ$, $R^{\mathrm{H}} = R$, $Q^{\mathrm{H}} = Q$, and if $|\, R \,| \leq \rho ee^{\mathrm{T}}$, $|\, Q \,| \leq \sigma ee^{\mathrm{T}}$, $|\, A \,| \leq \alpha ee^{\mathrm{T}}$, then

$$|\, \lambda_i(A) \,| \leq n\alpha, \qquad |\, \mathrm{Re}\, \lambda_i(A) \,| \leq n\rho, \qquad |\, \mathrm{Im}\, \lambda_i(A) \,| \leq (n - 1)\sigma.$$

45. Let

$$\Gamma = \begin{pmatrix} \gamma & \beta & \beta & \beta & \cdots \\ \alpha & \gamma & \beta & \beta & \cdots \\ \alpha & \alpha & \gamma & \beta & \cdots \\ \alpha & \alpha & \alpha & \gamma & \cdots \\ & & \cdot & \cdot & \cdot & \cdot & \cdot \end{pmatrix}.$$

Then $\delta(\Gamma) = (\beta - \alpha)^{-1}[\beta(\gamma - \alpha)^n - \alpha(\gamma - \beta)^n]$. Hence if $\alpha > 0$, $\beta > 0$, $\gamma > 0$,

$$\rho(\Gamma) = \gamma + (\beta\alpha^{1/n} - \alpha\beta^{1/n})/(\beta^{1/n} - \alpha^{1/n}).$$

46. If D is a real, nonnegative diagonal matrix, and U and V are unitary matrices, then the trace

$$\tau(D) \geq \operatorname{Re} \tau(UDV^{\mathrm{H}})$$

equality holding if and only if $U = V$. Hence for fixed A and arbitrary unitary V,

$$\max \operatorname{Re} \tau(AV)$$

is achieved when and only when AV is nonnegative semidefinite and Hermitian (von Neumann).

47. For fixed matrices A and B, with singular values

$$\alpha_1 \geq \cdots \geq \alpha_n \geq 0, \qquad \beta_1 \geq \cdots \geq \beta_n \geq 0,$$

and unitary U and V,

$$\max \operatorname{Re} \tau(AUBV) = \Sigma \, \alpha_i\beta_i.$$

48. Let $\| a \| = \phi(\alpha_1, \cdots, \alpha_n)$, considered as a function of its elements. The function ϕ is then, by definition, a *gauge function*. It is *symmetric* in case the norm is equilibrated and $\| Pa \| = \| a \|$ for any permutation matrix P. A generalized matrix norm $\nu(A)$ is said to be *unitarily invariant* in case $\nu(A) = \nu(AV) = \nu(VA)$ for any A and all unitary matrices V. If $\nu(A)$ is unitarily invariant, it is expressible as a symmetric gauge function $\phi(\alpha_1, \cdots, \alpha_n)$ of the singular values $\alpha_1, \cdots, \alpha_n$ of A (von Neumann). Examples are the spectral norm and the Euclidean norm of a matrix.

49. If $\phi(\alpha_1, \cdots, \alpha_n)$ is any symmetric gauge function; define $\nu_\phi(A) = \phi(\alpha_1, \cdots, \alpha_n)$ where the α_i are the singular values of A. Let $\psi(\alpha_1, \cdots, \alpha_n)$ be the gauge function associated with the dual norm. Then (von Neumann)

$$\max_{\nu_\phi(X)=1} \operatorname{Re} \tau(AX) = \nu_\psi(A).$$

50. If $\phi(\alpha_1, \cdots, \alpha_n)$ is a symmetric gauge function, then $\nu_\phi(A) = \phi(\alpha_1, \cdots, \alpha_n)$ is a unitarily invariant generalized norm (von Neumann).

51. Show that

$$\operatorname{lub} (A + B) \geq \operatorname{glb} (A) - \operatorname{glb} (B).$$

52. Show that

$$\text{glb } (A) - \text{lub } (B) \leq |\ \delta(A + B)\ |^{1/n} \leq \text{lub } (A) + \text{lub } (B).$$

53. The matrix B of (2.4.7) is said to be cyclic of index p. Apply the natural generalization of Exercise 48 to show that, for any $\lambda_j = \lambda_j(B)$,

$$\exp (2\pi i k/p)\lambda_j, \qquad k = 0, 1, \cdots, p - 1$$

is also a root of B.

54. If $A \geq 0$ is irreducible and P is the positive orthant, then (Birkhoff and Varga)

$$\max_{x \in P} \left[\min_{y \in P} y^{\mathrm{T}} A x / y^{\mathrm{T}} x \right] = \rho(A) = \min_{y \in P} \left[\max_{x \in P} y^{\mathrm{T}} A x / y^{\mathrm{T}} x \right].$$

55. A *stable matrix* is one each of whose roots has a negative real part. Use Exercise 8 to show that $-B$ is stable if and only if there exists a positive definite matrix M such that $BM + MB^{\mathrm{H}}$ is positive definite (Lyapunov).

56. Given any vector norm $\|\ x\ \|$ and any generalized matrix norm $\nu(A)$, there exists a scalar σ such that $\|\ A\ \| - \sigma\nu(A)$ is not only multiplicative, but consistent with the vector norm.

CHAPTER 3

Localization Theorems and Other Inequalities

3.0. Basic Definitions. By an *inclusion region* for a given matrix A will be meant a region of the complex plane that contains at least one of its characteristic roots. By an *exclusion region* will be meant one not containing any of its roots. Either type of region may or may not be connected. An *inclusion theorem* will be a theorem that exhibits an inclusion region, an exclusion theorem one that exhibits an exclusion region, and either will be called a *localization theorem*. Generally inclusion theorems will be stated only for a matrix A that is *normalizable*, that is, similar to a normal matrix, and hence similar to a diagonal matrix. For other matrices the case is much more complicated, and, in practice, less important.

For a normal matrix A, it is possible to state separation theorems. A *separation theorem* exhibits a locus that separates the complex plane into two or more regions, each of which is an inclusion region. These may be closed, the boundary between two regions being contained in each, or open, the boundary being excluded from both. Separation theorems do not hold in general for nonnormal matrices.

An exclusion theorem can always be regarded as a theorem which asserts the nonvanishing of certain determinants, since to say that α lies in an exclusion region for A is to assert that α is not a root of A, and hence that the matrix $\alpha I - A$ is nonsingular. There is a considerable amount of literature establishing conditions for the nonvanishing of certain determinants, and all such results define exclusion regions when properly interpreted.

Ideally, any method of "evaluating" the roots of a matrix is a method of obtaining smaller and smaller inclusion regions, or larger and larger exclusion regions. Even in the trivial case of a matrix whose roots are known exactly, it can be said that the entire complex plane from which these points are deleted is an exclusion region. Less trivially, however, it is not sufficient to assert that such and such a number is approximately equal to a root, since one wishes to know how close is the approximation, hence to have an inclusion region associated with and containing the approximation. Accordingly, the subject of this chapter may be considered central to the problem of evaluating characteristic roots. However, the theorems give also some information with only very simple computations, and sometimes even crude information may be sufficient.

Very simple exclusion theorems state that the roots of a Hermitian matrix can lie only on the real axis; of a skew-Hermitian matrix only on the imaginary axis; and of a unitary matrix only on the unit circle. Somewhat more sophisticated ones arose in the last chapter: (2.3.1) asserts that with any norm $|\lambda| > \|A\|$ is an exclusion region for A; (2.3.9) that for any K, $|\lambda| < \text{glb}_K(A)$ is also an exclusion region for A; Exercise 2.14 states conditions upon D and C sufficient for $\text{Re}\,\lambda \leq 0$ to be an exclusion region for $D - C$. These theorems come from simple application of the principle of norms, and the same will be true of the other localization theorems considered here.

From a general point of view, an exclusion theorem asserts an inequality satisfied by all proper values, and an inclusion theorem asserts an inequality satisfied by at least one. There is also a class of inequalities, especially for Hermitian matrices, relating the roots of the matrix to those of its principal minors. These are the classical minimax theorems, which will be considered next. After this, some of the simpler properties of the field of values of a matrix will be discussed. And finally consideration will be given to a group of inequalities due to Kantorovich, Wielandt, and Bauer, which are of basic importance both in discussions of rates of convergence of certain iterative methods of solving systems of equations, and also in obtaining error bounds in the use of direct methods.

3.1. Exclusion Theorems. These are the simplest to state and derive, and they have the longest history. Classical results are due to Gershgorin, Perron, A. T. Brauer, Ostrowski, and others. Any exclusion theorem, at least among those considered here, can also be considered a perturbation theorem, since they place bounds upon the variation of the roots as the matrix itself varies. These will be based upon the following lemma (Exercise 2.33):

Given any two matrices R and S, if there exists a vector $x \neq 0$ such that $Rx = Sx$, then for any K,

(1) $$\text{glb}_K(R) \leq \text{lub}_K(S), \qquad \text{glb}_K(S) \leq \text{lub}_K(R).$$

The two inequalities are, of course, entirely equivalent. In particular, let

(2) $$A = B + C,$$

and let λ be any root of A and x a proper vector belonging to it. Then

$$Bx = (\lambda I - C)x,$$

and, in fact, for any scalar μ,

$$(B - \mu I)x = [(\lambda - \mu)I - C]x.$$

It follows that *if λ is a proper value of $B + C$, then for any scalar μ, and* *arbitrary* K,

(3) $$\text{glb}_K [(\lambda - \mu)I - C] \leq \text{lub}_K (B - \mu I).$$

If $(\lambda - \mu)I - C$ is nonsingular, then

$$x = [(\lambda - \mu)I - C]^{-1}(B - \mu I)x.$$

Hence (3) can be sharpened somewhat: *If λ is a proper value of $B + C$, and μ is any scalar, then for every K*

(4) $$1 \leq \text{lub}_K \{[(\lambda - \mu)I - C]^{-1}(B - \mu I)\}$$

[if $(\lambda - \mu)I - C$ were singular the right member would be infinite].[*]

For any K, the relations (3) and (4) define regions, not necessarily connected, within which every root of $B + C$ must lie. Hence the complement of either region is an exclusion region. However, it is simpler to fix attention upon the regions (3) and (4) themselves.

Let $\mu = 0$, and $C = D$ the diagonal of A; then on taking $K = GE$, $g = Ge > 0$, where G is diagonal, the result is a Gershgorin theorem: *Any root of $D + B$ satisfies*

(5) $$1 \leq \text{lub}_g [(\lambda I - D)^{-1}B].$$

But if

(6) $$p = |B| g,$$

then

$$\text{lub}_g [(\lambda I - D)^{-1}B] = \max_i |\lambda - \delta_i|^{-1} \pi_i/\gamma_i,$$

where π_i and γ_i are elements of p and g respectively. Hence *every proper value of $D + B$ satisfies at least one of the inequalities*

(7) $$|\lambda - \delta_i| \leq \pi_i/\gamma_i.$$

Suppose B is irreducible, and consider a λ satisfying at least one relation (7) with equality but none with strict inequality. This is to say that for every i,

$$|\lambda - \delta_i| \geq \pi_i/\gamma_i,$$

hence

$$|\lambda I - D|^{-1} |B| g \leq g.$$

It follows, then, that

(8) $$\rho(|\lambda I - D|^{-1} |B|) \leq 1.$$

Moreover, if any of the inequalities is strict, then, since $|B|$ is irreducible, it follows that the inequality (8) becomes strict since g can be replaced by a g' where all inequalities are strict. In that event, $I - |\lambda I - D|^{-1} |B|$ is nonsingular, and, *a fortiori*, $I - (\lambda I - D)^{-1}B$, and hence $\lambda I - D - B$ is nonsingular, and λ cannot be a root of $D + B$. Hence *when B is irreducible,*

*This need not be so. See Numer. Math. **16** (1970) 141–144.

a root of $D + B$ cannot lie on the boundary of the union of the circular disks (7)
unless it lies on the boundary of each disk (Taussky, 1948).

Suppose the δ_i are all distinct, and consider

$$A(\tau) = D + \tau B.$$

Then $A(0) = D$, $A(1) = A$. For $\tau = 0$, the circular disks degenerate to the
points δ_i. But the roots are continuous functions of τ. Hence as τ increases,
the disks increase continuously in size, but while they remain disjoint, each
contains one and only one root $\lambda_i(\tau)$. By allowing τ to increase further it can
be seen that *if exactly ν of the disks* (7) *overlap but remain disjoint from the
others, then their union contains exactly ν roots of $D + B$, counting multiplicity.*
This remains true even when some or all of the δ_i coincide.

The same argument applied to B^H shows that each root of $D + B$ must
lie in at least one of the circular disks

$$(9) \qquad | \lambda - \delta_j | \leq \rho_j/\alpha_j, \qquad | B^H | a = r.$$

Special methods can be applied to strengthen (7), and mention will be
made of one result, due essentially to Ostrowski: *Every proper value of
$D + B$ satisfies at least one of the inequalities*

$$(10) \qquad | \lambda - \delta_i | \, | \lambda - \delta_j | \leq (\pi_i/\gamma_i)(\pi_j/\gamma_j).$$

Since B can be replaced by $\Gamma^{-1}B\Gamma$, where

$$\Gamma = \mathrm{diag}\,(\gamma_1, \gamma_2, \cdots, \gamma_n),$$

it is sufficient to prove the case for $\gamma_i = 1$, $g = e$. Suppose

$$(\lambda I - D)x = Bx.$$

By applying a permutational transformation, it can be arranged that

$$| \xi_1 | \geq | \xi_2 | \geq | \xi_j |, \qquad j = 3, 4, \cdots, n.$$

Then

$$(\lambda - \delta_1)\xi_1 = \beta_{12}\xi_2 + \cdots + \beta_{1n}\xi_n,$$
$$(\lambda - \delta_2)\xi_2 = \beta_{21}\xi_1 + \beta_{23}\xi_3 + \cdots + \beta_{2n}\xi_n,$$

hence

$$(\lambda - \delta_1)(\lambda - \delta_2) = (\beta_{12} + \beta_{13}\xi_3/\xi_2 + \cdots + \beta_{1n}\xi_n/\xi_2)$$
$$\times (\beta_{21} + \beta_{23}\xi_3/\xi_1 + \cdots + \beta_{2n}\xi_n/\xi_1),$$

and (10) follows from this for $i = 1$, $j = 2$.

In computing proper values and vectors, suppose an approximate set
has been found, so that

$$(11) \qquad X^{-1}AX = \Lambda + R,$$

where Λ is diagonal, and R is a matrix of small elements. Then the above
theorems can be applied to $\Lambda + R$ to obtain bounds for the deviation of each

diagonal element of Λ from a proper value. The most convenient vector g is naturally e. On the other hand, by making a particular γ_i large, the associated circle (7) may be made small, and also disjoint from all the others. If disjoint, then that circle can contain only a single proper value, and the smaller the value π_i/γ_i can be made by suitable choice of g without producing overlap with one of the other disks, the better will be the estimate of error.

Now let $K = S$, the unit sphere. Hereafter, in the present chapter, this will be understood unless otherwise indicated explicitly. Suppose C is normalizable and

$$(12) \qquad C = P_C \Lambda_C P_C^{-1}.$$

If

$$(C + B)x = \lambda x,$$

then

$$(\lambda I - \Lambda_C)y = P_C^{-1} B P_C y, \qquad x = P_C y.$$

Hence

$$\text{glb } (\lambda I - \Lambda_C) \leq \text{lub } (P_C^{-1} B P_C) \leq \text{lub } (B) \text{ lub } (P_C^{-1}) \text{ lub } (P_C).$$

The matrix P_C which diagonalizes C is not unique, since, in particular, it could be replaced by $P_C D$, where D is any nonsingular diagonal. Let

$$(13) \qquad \sigma(C) = \min_{P_C} \text{lub } (P_C^{-1}) \text{ lub } (P_C),$$

where the minimum is taken with respect to all matrices P_C that satisfy (12). Then

$$\text{glb } (\lambda I - \Lambda_C) \leq \text{lub } (B)\sigma(C).$$

Hence, *if $A = B + C$, where C is normalizable, then any root of A lies in one of the circular disks*

$$(14) \qquad |\, \lambda - \lambda_i(C) \,| \leq \text{lub } (B)\sigma(C),$$

where $\sigma(C)$ is defined by (13).

If B is also normalizable,

$$(15) \qquad B = P_B \Lambda_B P_B^{-1},$$

then

$$(\lambda I - \Lambda_C)y = P_C^{-1} P_B \Lambda_B P_B^{-1} P_C y;$$

hence if

$$(16) \qquad \sigma(B, C) = \min_{P_B, P_C} \text{lub } (P_B^{-1} P_C) \text{ lub } (P_C^{-1} P_B),$$

and it follows that *if $A = B + C$ where B and C are both normalizable, then any root of A lies on one of the circular disks*

$$(17) \qquad |\, \lambda - \lambda_i(C) \,| \leq \sigma(B, C) \, \rho(B).$$

Since it is also true that

$$A = (B - \mu I) + (C + \mu I),$$

for *any scalar* μ, it follows that *any root of A lies in one of the circular disks*

(18) $| \lambda - \mu - \lambda_i(C) | \leq \sigma(B, C) \, \rho(B - \mu I).$

The theorem just formulated has a rather simple geometric interpretation, especially in case B and C are both normal matrices. When B and C are normal, P_B and P_C can be taken to be unitary, and $\sigma(B, C) = 1$. The right member represents the radius of a circular disk with center μ which encloses all proper values of B, within or on the boundary. The interpretation is then as follows: *Let B and C be normal matrices, and let Σ be any circular disk which contains all proper values of B. For each i, form the circular disk Σ_i by translating Σ along the vector $\lambda_i(C)$. Then the union of all disks Σ_i contains all proper values of $A = C + B$.* This theorem is due to Wielandt. The interpretation in the more general case differs only in that the translation must be accompanied by an expansion in the ratio $\sigma(B, C)$.

It might be remarked that any K defining an absolute norm could have been used in place of S to derive (18), provided $\sigma(B, C)$ is defined accordingly. Generally speaking, however, σ is least when defined with reference to S.

Two corollaries follow almost immediately from Wielandt's case of normal matrices. Let

$$2B = A + A^{\mathrm{H}}, \qquad 2C = A - A^{\mathrm{H}}.$$

Then B is Hermitian and C is skew-Hermitian, hence both are normal. All roots of B are real, all roots of C are pure imaginary. The smallest circle containing all roots of B is one whose center is on the real axis, and when this circle is translated it remains within a strip bounded by the vertical lines through the least and the greatest of the $\lambda_i(B)$. Hence all roots of A must be contained in this strip. On interchanging B and C, it can be concluded that *all roots of A lie in or on the least rectangle with sides parallel to the real and imaginary axes that contains all roots of B and all roots of C.* This is the Bendixson theorem. By further application of the theorem, a still further refinement is possible (Exercise 3).

The second corollary is a perturbation theorem for Hermitian matrices. *Suppose B and C are both Hermitian, and let*

$$\beta_i = \lambda_i(B), \qquad \beta_1 \geq \beta_2 \geq \cdots \geq \beta_n,$$
$$\gamma_i = \lambda_i(C), \qquad \gamma_1 \geq \gamma_2 \geq \cdots \geq \gamma_n,$$
$$2\mu = \beta_1 + \beta_n.$$

Then for every $\lambda = \lambda(B + C)$, there is a γ_i such that

(19) $\beta_n \leq \lambda - \gamma_i \leq \beta_1.$

In particular let

$$A(\tau) = \tau B + C, \qquad \alpha_i(\tau) = \lambda_i[A(\tau)],$$
$$\alpha_1(\tau) \geq \cdots \geq \alpha_n(\tau).$$

Since $\alpha_i(0) = \gamma_i$, application of (19) with τB replacing B shows that for $\tau = 1$,

(20) $$\beta_n \leq \alpha_i - \gamma_i \leq \beta_1.$$

3.2. Inclusion and Separation Theorems.

Suppose A is normalizable,

(1) $$A = P\Lambda P^{-1}, \qquad \Lambda = \operatorname{diag}(\lambda_1, \cdots, \lambda_n),$$

and let

(2) $$\sigma = \sigma(A),$$

where $\sigma(C)$ is defined in (3.1.13). For any $y \neq 0$,

$$\sigma \operatorname{lub}(\Lambda) \geq \operatorname{lub}(A) \geq \| Ay \| / \| y \|.$$

Moreover, if $\alpha(\lambda)$ and $\beta(\lambda)$ are polynomials and

$$\gamma(\lambda) = \alpha(\lambda)/\beta(\lambda),$$

then

$$\sigma \operatorname{lub}[\gamma(\Lambda)] \geq \| \gamma(A)y \| / \| y \|.$$

Hence, if

$$y = \beta(A)v,$$

then

$$\operatorname{lub}[\alpha(A)/\beta(A)] \geq \| \alpha(A)v \| / \| \beta(A)v \|,$$

and therefore

(3) $$\operatorname{lub}[\alpha(\Lambda)/\beta(\Lambda)] \geq \sigma^{-1} \| \alpha(A)v \| / \| \beta(A)v \|.$$

The left member, however, has the value

$$\max | \alpha(\lambda_i)/\beta(\lambda_i) |.$$

It follows that *the region defined by*

(4) $$| \alpha(\lambda)/\beta(\lambda) | \geq \sigma^{-1} \| \alpha(A)v \| / \| \beta(A)v \|$$

is an inclusion region for any polynomials $\alpha(\lambda)$ *and* $\beta(\lambda)$, *and any vector* $v \neq 0$. But since α and β can be any polynomials, it follows that

(5) $$| \alpha(\lambda)/\beta(\lambda) | \leq \sigma \| \alpha(A)v \| / \| \beta(A)v \|$$

is also an inclusion region for any $\alpha(\lambda)$, $\beta(\lambda)$, and $v \neq 0$.

It is convenient to say that the *norm* (with respect to A and v) *of a polynomial* $\alpha(\lambda)$ is $\| \alpha(A)v \|$, and that *polynomials* $\alpha(\lambda)$ *and* $\beta(\lambda)$ *are orthogonal* in case the vectors $\alpha(A)v$ and $\beta(A)v$ are orthogonal.

Hence suppose $\alpha(\lambda)$ and $\beta(\lambda)$ are mutually orthogonal and normalized with respect to A and v, and η is any scalar. Then

$$\| [\alpha(A) - \eta\beta(A)]v \|^2 = \| [\bar{\eta}\alpha(A) + \beta(A)]v \|^2 = 1 + | \eta |^2.$$

Therefore, *for any two orthonormal polynomials $\alpha(\lambda)$ and $\beta(\lambda)$, and any scalar η, the relation*

(6) $$| \alpha(\lambda) - \eta\beta(\lambda) |/| \bar{\eta}\alpha(\lambda) + \beta(\lambda) | \le \sigma$$

defines an inclusion region for A.

Finally, let

$$\tau = -\bar{\eta}\sigma = \exp i\theta$$

for any real θ. Then (6) becomes

(7) $$\mathrm{Re}\,[\tau\alpha(\lambda)/\beta(\lambda)] \le (\sigma - \sigma^{-1})/2.$$

Hence, (7) *is an inclusion region, where $\alpha(\lambda)$ and $\beta(\lambda)$ are orthonormal with respect to A and v, and $| \tau | = 1$.*

It is to be observed that in all cases the scalar $\sigma \ge 1$, and $\sigma = 1$ only for a normal matrix A. In this case the regions (4) and (5) are complementary sets in the complex plane with the boundary common to both. Hence the inclusion theorem becomes a separation theorem. It is easy to see also that the complement of (6) is another region of the form (6), and that of (7) another region of the form (7), when $\sigma = 1$, and again separation theorems hold.

In Section 1.6 a sequence of orthogonal polynomials $\phi_\nu(\lambda)$ were defined. Let

$$\psi_\nu^*(\lambda) = \phi_\nu(\lambda)/\| \phi_\nu(A)v \|$$

be the same polynomials normalized. The most important special case of (6) is then provided by the polynomials ψ_0 and ψ_1:

(8) $$\begin{aligned} \beta(\lambda) &= \psi_0(\lambda) = \mu_{00}^{-\frac{1}{2}} \\ \alpha(\lambda) &= \psi_1(\lambda) = (\lambda - \mu_{01}/\mu_{00})/(\mu_{11} - \mu_{01}\mu_{10}/\mu_{00})^{\frac{1}{2}} \end{aligned}$$

This gives

(9) $$| \psi_1(\lambda) - \eta\,\psi_0 |/| \bar{\eta}\psi_1(\lambda) - \psi_0 | \le \sigma,$$

and for any η the region is a circle. For $\eta = 0$, it is a circle with radius

$$\sigma(\mu_{11} - \mu_{01}\mu_{10}/\mu_{00})^{\frac{1}{2}}/\mu_{00}^{\frac{1}{2}}$$

and center at μ_{01}/μ_{00}. The case $\eta = 0$ for A Hermitian is Weinstein's theorem.

The region (7) when $\alpha(\lambda)$ and $\beta(\lambda)$ are given by (8) is

$$\mathrm{Re}\,\tau\psi_1(\lambda) \le (\sigma - \sigma^{-1})/2, \qquad | \tau | = 1.$$

For a normal matrix the right member vanishes and the normalization of the polynomials is irrelevant. The theorem then states that *for a normal matrix any line through the Rayleigh quotient* μ_{01}/μ_{00} *separates the complex plane into two inclusion regions, the line itself being common to both.* The result is otherwise known. Geometrically it means that any Rayleigh quotient of a normal matrix lies in the convex hull of the proper values of the matrix. *For an arbitrary normalizable matrix A* the result is that *a line shifted a certain distance* from μ_{01}/μ_{00} *divides the complex plane into two half-planes, of which the one that contains the Rayleigh quotient is an inclusion region.*

The inequality (6) requires that $\alpha(\lambda)$ and $\beta(\lambda)$ be orthogonal and have equal norms. If $\alpha(\lambda)$ is one of the normed orthogonal polynomials $\psi_\rho(\lambda)$, then any of its divisors is orthogonal to it. In particular, let ω be any zero of $\psi_\rho(\lambda)$ and let $\beta(\lambda)$ be the normalized quotient $\psi_\rho(\lambda)/(\lambda - \omega)$. Then, for a suitable constant ϵ, the inequality (6) becomes

$$| (\lambda - \omega) - \eta\epsilon |/| \bar{\eta}(\lambda - \omega) + \epsilon | \le \sigma,$$

or, for $\eta = 0$,

$$| \lambda - \omega | \le | \epsilon | \sigma.$$

Hence inclusion circles about the zeros of the basic orthogonal polynomials $\psi_\rho(\lambda)$ or $\phi_\rho(\lambda)$ can always be found. In case $\rho = \nu$, the degree of v, then $\epsilon = 0$, and the circles reduce to points.

3.2.1. *The Matrix A is Normal.* For this case the right member of (7) vanishes since $\sigma = 1$, the norms of α and β are irrelevant, and the theorem can be restated by saying that *for any normal matrix A, if $\alpha(\lambda)$ and $\beta(\lambda)$ are orthogonal, then for any scalar τ, the locus*

$$\text{Re } \tau\alpha(\lambda)/\beta(\lambda) = 0$$

separates the complex plane into two regions, each of which is an inclusion region. For $\beta = 1$, and α linear,

$$\alpha = \lambda - \mu_{01}/\mu_{00},$$

this states that for any vector v, the locus

$$\text{Re } \tau[\lambda - v^{\text{H}}Av/v^{\text{H}}v] = 0$$

separates two inclusion regions. If $\theta(\bar{\lambda}, \lambda)$ is any polynomial in $\bar{\lambda}$ and λ, then for normal A, the matrix $\theta(A^{\text{H}}, A)$ is defined and normal, hence the locus

$$\text{Re } \tau[\theta(\bar{\lambda}, \lambda) - v^{\text{H}}\theta(A^{\text{H}}, A)v/v^{\text{H}}v] = 0$$

separates two inclusion regions. Thus

$$\theta(\bar{\lambda}, \lambda) - v^{\text{H}}\theta(A^{\text{H}}, A)v/v^{\text{H}}v$$

will be called a *separation polynomial.* Hence *if A is normal the locus*

(10) $$\text{Re } \tau\theta(\bar{\lambda}, \lambda) = 0$$

separates the complex plane into two inclusion regions if $\theta(\bar{\lambda}, \lambda)$ is any separation polynomial. Hence (10) will be called a *separation locus.*

The polynomials $\phi_\rho(\lambda)$ that were exhibited in Section 1.6 are separation polynomials, and so is the product $\overline{\alpha(\lambda)}\phi_\rho(\lambda)$ where $\alpha(\lambda)$ is any polynomial of degree less than ρ. For $\rho > 1$, let

$$\phi_\rho(\lambda) = \alpha(\lambda)\beta(\lambda),$$

with neither α nor β constant. Then

$$\overline{\alpha(\lambda)}\phi(\lambda) = \mid \alpha(\lambda) \mid^2 \beta(\lambda)$$

is a separation polynomial. Since $\mid \alpha(\lambda) \mid^2$ is real and nonnegative, the loci

$$\operatorname{Re} \tau \mid \alpha(\lambda) \mid^2 \beta(\lambda) = 0$$

and

(11) $$\operatorname{Re} \tau\beta(\lambda) = 0$$

are identical except at the zeros of $\alpha(\lambda)$. Hence (11) is a separation locus, and this is true when $\beta(\lambda)$ is any nonconstant divisor of $\phi_\rho(\lambda)$. By taking $\beta(\lambda)$ to be a linear factor, it can be concluded that any line through any zero of $\phi_\rho(\lambda)$ is a separation locus, hence that *all zeros of the* $\phi_\rho(\lambda)$ *lie in the convex hull of the characteristic roots of a normal matrix A*. If $\beta(\lambda)$ is a quadratic factor, the locus is a hyperbola through its zeros. Applied to a Hermitian matrix, whose roots lie on the real axis, or to a unitary matrix, whose roots lie on the unit circle, this implies that a segment or arc joining any two zeros of $\phi_\rho(\lambda)$ is an inclusion segment or arc (Bueckner).

Define the kernel polynomials

(12) $$\kappa_\sigma(\lambda, \mu) = -\mu_{00} \, \delta\begin{pmatrix} \mu_{00} & \cdots & \mu_{0\sigma} & 1 \\ \cdots\cdots\cdots\cdots\cdots \\ \mu_{\sigma 0} & \cdots & \mu_{\sigma\sigma} & \bar\mu^\sigma \\ 1 & \cdots & \lambda^\sigma & 0 \end{pmatrix} \delta^{-1}(M_\sigma)$$

$$= \mu_{00}(1, \lambda, \cdots, \lambda^\sigma)M_\sigma^{-1}(1, \mu, \cdots, \mu^\sigma)^{\mathrm{H}}$$

where the latter form is a consequence of (1.4.16′). Then it can be verified that

$$\kappa_\sigma(\lambda, \mu)\overline{\gamma(\lambda)(\lambda - \mu)}$$

is a separation polynomial if $\gamma(\lambda)$ is any polynomial of degree $\sigma - 1$ or less. For fixed μ, let

$$\kappa_\sigma(\lambda, \mu) = \alpha(\lambda)\beta(\lambda)$$

where neither α nor β is constant. Then an argument parallel to that given above shows that $\overline{\beta(\lambda)}(\lambda - \mu)$ is a separation polynomial, hence

$$\operatorname{Re} \tau(\bar\lambda - \bar\mu)\beta(\lambda) = 0$$

is a separation locus. If β is linear, the locus is a circle through μ and a zero of κ_σ. As τ varies, all circles through these two points are obtained. Hence any circle through μ and a zero of $\kappa_\sigma(\lambda, \mu)$ is a separation locus. For A Hermitian the theorem is due to Wielandt, for $\sigma = 1$ it is Temple's theorem.

Hitherto the boundaries of an inclusion region have been assumed to be a part of the region, in order to avoid complicating the discussion, but for normal matrices the boundary can generally be excluded. This is of very little practical significance, since in computation rounding errors will make it impossible, ordinarily, to determine whether or not a particular point lies exactly on a boundary. However, if $\theta(\bar{\lambda}, \lambda)$ is any separation polynomial, and if

$$\theta(A^{\mathrm{H}}, A)v \neq 0,$$

then v is not a characteristic vector of $\theta(A^{\mathrm{H}}, A)$, since for a characteristic vector the Rayleigh quotient, in this case 0, is the characteristic root to which it belongs. Therefore 0 is not a corner of the polygon which bounds the field of values (cf. Section 3.3), and hence lies either in the interior or on one of the sides. Hence, *barring at most a single exceptional amplitude for τ, all inclusion regions can be defined by strict inequalities with the boundary excluded.*

3.3. Minimax Theorems and the Field of Values.

Let X be a matrix of orthonormal columns:

(1) $$X^{\mathrm{H}}X = I.$$

Then XX^{H} is a projector (Section 1.3) and $XX^{\mathrm{H}}x$ is the orthogonal projection of any vector x on the space of X. Let $u = X^{\mathrm{H}}x$. If the columns of X are thought of as representing an orthogonal coordinate system in this subspace, then, since the projection is Xu, the elements of the vector u are the coordinates of the projection when referred to these axes. For any matrix A, the vector AXu, obtained by multiplying A by the projection, is not necessarily in the same subspace, but $XX^{\mathrm{H}}AXu$ is the orthogonal projection upon the subspace, and $X^{\mathrm{H}}AXu$ provides the coordinates of the projection referred to X. Hence $X^{\mathrm{H}}AX$ will be called the *section* of the matrix A in the space of X. In particular, when the columns of X are selected columns of I, then $X^{\mathrm{H}}AX$ is a principal submatrix of A. In Section 1.6 some consideration was given to sections defined by the vectors of a Krylov sequence. The primary interest here, however, will be in arbitrary sections of a Hermitian matrix.

First consider the one-dimensional section of an arbitrary matrix A. This is the set of all (complex) scalars of the form

(2) $$\lambda(x) = \lambda(A, x) = x^{\mathrm{H}}Ax/x^{\mathrm{H}}x,$$

as x varies over all possible nonnull vectors. This set is known as the *field of values* of A, and will be designated $F(A)$. It will be shown later that $F(A)$ is a convex set, and it contains all proper values since

$$Ax = \lambda x \Rightarrow \lambda = \lambda(A, x).$$

The field of values is the set of all possible Rayleigh quotients; if A is normal it is known that any line through a Rayleigh quotient is a separation locus (Section 3.2); hence any point in $F(A)$ lies in the convex hull of the proper values of A (when A is normal). But note that for any unitary matrix U, $F(U^H A U) = F(A)$; consequently, there is no restriction in supposing A to be triangular (Exercise 1.3), and, if normal, then diagonal. Hence for a normal matrix,

$$\lambda(x) = \Sigma \, \lambda_i \mid \xi_i \mid^2 / \Sigma \mid \xi_i \mid^2,$$

that is any point in $F(A)$ is a weighted mean of the proper values of a normal matrix, and conversely. Hence *for a normal matrix the field of values coincides with the convex hull of the proper values.* In the special case of a Hermitian matrix, the field of values coincides with the segment of the real axis bounded by the two extreme roots.

3.3.1. *Hermitian Matrices.* The problem in this section will be that of relating the proper values of a section of a Hermitian matrix A with those of A itself. Hence, suppose A is Hermitian, and let

(3) $$\lambda_1(A) \geq \lambda_2(A) \geq \cdots \geq \lambda_n(A).$$

Then

$$\lambda_1(A) \geq \lambda(A, x) \geq \lambda_n(A),$$

and in particular

(4) $$\lambda_1(A) = \max_{x \neq 0} \lambda(A, x), \qquad \lambda_n(A) = \min_{x \neq 0} \lambda(A, x).$$

Let E_m be any subspace of the entire space E_n. An orthonormal basis for E_m is any set of m orthonormal vectors in E_m, and if U_m is the matrix whose columns are these vectors in some order, then

(5) $$U_m{}^H U_m = I,$$

and conversely, if U_m is any matrix whose columns are in E_m and which satisfies (5), then these columns provide an orthonormal basis for E_m. If the columns of U_{n-m} are an orthonormal basis for the space E_{n-m} complementary to E_m, then the matrix $U = (U_m, U_{n-m})$ is unitary.

Let x be constrained to lie in E_m. Then

$$x = U_m u$$

for some vector u. Let

(6) $$A = V \Lambda V^H, \qquad \Lambda = \text{diag} \, (\lambda_1, \cdots, \lambda_n)$$
$$V^H V = I,$$

and let

(7) $$V = (V_m, V_{n-m}),$$

where the columns of V_m are proper vectors belonging to $\lambda_1, \lambda_2, \cdots, \lambda_m$, and columns of V_{n-m} are proper vectors belonging to $\lambda_{m+1}, \cdots, \lambda_n$. The equations

$$U_m u = \phi v_m + V_{n-m} v$$

are n homogeneous equations in the $n + 1$ unknowns which are the scalar ϕ, the elements of the m-dimensional vector u and the elements of the $(n - m)$-dimensional vector v. Hence they always have a nontrivial solution. Moreover, the columns of U_m are linearly independent, as are v_m, the last column of V_m, and the columns of V_{n-m}. Hence $U_m u = 0$ implies $\phi = 0$ and $v = 0$. Hence there is at least one vector $x = U_m u \neq 0$ common to E_m and the space of v_m and V_{n-m}. For such a vector $\lambda(x)$ is a weighted mean of $\lambda_m, \cdots, \lambda_n$, therefore $\lambda(x) \leq \lambda_m$. It follows that for any E_m

$$\min_{x \in E_m} \lambda(x) \leq \lambda_m.$$

But for $U_m = V_m$ the equality holds. Consequently

$$(8) \qquad \max_{E_m} \min_{x \in E_m} \lambda(x) = \lambda_m.$$

In words, this states the following: Given any Hermitian matrix A whose roots are ordered as in (3), *the least root of an m-dimensional section of A cannot exceed λ_m, but there exists a particular section for which it is equal to λ_m.* Analogously

$$(9) \qquad \min_{E_m} \max_{x \in E_m} \lambda(x) = \lambda_{n-m+1}.$$

These are the classical *minimax* and *maximin* theorems. For $m = n - 1$, these imply that for any U_{n-1} with

$$(10) \qquad U_{n-1}^H U_{n-1} = I,$$

$$\lambda_1(A) \geq \lambda_1(U_{n-1}^H A U_{n-1}) \geq \lambda_2(A) \geq \cdots \geq \lambda_{n-1}(A)$$
$$\geq \lambda_{n-1}(U_{n-1}^H A U_{n-1}) \geq \lambda_n(A).$$

This is known as a *separation theorem*, and it states that *the proper values of any $(n - 1)$-dimensional section of a Hermitian matrix A separate those of A.* The usual statement speaks only of a principal submatrix of order $n - 1$, which is a particular section.

Let U_m satisfy (5), and consider the trace

$$\tau(U_m^H A U_m).$$

This is equal to the sum of the roots of $U_m^H A U_m$. Repeated application of the separation theorem shows that this cannot exceed $\lambda_1 + \lambda_2 + \cdots + \lambda_m$, nor can it be exceeded by $\lambda_{n-m+1} + \cdots + \lambda_n$. On the other hand these bounds can be attained by taking the columns of U_m to be proper vectors. Hence

$$(11) \qquad \lambda_1 + \lambda_2 + \cdots + \lambda_m = \max_{U_m} \tau(U_m^H A U_m),$$

$$(12) \qquad \lambda_{n-m+1} + \cdots + \lambda_n = \min_{U_m} \tau(U_m^H A U_m).$$

Thus *the trace of any m-dimensional section of a Hermitian matrix A is bounded by the sum of the m largest and that of the m smallest of the proper values of A. Moreover, either bound can be attained.*

Let A be nonnegative semidefinite. Then its mth compound $A^{(m)}$ is also nonnegative semidefinite, and its proper vectors are the columns of the compound $V^{(m)}$. Let $U = (U_m, U_{n-m})$ be unitary. Then $U^{(m)}$ is unitary and application of the theorem on compound matrices shows that the element at the upper left of $U^{\mathrm{H}(m)}A^{(m)}U^{(m)}$ is the determinant $\delta(U_m{}^{\mathrm{H}}AU_m)$. Hence, for a nonnegative semidefinite matrix A,

$$
\begin{aligned}
\lambda_1\lambda_2\cdots\lambda_m &= \max_{U_m}\,\delta(U_m{}^{\mathrm{H}}AU_m),\\
\lambda_{n-m+1}\cdots\lambda_n &= \min_{U_m}\,\delta(U_m{}^{\mathrm{H}}AU_m).
\end{aligned}
\tag{13}
$$

Thus *the determinant of an m-dimensional section of a nonnegative, semidefinite matrix A is bounded by the product of the m greatest and that of the m least of the proper values of A. Moreover, either bound can be attained.*

The maximin theorem has a generalization due to Wielandt that includes both (8) and (11). As a first step toward the development of this generalization it will be shown that given any vector $x_1 \neq 0$, there exists a vector $x_n \neq 0$ orthogonal to it such that

$$\lambda(x_1) + \lambda(x_n) \le \lambda_1 + \lambda_n.$$

Let $v = v_n$ be the proper vector belonging to λ_n, and let w be orthogonal to v in the plane of x_1 and v. There is no restriction in supposing x_1, v, and w to be unit vectors and it will be shown that a unit vector x_n orthogonal to x_1 in the same plane is effective. Let both x_1 and x_n be resolved along the vectors w and v. Then x_1 and x_n are representable

$$
\begin{aligned}
x_1 &= \alpha w + \beta v,\\
x_n &= \bar{\beta}w - \bar{\alpha}v,\\
\alpha\bar{\alpha} + \beta\bar{\beta} &= 1,
\end{aligned}
$$

and hence

$$
\begin{aligned}
\lambda(v) &= \lambda_n,\\
\lambda(x_1) &= \alpha\bar{\alpha}\lambda(w) + \beta\bar{\beta}\lambda_n,\\
\lambda(x_n) &= \beta\bar{\beta}\lambda(w) + \alpha\bar{\alpha}\lambda_n,\\
\lambda(x_1) + \lambda(x_n) &= \lambda(w) + \lambda_n \le \lambda_1 + \lambda_n,
\end{aligned}
$$

which proves the assertion.

Next, let E_m be any subspace of dimension m. There exists a vector $x_m \in E_m$, and a vector x_n orthogonal to x_m such that

$$\lambda(x_m) + \lambda(x_n) \le \lambda_m + \lambda_n.$$

For there is at least one nonnull vector in the intersection of E_m with the space of the proper vectors $v_m, \cdots, v_n$. Let x_m be such a vector. If

$$V_{n-m+1} = (v_m, \cdots, v_n),$$

then the result just obtained can be applied to the section

$$V_{n-m+1}^{\mathrm{H}} A V_{n-m+1},$$

whose greatest proper value is λ_m, and the required result follows since x_m is expressible as a linear combination of the columns of V_{n-m+1}.

Finally, if $m < p \leq n$, and if E_p and $E_m \subset E_p$ are spaces of dimension p and m, respectively, then there exist orthogonal vectors $x_m \in E_m$ and $x_p \in E_p$ such that

$$\lambda(x_m) + \lambda(x_p) \leq \lambda_m + \lambda_p.$$

For let the columns of U_p form an orthonormal basis for E_p, and consider the section

$$U_p^{\mathrm{H}} A U_p.$$

Its proper values λ_i' satisfy

$$\lambda_i' \leq \lambda_i, \qquad I = 1, 2, \cdots, p.$$

But the results so far obtained, when applied to the section, imply that there exist vectors $x_m \in E_m$ and $x_p \in E_p$, nonnull and mutually orthogonal, such that

$$\lambda(x_m) + \lambda(x_p) \leq \lambda_m' + \lambda_p'.$$

Hence the result follows *a fortiori*.

Now for any p and $m < p$ there exist spaces E_p and $E_m \subset E_p$ such that for any $x_m \in E_m$ and some $x_p \in E_p$ orthogonal to it,

$$\lambda(x_m) + \lambda(x_p) \geq \lambda_m + \lambda_p.$$

Such spaces are, in fact, the spaces defined by the proper vectors. It follows, therefore, that for $m < p \leq n$,

(14)
$$\max_{E_m \subset E_p} \min_{x_m \in E_m,\, x_p \in E_p} [\lambda(x_m) + \lambda(x_p)] = \lambda_m + \lambda_p$$
$$x_m^{\mathrm{H}} x_p = 0.$$

In words, for any p, and $m < p$, consider all possible subspaces of dimension p, and all possible subspaces of dimension m contained in that subspace. Form all possible Rayleigh quotients of two mutually orthogonal vectors, both lying in the larger subspace, and at least one in the smaller. The sum of these Rayleigh quotients has a minimum which cannot exceed $\lambda_m + \lambda_p$, but which can be equal to it for at least one special choice of E_p and E_m.

Wielandt's theorem is more general than this and applies to an arbitrary number of nested subspaces, but this general form will not be needed here.

Another theorem of Wielandt generalizes the inequality (3.1.19) as follows: Let

(15) $$A = B + C,$$

(16)
$$\alpha_i = \lambda_i(A), \qquad \beta_i = \lambda_i(B), \qquad \gamma_i = \lambda_i(C),$$
$$\alpha_1 \geq \cdots \geq \alpha_n, \qquad \beta_1 \geq \cdots \geq \beta_n, \qquad \gamma_1 \geq \cdots \geq \gamma_n.$$

Then

(17) $$\alpha_i + \alpha_j + \cdots \leq \gamma_i + \gamma_j + \cdots + \beta_1 + \beta_2 + \cdots,$$

where the same number of roots appear for each matrix. While the proof will not be given here, the method of proof can be illustrated for the special case

(18) $$\alpha_m \leq \gamma_m + \beta_1.$$

The method of Section 3.1 seems not to extend.

Let E_m be the invariant subspace of A belonging to the roots $\alpha_1, \alpha_2, \cdots, \alpha_m$. Then for any $x \in E_m$,

$$\alpha_m \leq \lambda(A, x) = \lambda(B, x) + \lambda(C, x) \leq \beta_1 + \lambda(C, x).$$

But

$$\min_{x \in E_m} \lambda(C, x) \leq \gamma_m,$$

and this completes the proof.

3.3.2. *General Matrices.* For any matrix A let

(19) $$2R = A + A^H, \qquad 2iS = A - A^H.$$

Then R and S are Hermitian and

(20) $$A = R + iS.$$

Moreover, if

$$\rho(x) = \lambda(R, x), \qquad \sigma(x) = \lambda(S, x),$$

then ρ and σ are real, and

(21) $$\lambda(A, x) = \rho(x) + i\sigma(x).$$

Consequently if R and S have proper values of ρ_i and σ_i with

$$\rho_i \geq \rho_{i+1}, \qquad \sigma_i \geq \sigma_{i+1},$$

then $F(A)$ is contained in the rectangle with vertical sides at ρ_n and ρ_1, and horizontal sides at σ_n and σ_1. Moreover each of these sides contains at least one point of $F(A)$ since for x one could take a proper vector of R or of S. Hence each of these lines is a support line of $F(A)$.

Let

$$A_\theta = (\exp i\theta)A.$$

Then $F(A_\theta) = (\exp i\theta)F(A)$, which is to say that $F(A_\theta)$ is obtained by rotating $F(A)$ through an angle θ about the origin. Hence any support line of $F(A_\theta)$ rotated through an angle of $-\theta$ becomes a support line of $F(A)$.

It will now be shown that $F(A)$ *is convex*, and this will imply that $F(A)$ is the intersection of all those half-planes containing $F(A)$ and bounded by its support lines. It is sufficient to prove the theorem only for matrices of second order, for the following reason. Suppose μ_1 and μ_2 are any two distinct points in $F(A)$, and let x_1 and x_2 be vectors for which

$$\mu_1 = \lambda(A, x_1), \qquad \mu_2 = \lambda(A, x_2).$$

Then x_1 and x_2 are linearly independent and define a plane. Let y_1 and y_2 be any two orthonormal vectors in the plane of x_1 and x_2:

$$y_1{}^H y_1 = y_2{}^H y_2 = 1, \qquad y_1{}^H y_2 = 0.$$

That being the case, μ_1 and μ_2 are points in the field of values of the second-order matrix

$$A' = (y_1, y_2)^H A(y_1, y_2),$$

and clearly $F(A') \subset F(A)$. Hence if the field of values of any second-order matrix is known to be convex, it follows that every point on the segment joining μ_1 and μ_2 is contained in $F(A')$, and hence in $F(A)$. But μ_1 and μ_2 were arbitrary distinct points in $F(A)$ and therefore it follows that $F(A)$ is convex.

Now suppose A is of second order. If U is any unitary matrix, then $F(U^H A U) = F(A)$; consequently there is no restriction in supposing A to be upper triangular. For any scalar μ, $F(A - \mu I)$ is obtained from $F(A)$ by a translation only, hence it may be supposed that where $\beta \geq 0$, $|\epsilon| = 1$,

$$A = \begin{pmatrix} \alpha & 2\beta\,\epsilon \\ 0 & -\alpha \end{pmatrix}.$$

If $\beta = 0$, A is normal and the field of values of a normal matrix is known to be convex. Hence suppose $\beta \neq 0$. Then a similarity transformation with the matrix ve $=$ diag $(1, \epsilon)$ followed by division by β replaces the matrix A by one of the same form in which $\beta\epsilon = 1$.

Except for a scalar factor of unit modulus any unit vector in 2-space has the form of x where

$$x^H = (\cos \theta/2, e^{-i\omega} \sin \theta/2),$$

so that

$$\lambda(A, x) = \alpha \cos \theta + \cos \omega \sin \theta + i \sin \omega \sin \theta$$
$$= \rho + i\sigma.$$

Hence

$$(22) \qquad (\rho - \alpha \cos \theta)^2 + \sigma^2 = \sin^2 \theta.$$

Let

$$\tau = \cos \theta.$$

Then the roots of (22), considered as an equation in τ, are real if and only if

(23) $$\rho^2 + (1 + \alpha^2)\sigma^2 \leq 1 + \alpha^2.$$

Moreover, by setting first $\tau = 1 + \nu$ and then $\tau = -1 - \nu$, it is readily verified that neither equation in ν has positive real roots. Consequently any point $\rho + i\sigma$ with ρ and σ satisfying (23) is a point in the field of values of the second-order matrix A and there are no others. But this is a region bounded by an ellipse and is certainly convex. This completes the proof that $F(A)$ is convex for arbitrary matrices A.

If A is normal, then $F(A)$ coincides with the convex hull of the roots of A, but the converse does not hold, so the property is not characteristic for normal matrices. However, a partial result can be stated as follows (Kippenhahn): *If λ is a root of A lying on the boundary of $F(A)$, and if x is a proper vector belonging to λ, then x is a proper vector also of A^H.*

If A is replaced by a suitable A_θ, and

$$\lambda = \rho + i\sigma,$$

then ρ will be a maximal root of R. Hence

$$Ax = Rx + iSx = (\rho + i\sigma)x,$$

and

$$Rx = \rho x.$$

Therefore

$$Sx = \sigma x.$$

But

$$A^H = R - iS,$$
$$A^H x = (\rho - i\sigma)x,$$

which establishes the result.

If A is normalizable (3.2.7) with α linear and β constant imply that *no point of the boundary of $F(A)$ can be farther than $(\sigma - \sigma^{-1})/2$ from the convex hull of the proper values of A*, where $\sigma = \sigma(A)$ is defined in Section 3.1.

3.4. Inequalities of Wielandt and Kantorovich. The spectral condition number $\kappa(A)$ of an arbitrary matrix A is defined in terms of the spectral norm:

(1) $$\kappa(A) = \| A \| \| A^{-1} \|.$$

For a diagonalizable matrix A the condition number of a diagonalizing matrix P has figured prominently in the discussion of inclusion theorems (Section 3.2). The condition number of the matrix itself will arise in connection with methods of inverting it. The inequalities of Wielandt and of Kantorovich exhibit important geometric properties of the number.

It will be assumed that A is nonsingular, hence $\kappa(A)$ finite. Let

(2)
$$M = A^{\mathrm{H}}A.$$

Then
$$\kappa(M) = \kappa^2(A).$$

Let x and y be any orthonormal pair of vectors, and form the two-dimensional section

(3)
$$G = (x, y)^{\mathrm{H}}M(x, y).$$

If γ_1 and $\gamma_2 \leq \gamma_1$ are the proper values of G, and if the singular values of A are

$$\sigma_1 \geq \sigma_2 \geq \cdots \geq \sigma_n,$$

then $\sigma_i{}^2 = \lambda_i(M)$, and

$$\sigma_1{}^2 \geq \gamma_1 \geq \gamma_2 \geq \sigma_n{}^2.$$

Consider

$$1 - \frac{|\,x^{\mathrm{H}}My\,|^2}{x^{\mathrm{H}}Mx\,y^{\mathrm{H}}My} = \frac{4\delta(G)}{(x^{\mathrm{H}}Mx + y^{\mathrm{H}}My)^2 - (x^{\mathrm{H}}Mx - y^{\mathrm{H}}My)^2}$$
$$= \frac{4\gamma_1\gamma_2}{(\gamma_1 + \gamma_2)^2 - (x^{\mathrm{H}}Mx - y^{\mathrm{H}}My)^2}.$$

Hence if x and y are allowed to vary while γ_1 and γ_2 remain fixed, the right member is minimized and

$$\frac{|\,x^{\mathrm{H}}My\,|^2}{x^{\mathrm{H}}Mx\,y^{\mathrm{H}}My}$$

is maximized when $x^{\mathrm{H}}Mx = y^{\mathrm{H}}My$. When this is true,

$$\frac{|\,x^{\mathrm{H}}My\,|^2}{x^{\mathrm{H}}Mx\,y^{\mathrm{H}}My} = \frac{(\gamma_1/\gamma_2 - 1)^2}{(\gamma_1/\gamma_2 + 1)^2}.$$

But this is monotonically increasing in γ_1/γ_2, hence is greatest when

$$\gamma_1 = \sigma_1{}^2, \qquad \gamma_2 = \sigma_n{}^2.$$

To maximize (4), therefore, with respect to all orthonormal pairs x and y, these vectors must be taken in the planes of v_1 and v_n, the proper vectors of M belonging to its roots $\sigma_1{}^2$ and $\sigma_n{}^2$, and unless $\sigma_1 = \sigma_n$,

$$x = (v_1 + v_n)/\sqrt{2},$$
$$y = (v_1 - v_n)/\sqrt{2}.$$

Evidently

(4)
$$\kappa = \kappa(A) = \sigma_1/\sigma_n.$$

Hence if the angle θ is defined by

(5) $$(\kappa - \kappa^{-1})/(\kappa + \kappa^{-1}) = \cos \theta,$$

the result is *the inequality of Wielandt,*

(6) $$\frac{|x^H My|^2}{x^H Mxy^H My} \leq \cos^2 \theta.$$

From (2) it is clear that the left member represents the squared cosine of the angle between Ax and Ay. Hence, *if A is considered to represent a transformation, θ as defined by (5) is the minimal angle between the transforms of any orthogonal pair of vectors.* Note that

(7) $$\kappa = \cot \theta/2.$$

The Kantorovich inequality can be obtained as a special case by setting

$$y = (x^H x)M^{-1}x - (x^H M^{-1}x)x.$$

Let

(8) $$\mu_\nu = x^H M^\nu x.$$

Then

$$y = \mu_0 M^{-1}x - \mu_{-1}x,$$
$$x^H My = \mu_0{}^2 - \mu_1\mu_{-1},$$
$$y^H My = \mu_{-1}(\mu_1\mu_{-1} - \mu_0{}^2).$$

Hence

$$\frac{\delta[(x, y)^H M(x, y)]}{x^H Mxy^H My} = \frac{\mu_0{}^2}{\mu_1\mu_{-1}}.$$

But (6) implies that the left member of this is not less than $\sin^2 \theta$. From this follows the Kantorovich inequality,

(9) $$(x^H x)^2 \geq x^H Mxx^H M^{-1}x \sin^2 \theta.$$

Evidently also

(10) $$\mu_\nu{}^2 \geq \mu_{\nu-1}\mu_{\nu+1} \sin^2 \theta$$

for any ν.

A generalization of the Wielandt inequality can be made by considering vectors x and y at an arbitrary but fixed angle ϕ. It is easily verified that the characteristic roots of the matrix

$$\begin{pmatrix} \cos \phi/2 & -\sin \phi/2 \\ \cos \phi/2 & \sin \phi/2 \end{pmatrix} \begin{pmatrix} \gamma_1 & 0 \\ 0 & \gamma_2 \end{pmatrix} \begin{pmatrix} \cos \phi/2 & \cos \phi/2 \\ -\sin \phi/2 & \sin \phi/2 \end{pmatrix}$$

are $2\gamma_1 \cos^2 \phi/2$ and $2\gamma_2 \sin^2 \phi/2$. Therefore, if x and y are at an angle ϕ, these quantities replace γ_1 and γ_2 in (4), and hence in (5), and

(11) $\dfrac{|\,x^H M y\,|^2}{x^H M x y^H M y} \leq (\kappa \cot \phi/2 - \kappa^{-1} \tan \phi/2)^2/(\kappa \cot \phi/2 + \kappa^{-1} \tan \phi/2)^2.$

Denote the left member by $\cos^2 \phi_A$. By a standard trigonometric identity, this can be expressed in terms of $\cot \phi_A/2$, and the following conclusion results: *if x and y enclose an angle of not less than ϕ, then Ax and Ay enclose an angle ϕ_A satisfying*

(12) $\cot \phi_A/2 \leq \kappa(A) \cot \phi/2,$

where ϕ is defined by (1) with (5) or (7) and all angles are taken in the first quadrant.

REFERENCES

There are rather obvious relations between exclusion theorems and determinantal bounds, especially conditions for the nonvanishing of the determinant, as mentioned at the end of Chapter 2, and although the treatment here has stressed the exclusion theorems, it seems appropriate to list also treatments of determinantal bounds, of which there are many. Taussky (1949) summarizes the literature on the latter formulation. The basic theorem for real determinants, that if the diagonal is dominant the determinant is nonvanishing, seems to be due to Levy (1881), although the names of Hadamard (1903) and of Minkowski (1900) are often associated. Gerschgorin (1931) applied the theorem to the determination of exclusion sets; Brauer and his students, and Parodi, in a number of publications of which a few are listed, have further elaborated the basic idea; Householder (1956) showed the theorem to be a consequence of the norm principle. Returning to determinantal bounds, reference may be made to Price (1951), and to papers by Ostrowski and by Haynsworth. More general exclusion theorems have been obtained by Fiedler and Pták (1960) and Fiedler (1960). The perturbation theorems of Section 3.1 are due to Householder and Bauer (1961), and (3.1.18) generalizes a theorem of Wielandt (1955). See also Bauer and Fike (1960).

Browne (1930) summarizes earlier results giving bounds for the real and imaginary parts, separately, of the proper values, notable contributions being by Hirsch (1901), Bendixson (1902), Bromwich (1906), Pick (1922), Browne himself, and, later Rohrbach (1931) and Barankin (1945). These considerations lead naturally into the field of values, and here mention may be made of a series of papers by Parker, and of Toeplitz (1918), Hausdorf (1919), Kippenhahn (1951), Moyles and Marcus (1955), and Givens (1952). The proof given here of the convexity of the field of values is due to Hans Schneider (unpublished), but the fact was first established by Toeplitz (1918).

For a survey of earlier results on inclusion theorems see Bauer and

Householder (1960a). Particular mention may be made here of Temple (1929), Weinstein (1934), and the several papers by Wielandt, and Bückner (1952).

That the roots of a principal minor separate those of a symmetric matrix was shown by Cauchy (1829). The minimax theorem is to be found in Courant-Hilbert, and the more elaborate generalization is given by Wielandt (1955). Special attention should be called also to Lidskiĭ (1950). A number of other papers dealing with somewhat related topics may be recognized from their titles.

Kantorovič (1948) and Wielandt (1953) state the inequalities given here, and the generalization is due to F. L. Bauer (1961), who also showed that Wielandt's inequality implies that of Kantorovič. See also Greub and Rheinboldt (1959). Schopf (1960) generalized the Kantorovič inequality in a different direction, and provided a new and elegant proof of the inequality itself.

PROBLEMS AND EXERCISES

1. Let D be the diagonal of $A = D + B$, and let

$$M = |A| = |D| + |B|, \qquad \mu = \rho(M).$$

Then any proper value of A lies in at least one of the circular disks centered at δ_i and passing through μ (Kotelyanski):

$$|\lambda - \delta_i| \leq |\mu - \delta_i|.$$

2. Let $P \geq 0$ be nonnegative with a null diagonal, and let

$$R = \mathrm{diag}\,(\rho_1, \cdots, \rho_n) > 0$$

be nonsingular. If for every matrix B such that $|B| = P$ and diagonal D, every proper value of $A = D + B$ lies in at least one of the circular disks

$$|\lambda - \delta_i| \leq \rho_i,$$

then (Fan)

$$\rho(R^{-1} P) \leq 1.$$

3. Prove Wielandt's refinement on Bendixson's theorem as follows: Let $A = H + S$, where H is Hermitian and S skew-Hermitian. Let $\eta_j = \lambda_j(H)$, $\sigma_k = \lambda_k(S)$. By a hyperbolic region will be meant either of the two regions into which the complex plane is separated by a rectangular hyperbola with axes parallel to the real and imaginary axes. Then any hyperbolic region containing all points $\eta_j + \sigma_k$ contains all roots of A.

4. Let A be normal; let $Ax = Dx \neq 0$, where D is diagonal, and if $e_i^T x = e_i^T A x = 0$, then $\delta_i = 0$. Then any circular disk that contains every δ_i contains at least one root of A (Walker and Weston; for A Hermitian the theorem is due to Collatz).

5. Let A and B be Hermitian and B positive definite. For any x, let $y = Ax$, $z = Bx$, and $y = Dz$, where D is diagonal, and if $\eta_i = \zeta_i = 0$ then $\delta_i = 0$. If $|\lambda - \mu| \leq \rho$ is a disk that contains every δ_i, then the disk $|\lambda - \mu| \leq \rho \,[\text{lub }(B)/\text{glb }(B)]^{\frac{1}{2}}$ contains at least one root κ of $\delta(A - \kappa B) = 0$ (Bartsch).

6. For the normal matrix

$$\begin{pmatrix} A & a \\ a^{\mathrm{H}} & \alpha \end{pmatrix}$$

the disk $|\lambda - \alpha| \leq \| a \|_s$ is an inclusion region.

7. The inclusion regions defined by (3.2.9) are circular disks, and the disk of minimal radius is given by $\eta = 0$. The center of the minimal disk is μ_{12}/μ_{11}. Let the Riemann sphere be constructed with this center and with the radius $(\mu_{22} - \mu_{12}\mu_{21}/\mu_{11})^{\frac{1}{2}}$, and let the minimal disk be projected from the north pole upon the sphere, the projection forming a spherical cap. Then any other disk of the family can be obtained by sliding the cap over the surface of the sphere and projecting back upon the plane (Wielandt). A similar transformation relates the regions of any other of the families (3.2.6) [for variable η and fixed polynomials $\alpha(\lambda)$ and $\beta(\lambda)$].

8. By considering the νth compounds of the matrix $M = A^{\mathrm{H}}A$, and of an arbitrary matrix V, obtain the following generalization of the Kantorovich inequality (Fan and Schopf): If V_ν is a matrix of ν columns, then

$$\delta^2(V_\nu{}^{\mathrm{H}}V_\nu) \geq \delta(V_\nu{}^{\mathrm{H}}MV_\nu)\delta(V_\nu{}^{\mathrm{H}}M^{-1}V_\nu)\sin^2 \Theta_\nu,$$

where

$$\cos \Theta_\nu = (\sigma_1{}^2 \cdots \sigma_\nu{}^2 - \sigma_{n-\nu+1}^2 \cdots \sigma_n{}^2)/(\sigma_1{}^2 \cdots \sigma_\nu{}^2 + \sigma_{n-\nu+1}^2 \cdots \sigma_n{}^2).$$

9. If a Hermitian matrix has a root λ of multiplicity ν, then λ is a root of multiplicity $\nu - 1$ of every $(n - 1)$-dimensional section.

10. A Jacobi matrix is of the form

$$A = \begin{pmatrix} \alpha_1 & \beta_1 & 0 & 0 & \cdots \\ \gamma_1 & \alpha_2 & \beta_2 & 0 & \cdots \\ 0 & \gamma_2 & \alpha_3 & \beta_3 & \cdots \\ 0 & 0 & \gamma_3 & \alpha_4 & \cdots \\ & \cdots & \cdots & \cdots & \end{pmatrix}$$

where every α_i is real and every $\gamma_i\beta_i \geq 0$. If $\gamma_i\beta_i = 0$ only when $\beta_i = \gamma_i = 0$, then there exists a nonsingular diagonal matrix D such that DAD^{-1} is symmetric. Hence any Jacobi matrix has only real roots.

11. Show that the characteristic polynomial of the Jacobi matrix of the last problem can be obtained from the recursion

$$\phi_0 = 1,$$
$$\phi_1 = \lambda - \alpha_1,$$
$$\phi_i = (\lambda - \alpha_i)\phi_{i-1} - \beta_{i-1}\gamma_{i-1}\phi_{i-2}, \quad i > 1.$$

Hence show that A can have a multiple root only if some $\beta_i = 0$ or some $\gamma_i = 0$. Moreover, the number of roots exceeding any number λ' is equal to the number of sign variations in the sequence $\phi_0, \phi_1(\lambda'), \phi_2(\lambda'), \cdots, \phi_n(\lambda')$ (hence the ϕ_i form a Sturm sequence).

12. For a given normalizable matrix A and vector $v \neq u$, among all monic polynomials $\alpha(\lambda)$ of degree ν (less than the degree of v), the polynomial $\phi_\nu(\lambda)$ has the least norm, and hence minimizes the right member of (3.2.5) for given $\beta(\lambda)$.

13. For a normal matrix A and a vector $v \neq 0$, any circle through μ and $(\mu_{22} - \mu_{12}\bar{\mu})/(\mu_{21} - \mu_{11}\bar{\mu})$ is a separation locus (for A Hermitian, the theorem is due to Temple).

14. The Lanczos polynomials $\chi_\nu(\lambda)$ are separation polynomials, and, in fact, if $\alpha(\lambda)$ is any polynomial of degree less than ν, then $\alpha(\lambda)\chi_\nu(\lambda)$ is a separation polynomial.

15. Among all polynomials of degree ν which assume a specified value at a given point μ_1, the polynomial of least norm has the form $\gamma\kappa_\nu(\lambda, \mu)$ with γ a constant (Szego).

16. Let $A = D + B$, $D = \text{diag}(1, 0, \cdots)$ where remaining elements are nonpositive, and

$$| B | \leq (ee^T - I)\epsilon.$$

Show by Gershgorin's theorem that, if

$$\epsilon^{-1} > (n - 2) + 2\sqrt{n - 1},$$

there is exactly one root in the circle with center unity and radius

$$\epsilon(n - 1)^{\frac{1}{2}}.$$

17. If

$$B = \begin{pmatrix} \alpha & a^H \\ a & A \end{pmatrix}$$

is Hermitian, $\eta = \lambda_1(B) - \lambda_1(A)$, then (Fan)

$$\lambda_1(B) \leq \alpha + \| a \|_s^2/\eta.$$

18. If A is normal, with roots satisfying $| \alpha_1 | \geq \cdots \geq | \alpha_n |$, and U is unitary, then (Loewy and Brauer)

$$| \alpha_1 | \geq | \lambda_i(AU) | \geq | \alpha_n |.$$

Hence if

$$U = \begin{pmatrix} U_{11} & U_{12} \\ U_{21} & U_{22} \end{pmatrix},$$

then

$$\delta \left[\begin{pmatrix} U_{11} - \mu I & U_{12} \\ U_{21} & U_{22} \end{pmatrix} \right] = 0 \Rightarrow | \mu | \geq 1.$$

19. If A is positive definite with roots $\alpha_1 \geq \cdots \geq \alpha_n$, and X is any matrix of $\nu \leq n$ columns, then

$$\lambda_1 \cdots \lambda_\nu \delta(X^H X) \geq \delta(X^H A X) \geq \lambda_{n-\nu+1} \cdots \lambda_n \delta(X^H X).$$

20. Let A be normal and X as above and of rank ν. Then

$$\delta(X^H A X)/\delta(X^H X)$$

lies in the convex hull of the products ν at a time of the α_i.

21. Let $B = Q^H Q$ be positive definite, $A = Q^H C Q$ be Hermitian. If $V^H C V = \Gamma = \text{diag}(\gamma_1, \gamma_2, \cdots, \gamma_n)$, $V^H V = I$, then

$$X = Q^{-1} V$$

satisfies

$$X^H A X = \Gamma, \qquad X^H B X = I.$$

Moreover

$$A x_i = \gamma_i B x_i.$$

22. With A and B as above, let $\gamma_1 \geq \gamma_2 \geq \cdots \geq \gamma_n$ be solutions of $\delta(A - \gamma B) = 0$. Show that

$$\max_{E_m} \min_{x \in E_m} x^H A x / x^H B x = \gamma_m,$$

$$\min_{E_m} \max_{x \in E_m} x^H A x / x^H B x = \gamma_{n-m+1},$$

where E_m is any m-dimensional linear subspace of the space E_n.

23. With the γ_i as above, $\alpha_1 \geq \alpha_2 \geq \cdots \geq \alpha_n$ the roots of A, $\beta_1 \geq \cdots \geq \beta_n > 0$ the roots of B, then (Ostrowski)

$$\beta_1 \gamma_i \geq \alpha_i \geq \beta_n \gamma_i.$$

24. If B is positive definite and A Hermitian, then

$$x^H B x > | x^H A x | \qquad \text{for every } x$$

if and only if $\rho(B^{-1} A) < 1$. In this event

$$\beta_\nu > \alpha_\nu, \qquad \beta_\nu > \alpha_{n-\nu+1}.$$

25. If A is Hermitian, it is positive definite if and only if there is a complete nested sequence of principal submatrices each having a positive determinant.

26. Let A be positive definite

$$A = V \Lambda V^H, \qquad V^H V = I, \qquad \Lambda = \text{diag}\,(\lambda_1, \cdots, \lambda_n).$$

Let

$$\gamma_i = \log \lambda_i, \qquad \Gamma = \text{diag}\,(\gamma_i, \cdots, \gamma_n),$$
$$C = V \Gamma V^H.$$

Then

$$A = \exp C, \qquad C = \log A.$$

Hence obtain (3.3.11) and (3.3.12) from (3.3.13).

27. Let $0 \le \mu \le 1$, let C be nonnegative semidefinite, B positive definite, and let $A(\mu) = \mu B + (1 - \mu)C$. If

$$\alpha_1 \ge \cdots \ge \alpha_n, \qquad \beta_1 \ge \cdots \ge \beta_n, \qquad \gamma_1 \ge \cdots \ge \gamma_n$$

are the roots of A, B, and C, respectively, then

$$\prod_\nu^n \alpha_i \ge \prod_\nu^n \beta_i{}^\mu \prod_\nu^n \gamma_i^{1-\mu}.$$

[N.B. By applying Exercise 21 to the case $\nu = 1$, A, B, and C can be supposed diagonal. Thence consider sections in subspaces.]

28. Let $C = AB$, where A and B are arbitrary with singular values $\alpha_1 \ge \alpha_2 \ge \cdots \ge 0$, $\beta_1 \ge \beta_2 \ge \cdots \ge 0$, $\gamma_1 \ge \gamma_2 \ge \cdots \ge 0$. By forming $WH = AB$, $W^H W = I$, $H^H = H$, and applying the Cauchy inequality and the minimax theorems to $(x^H H x)^2$, show that if $i + j + 1 \ge n$, then

$$\gamma_{i+j+1} \le \alpha_{i+1}\beta_{j+1}.$$

29. If A has the singular values $\alpha_1 \ge \alpha_2 \ge \cdots \ge \alpha_n \ge 0$, then

$$\alpha_{i+1} = \max_{E_{n-i}} \min_{x \in E_{n-i}} \| Ax \|_s / \| x \|_s$$
$$\alpha_i = \min_{E_i} \max_{x \in E_i} \| Ax \|_s / \| x \|_s,$$

where E_{n-i} and E_i are arbitrary subspaces of dimension $n - i$ and i, respectively.

30. If $D = A + B$, where A and B are as in Exercise 28, and $\delta_1 \ge \cdots \ge \delta_n \ge 0$ are the singular values of D, then (Fan)

$$\delta_{i+j+1} \le \alpha_{i+1} + \beta_{j+1}.$$

31. Let $\sigma_1(A) \ge \sigma_2(A) \ge \cdots \ge \sigma_n(A) \ge 0$ be the singular values of A. Apply Exercise 28 to show that

$$\| A \|^2 = \sigma_1{}^2(A) + \cdots + \sigma_k{}^2(A)$$

defines a norm, consistent with the Euclidean vector norm.

32. For any matrix A, if

$$| \lambda_1(A) | \geq \cdots \geq | \lambda_n(A) |, \qquad \sigma_1(A) \geq \cdots \geq \sigma_n(A),$$

then

$$\prod_1^k \sigma_i(A) \geq \prod_1^k | \lambda_i(A) |.$$

33. For any matrices A and B, if $C = A + B$ and the singular values are in decreasing order, then

$$\sum_i^k \sigma_i(C) \leq \sum_1^k \sigma_i(A) + \sum_1^k \sigma_i(B).$$

34. Let A be of order n with

$$B = \begin{pmatrix} \alpha & a^{\mathrm{H}} \\ a & A \end{pmatrix}$$

Hermitian. Then if β is a simple root of A,

$$\delta(\beta I - B) \neq 0$$

unless the matrix $(a, \beta I - A)$ is of rank $n - 1$ or less. Hence if $(a, \beta I - A)$ is of rank n for every root β of A, then the roots of A strictly separate those of B.

35. If A is positive definite, $\delta = \delta(A)$, $\tau = \tau(A)$, $\lambda_\nu = \lambda_\nu(A)$, show that (Wegner, 1953)

$$\delta(n - 1)^{n-1} \leq \lambda_\nu(\tau - \lambda_\nu)^{n-1}.$$

Hence if $\lambda_1 \geq \lambda_2 \geq \cdots \geq \lambda_n$, then

$$\lambda_n \geq \tau - (\tau^n - \alpha)^{1/n}, \qquad \alpha = n\delta(n - 1)^{n-1},$$

and

$$\lambda_1 \leq \tau - (n - 1)(\delta/\mu)^{1/(n-1)},$$

if

$$\mu \geq \lambda_1.$$

36. For any nonsingular A, using spherical norms, show that (Kato, 1960)

$$\mathrm{lub}\,(A^{-1}) \leq \mathrm{lub}^{n-1}\,(A)/| \delta(A) |.$$

CHAPTER 4

The Solution of Linear Systems: Methods of Successive Approximation

4.0. Direct Methods and Others. The general problem to be considered in this and in the next chapter is that of finding X to satisfy

(1) $$AX = M,$$

where A will be *assumed nonsingular* unless singularity is explicitly permitted. If $M = I$, then $X = A^{-1}$, and the problem is to invert A. If $M = h$ is a single vector, the problem is the ordinary one of solving a system $Ax = h$ to find the vector solution x. Often the solution is required for several distinct vectors h but the same A, in which case the distinct vectors h can be considered as columns of a matrix M, whether or not they are all given in advance. The number of columns expected to be in M will affect the choice of a method, and since

(2) $$X = A^{-1}M,$$

if there are to be many columns in M there may be an advantage in finding A^{-1} and then proceeding by simple matrix multiplication. For error estimates it is advantageous to find at least an approximation to A^{-1} even when only a single vector x is required, as will be brought out later. In any event, attention will be directed primarily toward only the two cases, $M = I$, and $M = h$.

A *direct method* provides an algorithm guaranteed to lead to the exact solution after a finite number of mathematical operations. For most methods only arithmetic operations are required, but some also require square roots. In any event, the machine operations deviate from the mathematical operations, and the result of the machine operations will be an approximation only, which may or may not be adequate. If not adequate, the application of the direct method will need to be followed by a *self-correcting method of successive approximation*, which is guaranteed only to provide a sequence that converges to the true solution. Even this guarantee is, however, good only when the operations are the prescribed mathematical ones, and some analysis may be required to make sure how far terms actually computed with the machine operations do in fact improve the approximations.

Direct methods will be discussed in Chapter 5; here only methods of successive approximation will be considered. The simplest of these prescribe certain operations to be performed upon an arbitrary approximation in order to obtain a better one. These operations are often the same, or nearly so, from one step to the next, and hence methods of successive approximation are often called *iterative*. If successive steps are identical, the method is called *stationary*. However, there are direct methods, such as the method of conjugate gradients (Chapter 5), that are essentially iterative in character, hence this term does not properly describe the class. The term *relaxation* is also used with varying connotations in connection with methods of successive approximation. This term was introduced by Southwell to designate a method (Section 4.2) in which each step to be taken is selected according to a certain criterion from among a finite number of allowable alternatives. The term will be so used here. The selection is easy in hand computation but not in machine computation (at least on existing machines), hence a method of relaxation is not recommended for machine computation.

A method of successive approximation may be applied in order to improve an approximate solution already obtained by a direct method, but if it converges, it converges for any initial "approximation," however poor. For nonlinear problems the initial approximation must ordinarily lie in some neighborhood of a solution. Because of this freedom available in solving linear systems, the method may also be applied at the outset to an initial "approximation," obtained intuitively or arbitrarily, without the application of a direct method. The advantage of doing this is that for a method of successive approximation the programing is usually fairly simple, and the fast-storage requirements modest, even for very large systems. This is especially true of the large sparse matrices that usually arise in the solution of differential equations. The disadvantage is that convergence may be very slow, and modifications of the method for accelerating convergence will inevitably complicate the method and augment the storage requirements. Decisions about the use of a direct method at the outset, or about the use of accelerating devices, can only be made by considering both the system itself and the available facilities.

4.0.1. *Errors and Remainders.* Suppose an approximation x^* has been obtained by whatever method applied to the equations

$$(3) \qquad\qquad Ax = h,$$

with A nonsingular, and let

$$(4) \qquad\qquad s = x - x^*, \qquad r = h - Ax^* = As.$$

Then for any norm

$$(5) \qquad\qquad \| s \| \leq \| A^{-1} \| \, \| r \|.$$

Since r is available but s is not, relation (5) provides the only clue as to the magnitude of the actual error s. Suppose C is an approximation to A^{-1} and let

(6) $H = I - CA, \qquad \| H \| < 1.$

Then

(7)
$$A^{-1} = (I - H)^{-1}C,$$
$$\| A^{-1} \| \leq \| C \|/(1 - \| H \|).$$

Hence (5) can be applied provided any reasonable approximation C of A^{-1} is known. This bears out the remark made above about the advantage of an approximate inverse [see also (13) below].

The remainder of this chapter will be separated into three sections, corresponding to three broad classes of methods. The first applies only to inversion. All others apply in general to the system (3) and can be described in general terms as follows: *Given an initial approximation x_0, to form a sequence of approximations by the recursion*

(8) $x_{\nu+1} = x_\nu + C_\nu r_\nu,$

where

(9) $s_\nu = x - x_\nu, \qquad r_\nu = h - Ax_\nu = As_\nu.$

If

(10) $H_\nu = I - C_\nu A,$

then

(11) $s_{\nu+1} = H_\nu s_\nu, \qquad r_{\nu+1} = AH_\nu A^{-1}r_\nu.$

The several methods applicable to (3) differ in the formation of the matrices C_ν, or, equivalently, H_ν. It is certainly sufficient for convergence that

(12) $\| H_\nu \| \leq - \epsilon < 1$

in some norm for every ν. Such methods will be called *norm-reducing* and these and related methods are discussed in the last section. Any such method would converge whatever the initial vector, and the choice of the C_ν is independent of the x_ν. Such methods are available only for matrices A of certain special forms. In the *methods of projection*, however, each H_ν is a projector, hence $\| H_\nu \| \geq 1$, and such an H_ν will reduce only those vectors lying in a certain subspace. Necessarily, therefore, the H_ν must vary from step to step.

There is still another method, however, that can be disposed of rather briefly, and that may be called the method of *successive refinement*. According to (9), each s_ν satisfies with r_ν a system of the form (3), with the same matrix A. Hence suppose $x^* = x_0$ is the result of applying a direct method to (3).

The result of this computation is expressible

$$x_0 = Ch,$$

where C is the matrix representing the operations that have been performed upon h in the course of applying the given method. Presumably, then, C is some approximation to A^{-1}. Although it is not given explicitly, if the same method were applied to the system

$$As_0 = r_0,$$

many of the results of the previous calculation can be carried over, and the result will be the formation of a vector Cr_0, with the same matrix C, hence

$$x_1 = x_0 + Cr_0$$

should be an improvement upon the approximation x_0. This amounts to forming

$$s_1 = Hs_0,$$

and the process can be repeated. Here $C_\nu = C$ for every ν, and $H_\nu = H$.

In this method C and H are not given explicitly, although in most methods they are. From (9), (10), and (11), it follows that

$$x_{\nu+1} - x_\nu = C_\nu As_\nu,$$

$$s_\nu = (C_\nu A)^{-1}(x_{\nu+1} - x_\nu),$$

$$s_{\nu+1} = [(C_\nu A)^{-1} - I](x_{\nu+1} - x_\nu).$$

But

$$(C_\nu A)^{-1} - I = H_\nu(I - H_\nu)^{-1}.$$

Hence

(13) $\| s_{\nu+1} \| \leq \| x_{\nu+1} - x_\nu \| \, \| H_\nu \| / (1 - \| H_\nu \|),$

provided, of course, $\| H_\nu \| < 1$. Thus, in iterative methods, if H_ν, or some bound for $\| H_\nu \|$, is available, an estimate of error can be made by (13).

4.1. The Inversion of Matrices. Let X_ν be an approximation to A^{-1}, let

(1) $R_\nu = AS_\nu = I - AX_\nu$

represent the *residual*, while S_ν represents the *error* matrix,

(2) $S_\nu = A^{-1} - X_\nu,$

and consider the possibility of finding a sequence of matrices C_ν such that if

(3) $X_{\nu+1} = X_\nu + C_\nu R_\nu,$

the sequence of matrices X_ν will approach A^{-1} in the limit. This means that the matrices of each sequence

(4) $R_{\nu+1} = (I - AC_\nu)R_\nu, \qquad S_{\nu+1} = (I - C_\nu A)S_\nu,$

must vanish in the limit. Since A is assumed to be nonsingular, or the problem would have no meaning, the vanishing of one sequence is equivalent to the vanishing of the other. Moreover, since

$$I - AC_\nu = A(I - C_\nu A)A^{-1},$$

the multiplying matrices are similar, and have the same spectral radius.

Obviously the choice $C_\nu = A^{-1}$ would be optimal at any stage, since this would lead to $R_{\nu+1} = 0$, $X_{\nu+1} = A^{-1}$, exactly. Since the true inverse is not known, presumably the best current approximation would be the optimal available choice for C_ν, and, in fact, if

(5) $C_\nu = X_\nu,$

it follows immediately from (1) that

(6) $R_{\nu+1} = R_\nu^2, \qquad S_{\nu+1} = S_\nu A S_\nu.$

Hence, with any norm,

(7) $\| R_{\nu+1} \| \leq \| R_\nu \|^2, \qquad \| S_{\nu+1} \| \leq \| A \| \, \| S_\nu \|^2.$

This shows that if $\| R_\nu \| < 1$, the sequence converges quadratically. In fact, for convergence it is necessary and sufficient that

$$\rho(R_\nu) < 1$$

for any ν. The iteration may also be written in the form

(8) $X_{\nu+1} = X_\nu(2I - AX_\nu).$

Note that the method requires at each step two matrix multiplications, hence, in general, $2n^3$ scalar multiplications. From this point of view it is uneconomical, since only approximately n^3 or fewer scalar multiplications are required for a complete inversion by any of several standard direct methods. However, one can define

$$C_\nu = X_0$$

for every ν, thus forming two matrix products at the first step but only one for each following step. The process still converges, but only in the first order and not the second.

Conversely, it is possible to construct sequences with convergence of order three or higher, but at the expense of additional matrix multiplications per step. From (1) it follows that

$$A^{-1} = X_\nu(I + R_\nu + R_\nu^2 + \cdots),$$

assuming always that $\rho(R_\nu) < 1$, hence (8) is obtained by taking the first two terms on the right for the next approximation. If the first three terms are used, making

$$X_{v+1} = X_v(I + R_v + R_v^2),$$

then

$$R_{v+1} = R_v^3.$$

For this three multiplications are required, and four to obtain convergence of fourth order. But convergence of eighth order can be obtained with only two additional multiplications by the formula

$$X_{v+1} = X_v(I + R_v)(I + R_v^2)(I + R_v^4).$$

This amounts to summing the series, and by this type of factorization it is possible to double the number of terms after forming only two additional matrix products.

To return to the original second-order process it is, in fact, only Newton's method applied to the matrix equation

$$AX - I = 0.$$

It could equally well be applied to A^H which is equivalent to defining

$$R_v = I - X_v A,$$
$$X_{v+1} = (I + R_v)X_v,$$

thus reversing the order for each product. To form R_v, whether by this formula or by (1), and whether another step is to be taken or not, is quite natural as a means of testing the adequacy of X_v as an approximation.

It has been pointed out already that the two matrices, $I - AX_v$ and $I - X_v A$, are similar and hence have the same spectral radius. Nevertheless, it is possible that individual elements of one could be large even though all elements of the other are small. Thus, let

$$Au = \lambda u, \qquad v^T A = \mu v^T.$$

Then

$$A(A^{-1} - uv^T) = I - \lambda uv^T,$$
$$(A^{-1} - uv^T)A = I - \mu uv^T.$$

Hence if λ is large and μ small, and if

$$X_v = A^{-1} - uv^T$$

then

$$I - AX_v = \lambda uv^T, \qquad I - X_v A = \mu uv^T,$$

and the elements of the two matrices are in the ratio λ/μ. In this sense X_v may be considered a poor right-hand inverse but a good left-hand inverse. However, if $\lambda \neq \mu$, then $v^T u = 0$, hence uv^T is nilpotent, and therefore

$$\rho(\lambda uv^T) = \rho(\mu uv^T) = 0.$$

Either remainder, therefore, in this special case, has a spectral radius of zero.

If X_v is any approximation to A^{-1}, then

$$R_v = I - AX_v,$$
$$A^{-1} - X_v = X_v R_v (I - R_v)^{-1};$$

hence if $\| R_v \| < 1$, then

(9) $$\| A^{-1} - X_v \| \leq \| X_v R_v \| / (1 - \| R_v \|).$$

Likewise

(10) $$\| X_v^{-1} - A \| \leq \| R_v A \| / (1 - \| R_v \|).$$

In actual computation, naturally, the arithmetic operations called for by the formulas are replaced by the pseudo-operations of the machine. Hence if X_0 is the initial approximation to A^{-1} assumed to be the matrix actually stored in the machine, then the true result of the first iteration is

$$X_1 = X_0 + (X_0 R_0 *)*,$$

where

$$R_0 * = I - (AX_0)*,$$

the asterisks designating the results of the machine operations. From a consideration of the program it may be possible to obtain bounds δ and δ' such that, in terms of some norm, one is assured that

$$\| R_0 - R_0 * \| \leq \delta, \qquad \| X_0 R_0 * - (X_0 R_0 *)* \| \leq \delta'.$$

Then, since

$$X_0 R_0 - (X_0 R_0 *)* = X_0 (R_0 - R_0 *) + X_0 R_0 * - (X_0 R_0 *)*,$$

it follows that

$$\| X_0 R_0 - (X_0 R_0 *)* \| \leq \| X_0 \| \delta + \delta'.$$

But

$$R_1 = R_0^2 + A[X_0 R_0 - (X_0 R_0 *)*].$$

To be assured that there is actual improvement, at least in terms of the norm being employed, it is necessary that

$$\| R_0^2 \| + \| A \| (\| X_0 \| \delta + \delta') < \| R_0 \|,$$

which is true if

(11) $$\| A \| (\| X_0 \| \delta + \delta') < \| R_0 \| (1 - \| R_0 \|).$$

But only $R_0 *$, and not R_0, is available in fact, and

$$\| R_0 * \| - \delta \leq \| R_0 \| \leq \| R_0 * \| + \delta.$$

Hence the condition $\| R_0 \| < 1$ is assured only if

$$\| R_0 * \| + \delta < 1.$$

If $\| R_0^* \| < \frac{1}{2}$, then (11) is assured if

$$\| A \| (\| X_0 \| \delta + \delta') < (\| R_0^* \| - \delta)(1 + \delta - \| R_0^* \|).$$

If $\| R_0^* \| > \frac{1}{2}$, then (11) is assured if

$$\| A \| (\| X_0 \| \delta + \delta') < (\| R_0^* \| + \delta)(1 - \delta - \| R_0^* \|).$$

If these inequalities fail, the step is of questionable value.

4.2. Methods of Projection. A method of projection is one that assigns at any step a subspace, defined by the $\rho_\nu \geq 1$ linearly independent columns of a matrix Y_ν, and selects a u_ν in such a way that if

$$C_\nu r_\nu = Y_\nu u_\nu,$$

then

$$s_{\nu+1} = s_\nu - Y_\nu u_\nu$$

is reduced in some norm. The simplest norms to treat are the elliptic norms, defined by

$$(1) \qquad\qquad \| y \|^2 = y^H G y$$

where G is positive definite. In these terms, for given G and given Y_ν, it is required to minimize

$$(2) \qquad\qquad s_{\nu+1}^H G s_{\nu+1} = (s_\nu - Y_\nu u_\nu)^H G(s_\nu - Y_\nu u_\nu).$$

Let

$$(3) \qquad\qquad u_\nu = (Y_\nu^H G Y_\nu)^{-1} Y_\nu^H G s_\nu + w_\nu.$$

Then

$$(4) \quad s_{\nu+1}^H G s_{\nu+1} = s_\nu^H G s_\nu - s_\nu^H G Y_\nu (Y_\nu^H G Y_\nu)^{-1} Y_\nu^H G s_\nu + w_\nu^H Y_\nu^H G Y_\nu w_\nu.$$

The last term on the right is a quadratic form in w_ν with positive definite matrix $Y_\nu^H G Y_\nu$, and hence is nonnegative. Since w occurs in that term only, it follows that the right member of (4) is minimized by $w_\nu = 0$. When u_ν is given accordingly by (3) with $w_\nu = 0$, and the norm is defined by (1), it follows that

$$(5) \qquad \| s_\nu \|^2 - \| s_{\nu+1} \|^2 = s_\nu^H G Y_\nu (Y_\nu^H G Y_\nu)^{-1} Y_\nu^H G s_\nu.$$

The' matrix G that defines the norm is assumed to remain the same throughout. The matrices Y_ν, however, which are, ordinarily, single column vectors $(\rho_\nu = 1)$, must change from step to step. The choice for Y_ν is equivalent to the choice

$$(6) \qquad\qquad C_\nu A = Y_\nu (Y_\nu^H G Y_\nu)^{-1} Y_\nu^H G.$$

Evidently it is feasible only if

$$Y_v^H G = V_v^H A,$$

for some V_v, since otherwise C_v could be obtained only after inverting A.

In case A is itself positive definite, the simplest choice is evidently to take $G = A$. This will be considered later. Otherwise, two possibilities suggest themselves. The first is to take

$$(7) \qquad G = A^H A,$$

in which case

$$(8) \qquad C_v = Y_v (Y_v^H A^H A Y_v)^{-1} Y_v^H A^H,$$

$$(9) \qquad \| s_v \|^2 - \| s_{v+1} \|^2 = r_v^H A Y_v (Y_v^H A^H A Y_v)^{-1} Y_v^H A^H r_v.$$

If, as usually happens, Y_v is taken to be a single vector, y_v, so that u_v is a scalar μ_v, then the method is defined by

$$(10) \qquad \begin{aligned} x_{v+1} &= x_v + \mu_v y_v, \\ \mu_v &= y_v^H A^H r_v / y_v^H A^H A y_v, \end{aligned}$$

where

$$(11) \qquad \| s_v \|^2 - \| s_{v+1} \|^2 = | r_v^H A y_v |^2 / y_v^H A^H A y_v = \mu_v y_v^H A^H r_v,$$

the norm being that referred to the ellipsoid $A^{-1}S$, where S is the unit sphere. But this norm of s_v is precisely the spherical norm of r_v.

The right member of (11) can vanish for particular choices of y_v; also if $y_v = \eta s_v$, where η is any nonnull scalar, the right member has the value $\| s_v \|^2$, and the solution has been obtained. In practice the y_v are ordinarily taken to be the vectors e_i; hence $A y_v$ are taken to be the columns of A. These can be taken in rotation, or else at each step the particular e_i can be selected that maximizes the right member of (11). In either case it is seen from (10) that only a single element of x_v is adjusted at each step when the y_v are so chosen. The selection of the optimal e_i (in accordance with the method of relaxation) would formally accelerate convergence, but would necessitate computing the entire vector $r_v^H A$ at each step and selecting the one that gives the largest quotient.

The second obvious choice with a general, nonsingular matrix A, is to take

$$(12) \qquad G = I, \qquad Y_v^H = V_v^H A,$$

where V_v is selected and Y_v computed. For this,

$$(13) \qquad \begin{aligned} C_v &= A^H V_v (V_v^H A A^H V_v)^{-1} V_v^H, \\ \| s_v \|^2 - \| s_{v+1} \|^2 &= r_v^H V_v (V_v^H A A^H V_v)^{-1} V_v^H r_v, \end{aligned}$$

the norm being now the spherical norm. If each V_v is a single column v_v,

then

(14)
$$x_{\nu+1} = x_\nu + \mu_\nu A^H v_\nu,$$

$$\mu_\nu = v_\nu{}^H r_\nu / v_\nu{}^H A A^H v_\nu,$$

and

$$\| s_\nu \|^2 - \| s_{\nu+1} \|^2 = | r_\nu{}^H v_\nu |^2 / v_\nu{}^H A A^H v_\nu = \mu_\nu r_\nu{}^H v_\nu,$$

or

(15) $$s_{\nu+1}^H s_{\nu+1} / s_\nu{}^H s_\nu = 1 - | r_\nu{}^H v_\nu |^2 / [s_\nu{}^H s_\nu v_\nu{}^H A A^H v_\nu].$$

The generalized Wielandt inequality (Section 3.4) provides a bound for the right member of (15). Since

$$H_\nu = I - C_\nu A$$

is a projector, and hence idempotent, it is possible to write

$$\| s_{\nu+1} \|^2 = \| H_\nu s_\nu \|^2 = s_\nu{}^H H_\nu s_\nu = r_\nu{}^H M^{-1} w_\nu,$$

where

$$M = A A^H, \qquad w_\nu = A H_\nu s_\nu.$$

It can be verified that $w_\nu{}^H V_\nu = 0$; hence w_ν is orthogonal to the columns of V_ν, and therefore

$$w_\nu{}^H M^{-1} w_\nu = r_\nu{}^H M^{-1} w_\nu.$$

Hence

$$\| s_{\nu+1} \|^2 / \| s_\nu \|^2 = | r_\nu{}^H M^{-1} w_\nu |^2 / (r_\nu{}^H M^{-1} r_\nu w_\nu{}^H M^{-1} w_\nu).$$

If

$$\kappa(A) = \kappa(A^{-1}) = \kappa(A^H) = \| A \| \, \| A^{-1} \|,$$

it follows that if the angle between r_ν and w_ν is at least ϕ, then

(16)
$$\| s_{\nu+1} \| / \| s_\nu \| \leq \cos \psi,$$

$$\cot \psi/2 = \kappa(A) \cot \phi/2.$$

But w_ν is orthogonal to V_ν. Hence if ϕ is defined by

(17) $$\| V_\nu (V_\nu{}^H V_\nu)^{-1} V_\nu{}^H r_\nu \| / \| r_\nu \| = \sin \phi,$$

which is to say that $\pi/2 - \phi$ is the angle between r_ν and the space of V_ν, then the angle between r_ν and w_ν cannot be less than ϕ. Hence (F. L. Bauer) *if the first quadrant angle ϕ is defined by* (17), *then the iteration defined by* (13) *reduces the norm of the residual in the amount shown by* (16). The best estimate results when the space of V_ν contains r_ν, in which case

$$\| s_{\nu+1} \| / \| s_\nu \| \leq (\kappa + \kappa^{-1}) / (\kappa - \kappa^{-1}).$$

For a single vector v_ν, the simplest choice is an e_i, and these may be taken in rotation, or selection may be made at each step in accordance with some criterion [cf. Ostrowski, 1956]. From the above result, the natural choice is to maximize $|r_\nu^H e_i|$. If this is done, then

$$|r_\nu^H e_i| = \|r_\nu\|_e \geq \|r_\nu\| \, n^{-\frac{1}{2}}.$$

Also

$$e_i^T A A^H e_i \leq \|A\|^2{}_S.$$

But

$$\|r_\nu\| \geq \|A^{-1}\|^{-1}{}_S \|s_\nu\|,$$

and therefore it follows from (15) that

$$(18) \qquad \|s_{\nu+1}\|^2 / \|s_\nu\|^2 \leq 1 - n^{-1}\kappa^{-2}(A).$$

In place of a selected vector e_i, Gastinel (1958) uses a vector v_ν such that $|v_\nu| = e$, and

$$v_\nu^H r_\nu = \|r_\nu\|_{e'}.$$

But

$$\|r_\nu\|_{e'} \geq \|r_\nu\| \geq \|A^{-1}\|^{-1}{}_S \|s_\nu\|,$$

while

$$v_\nu^H A A^H v_\nu \leq \|A\|^2{}_S n.$$

Therefore this method leads to the same result (18) as the method of relaxation. Although the method avoids the need for a search, the number of arithmetic operations is greater.

The method of steepest descent takes

$$v_\nu = r_\nu.$$

This requires considerably more operations per step, but a much more favorable bound can be given for the right member of (15). In fact, by the inequality of Kantorovich in the form (3.4.10), it follows that

$$|r_\nu^H r_\nu|^2 / [s_\nu^H s_\nu r_\nu^H A A^H r_\nu] \geq \sin^2 \theta,$$

and therefore from (15),

$$(19) \qquad \|s_{\nu+1}\| / \|s_\nu\| \leq \cos \theta,$$

where θ is defined by (3.4.5).

It does not follow that the Euclidean norms of r_ν are monotonically decreasing, but (19) implies that

$$\|s_\nu\| \leq \|s_0\| \cos^\nu \theta,$$

and since

$$(20) \qquad \|s_0\| \leq \|A^{-1}\| \, \|r_0\|, \qquad \|r_\nu\| \leq \|A\| \, \|s_\nu\|,$$

therefore

$$(21) \qquad \|r_\nu\| / \|r_0\| \leq \kappa \cos^\nu \theta = \cot \theta / 2 \cos^\nu \theta.$$

Also, for a particular step, it follows from (14) that

$$\| r_{\nu+1} \|^2 / \| r_\nu \|^2 = r_\nu{}^H r_\nu r_\nu{}^H (AA^H)^2 r_\nu / (r_\nu{}^H AA^H r_\nu)^2 - 1 \leq \cot^2 \theta,$$

the latter inequality coming from a further application of the Kantorovich inequality. Hence

(22) $$\| r_{\nu+1} \| / \| r_\nu \| \leq \cot \theta.$$

Again attention is called to the fact that here the norms are Euclidean.

In case A is positive definite, the simplest choice for G is

(23) $$G = A.$$

Then $Y_\nu = V_\nu$, and

$$C_\nu = Y_\nu (Y_\nu{}^H A Y_\nu)^{-1} Y_\nu{}^H,$$

(24) $$u_\nu = (Y_\nu{}^H A Y_\nu)^{-1} Y_\nu{}^H r_\nu,$$

$$\| s_\nu \|^2 - \| s_{\nu+1} \|^2 = r_\nu{}^H Y_\nu u_\nu,$$

the norms being ellipsoidal.

In the method of block relaxation the matrices Y_ν are formed from columns of the identity matrix I, hence $Y_\nu{}^H A Y_\nu$ is a principal submatrix of A, one whose inverse is known or readily obtained. In the ordinary method of relaxation, each Y_ν is a single vector e_i selected by some criterion. If the choice is to maximize $| e_i{}^T r_\nu |$, then

$$| e_i{}^T r_\nu | = \| r_\nu \|_e \geq n^{-\frac{1}{2}} \| r_\nu \|_S$$

and

$$e_i{}^T A e_i \leq \| A \|_S.$$

In (24) the norms are ellipsoidal norms

(25) $$\| s \|^2 = s^H A s.$$

Let P be a matrix such that

$$A = P^H P.$$

Then

$$\| s \| = \| P s \|_S,$$

and

$$\| r_\nu \|_S = \| P^H P s_\nu \|_S \geq \| P^{-1} \|_S{}^{-1} \| s_\nu \|.$$

It is convenient to let

(26) $$\kappa = \kappa(P) = \cot \theta / 2, \qquad \kappa(A) = \kappa^2.$$

Then

$$\| s_\nu \|^2 - \| s_{\nu+1} \|^2 \geq \| s_\nu \|^2 / (n\kappa^2)$$

or

(27) $$\| s_{\nu+1} \|^2 / \| s_\nu \|^2 \leq 1 - (n\kappa^2)^{-1}.$$

An analogous argument applied to the method of Gastinel leads to the same result.

The method of steepest descent takes the single vector

$$v_\nu = r_\nu.$$

Application of the Kantorovich inequality leads again to (19), where, however, the norms are defined by (25), and θ by (26). The Euclidean norms of r_ν and of s_ν satisfy

(28) $$\| s_\nu \|_S / \| s_0 \|_S \leq \cot \theta/2 \cos^\nu \theta,$$

(29) $$\| r_\nu \|_S / \| r_0 \|_S \leq \cot 3\, \theta/2 \cos^\nu \theta.$$

Also, by application of the Kantorovich inequality,

(30) $$\| r_{\nu+1} \|_S / \| r_\nu \|_S \leq \cot \theta.$$

4.3. Norm-Reducing Methods. In the projective methods the matrices C_ν are ordinarily singular, but with unit spectral radius. Since they are singular they must be changed from step to step, and since they do not themselves lead to matrices H_ν with small norms, each is norm-reducing only for vectors having components in the subspace belonging to their small or vanishing proper values. Advantage is taken of the fact that such components will exist in general, even if they are not large.

The methods to be considered now are methods for developing matrices C_ν such that the multipliers H_ν, or products of them, have norms less than one, hence are generally norm-reducing wherever the residual may lie. Usually the matrices C_ν will be the same for every ν, and they must be nonsingular. But whereas a converging sequence of projections can be formed with any arbitrary nonsingular matrix A, the methods to be described now converge only for matrices of certain special classes. However, these classes are extremely important, and include most of the cases of interest. These methods differ from the one considered at the outset (Section 4.0.1) in that no preliminary steps are presumed to have been taken for obtaining an initial approximation either for the solution to the system or for the inverse of the matrix.

There are, in fact, just two main classes to which the methods apply. These are first, of positive definite matrices, and second, of matrices having a positive diagonal, and nonpositive off-diagonal elements. These will be treated separately.

The following theorem (Exercise 2.8) is due to Stein: *For any matrix B, $\rho(B) < 1$ if and only if there exists a positive definite matrix G such that $G - B^H G B$ is also positive definite.*

First, suppose that both G and $G - B^H G B$ are positive definite, and let

(1) $$G^{-1} = P P^H.$$

If

$$\beta = \| B \|_{PS},$$

then β^2 is the greatest zero of $\delta(\lambda G - B^H G B)$. Hence $\lambda G - B^H G B$ is semidefinite for $\lambda = \beta^2$ and positive definite for $\lambda > \beta^2$. Since it is, by hypothesis, positive definite for $\lambda = 1$, it follows that $\beta^2 < 1$.

Conversely, suppose $\rho(B) < 1 - \epsilon < 1$. Then (Section 2.3) there exists a nonsingular matrix P such that

$$\beta = \| P^{-1}BP \|_S < 1.$$

But then

$$\beta = \| B \|_{PS},$$

and β^2 is the greatest zero of $\delta(\lambda G - B^H G B)$, where G is defined by (1). Hence $\beta^2 G - B^H G B$ is semidefinite, and therefore $G - B^H G B$ is positive definite.

In the notation of Section 4.0.1, if

(2) $$C_\nu = C$$

for every ν, then

$$x_{\nu+1} = Ch + (I - CA)x_\nu,$$

or, if C is nonsingular,

$$C^{-1}x_{\nu+1} = h + (C^{-1} - A)x_\nu.$$

But this amounts to a decomposition of the matrix A into a sum

$$A = A_1 + A_2,$$

with A_1 nonsingular and readily inverted, and with

(3) $$C = A_1^{-1}.$$

Hence the process is defined by

(4) $$A_1 x_{\nu+1} = h - A_2 x_\nu.$$

The first special form to be considered will be

(5) $$A = I - B, \qquad B \geq 0.$$

There is no restriction in supposing A, hence also B, to be irreducible, since otherwise the problem could be reduced to that of solving in sequence two systems of lower order. Presumably B has a null diagonal, although this need not be assumed. The most elementary decomposition would be of the form

(6) $$A_1 = I = C, \qquad A_2 = -B = -H,$$

and the necessary and sufficient condition for convergence is that $\rho(B) < 1$.

Classical sufficient conditions are that one of the inequalities

$$\text{lub}_e\,(B) < 1, \qquad \text{lub}_{e'}\,(B) < 1$$

shall hold. More generally, it is sufficient that for any $g > 0$, either of the inequalities

$$\text{lub}_g\,(B) < 1, \qquad \text{lub}_{g'}\,(B) < 1,$$

shall hold. Moreover, since B is irreducible and nonnegative, in case of convergence there always exists a vector $g > 0$ for which the first (second) condition holds.

More generally, let

(7) $$A_1 = I - B_1 = C^{-1}, \qquad A_2 = -B_2,$$

where

(8) $$B_1 \geq 0, \qquad B_2 \geq 0, \qquad \rho(B_1) < 1.$$

It follows that

(9) $$C \geq 0.$$

Usually B_1 is nilpotent, $\rho(B_1) = 0$, and, in particular, B_1 is ordinarily the lower triangle, B_2 the upper triangle of B. Then

(10) $$H = I - CA = (I - B_1)^{-1}B_2,$$

and

(11) $$H \geq 0.$$

Stein and Rosenberg have shown that *if B is irreducible and $B_2 \neq 0$, and $\rho(B_1) < 1$, then one of the following three conditions holds:*

$$\rho(H) < \rho(B) < 1,$$
$$\rho(H) = \rho(B) = 1,$$
$$\rho(H) > \rho(B) > 1.$$

Since $\rho(B) > 0$, if $\rho(H) = 0$ and $\rho(B) < 1$, there is nothing to prove. Hence suppose $\gamma = \rho(H) > 0$, and

$$Hc = \gamma c \geq 0.$$

Then

$$(B_1 + \gamma^{-1}B_2)c = c.$$

If B is irreducible, so is $B_1 + \gamma^{-1}B_2$; hence $c > 0$. Let

(12) $$\phi(\delta) = \rho(B_1 + \delta^{-1}B_2).$$

Then $\phi(\delta) \geq 0$ and is strictly monotonically decreasing as δ goes from 0 to $+\infty$. Moreover

$$\phi(\gamma) = 1, \qquad \phi(1) = \rho(B), \qquad \gamma = \rho(H).$$

Hence
$$[1 - \rho(B)][1 - \rho(H)] = -(\gamma - 1)[\phi(\gamma) - \phi(1)] \geq 0.$$

The result now follows directly.

Thus if $\rho(B) \geq 1$, no method of this form can give a converging process. Hence suppose

$$\rho(B) < 1.$$

An almost immediate corollary is a theorem of Fiedler and Pták: *Let*

$$B_2 - P \geq 0, \qquad P \geq 0, \qquad P \neq 0,$$

and let

$$K = (I - B_1 - P)^{-1}(B_2 - P).$$

Then $\rho(B) < 1$ implies that

(13) $$\rho(K) < \rho(H).$$

Let

$$\omega(\delta) = \rho[B_1 + P + \delta^{-1}(B_2 - P)].$$

Then $\omega(\delta)$ is monotonically decreasing, and

$$\omega(\kappa) = 1, \qquad \kappa = \rho(K).$$

But $\gamma = \rho(H) < 1$, whence

$$B_1 + P + \gamma^{-1}(B_2 - P) \leq B_1 + \gamma^{-1}B_2.$$

Hence

$$\omega(\gamma) < 1, \qquad \kappa < \gamma.$$

The strict inequality in (13) permits a slight strengthening of the Stein-Rosenberg theorem: $\rho(H) = 0$ if and only if $B_2 = 0$.

A closely related theorem is due to Varga, in which the requirement

$$A^{-1} > 0$$

replaces that of irreducibility, in the last theorem. Note that H can be written

$$H = (I + E)^{-1}E, \qquad E = A^{-1}B_2.$$

Proper values of H and of E, when properly paired, are related by

$$\lambda_i(H) = \lambda_i(E)/[1 + \lambda_i(E)],$$

and, in particular

$$\rho(H) = \rho(E)/[1 + \rho(E)].$$

In fact, $\rho/(1 + \rho)$ is monotonic for $\rho \geq 0$. Since E and H are both nonnegative, $\rho(H)$ and $\rho(E)$ are proper values of H and E, respectively. Because of the monotonicity, the largest proper values of the two matrices must correspond. Also, because of the monotonicity $\rho(H)$, when considered as a function of $\rho(E)$, is properly monotonically increasing, and

$$0 < \rho(H) < 1.$$

Since $A^{-1} > 0$, whether E is reducible or not, a decrease in any element of B_2 would lead to a reduction in at least one element in an irreducible principal submatrix, and could cause an increase in none. Hence

$$\rho[A^{-1}(B_2 - P)] < \rho[A^{-1}B_2],$$

and hence

$$\rho(K) < \rho(H).$$

In this proof it is not required that A have the form (6), but only that $(A - B_2)^{-1} \geq 0$.

The foregoing results show that when the matrix A has the form (6), and $\rho(B) < 1$, then any decomposition of the form (7) and (8) provides a convergent sequence, and "smaller" matrices B_2 provide more rapid convergence. The comparison, however, is only qualitative, and not quantitative. A quantitative comparison can be made in certain special cases, the most important being that in which the matrix A has *property* (A) as defined by Young. A matrix A is said to have property (A) in case by some permutational transformation it can be put into the form

$$(14) \qquad \begin{pmatrix} D_1 & -M_2 \\ -M_1 & D_2 \end{pmatrix}$$

where D_1 and D_2 are diagonal matrices. A matrix in such a form is said to be *consistently ordered*. Property (A) often appears directly in the systems that arise from the numerical solution of partial differential equations. Also, it can often result from a feasible transformation. Thus, consider the product

$$\begin{pmatrix} A_{11}^{-1} & 0 \\ 0 & A_{22}^{-1} \end{pmatrix} \begin{pmatrix} A_{11} & A_{12} \\ A_{21} & A_{22} \end{pmatrix} = \begin{pmatrix} I & A_{11}^{-1}A_{12} \\ A_{22}^{-1}A_{21} & I \end{pmatrix},$$

which has property (A) and is consistently ordered. In many important cases matrices occur in the form

$$\begin{pmatrix} A_1 & -B_1 & 0 & 0 & \cdots \\ -C_1 & A_2 & -B_2 & 0 & \cdots \\ 0 & -C_2 & A_3 & -B_3 & \cdots \\ \cdot & \cdot & \cdot & \cdot & \cdot \end{pmatrix},$$

where each A_i is a square submatrix. If this matrix is multiplied by

$$\text{diag } (A_1^{-1}, A_2^{-1}, \cdots),$$

the product has property (A). To obtain a consistent ordering, first interchange the rows of A_1 with those of A_2, and make a corresponding interchange of columns. Now group together the second and third groups and repeat. In case each

$$A_i^{-1} \geq 0, \qquad B_i \geq 0, \qquad C_i \geq 0,$$

the resulting matrix is consistently ordered, and has the form (14).

Suppose the matrix A has the form (14), with

$$D_1 = I, \qquad D_2 = I.$$

Let

$$B_1 = \begin{pmatrix} 0 & 0 \\ M_1 & 0 \end{pmatrix}, \qquad B_2 = \begin{pmatrix} 0 & M_2 \\ 0 & 0 \end{pmatrix}.$$

Then

$$\delta \begin{pmatrix} \lambda I & -M_2 \\ -M_1 & \lambda I \end{pmatrix} = \lambda^\sigma \delta(\lambda^2 I - M_1 M_2),$$

where σ is some integer which depends upon the partitioning. But if H is defined by (10), then for some ν,

$$\delta(\lambda I - H) = \lambda^\nu \delta(\lambda I - M_1 M_2).$$

Hence *nonnull proper values of H are the squares of those of B, and in particular,*

$$\rho(H) = \rho^2(B).$$

Hence to take $C = (I - B_1)^{-1}$ gives convergence at twice the rate obtained with $C = I$.

Methods of over-relaxation (or under-relaxation), applied to a matrix A of the form (5), and a particular decomposition

(15) $$B = B_1 + B_2, \qquad B_1 \geq 0, \qquad B_2 \geq 0,$$

utilize a matrix

(16) $$C(\omega) = \omega(I - \omega B_1)^{-1},$$

hence lead to a matrix

(17) $$H(\omega) = (I - \omega B_1)^{-1}[(1 - \omega)I + \omega B_2].$$

This corresponds to the decomposition

$$A_1 = (\omega^{-1}I - B_1), \qquad A_2 = -[(\omega^{-1} - 1)I + B_2].$$

Let

(18) $$\eta(\omega) = \rho[H(\omega)].$$

Assuming, as before, that $\rho(B) < 1$, it follows that

$$\eta(1) = \rho(H) < 1, \qquad \eta(0) = 1.$$

Moreover, the theorem of Fiedler and Pták can be applied to show that for

$0 \leq \omega \leq 1$, $\eta(\omega)$ increases monotonically. To show this, observe first that

$$\omega A = I - \omega B_1 - [(1 - \omega)I + \omega B_2],$$

and that $\rho(\omega B_1) < 1$. Also the matrix within brackets is nonnegative. Hence to apply the theorem to ωA it is necessary only to show that

$$\rho[\omega B_1 + (1 - \omega)I + \omega B_2] < 1.$$

But

$$\omega B_1 + (1 - \omega)I + \omega B_2 = (1 - \omega)I + \omega B,$$

and since B has a norm less than 1, this matrix has a norm less than $(1 - \omega) + \omega = 1$. Hence the theorem applies and the assertion follows.

It follows from this, however, that *also $C(\omega)$ as defined by* (16) *provides a convergent sequence when $\omega \leq 1$.* But for $\omega < 1$ the convergence can only be slower, and it is natural to ask whether convergence would not be more rapid with $\omega > 1$. This is indeed true, although the analysis is complicated by the fact that H may not be nonnegative.

It will be assumed that H is noncyclic (Section 2.4). Let

$$H(\omega)y = \lambda y, \qquad \lambda \neq 0.$$

Then

$$(B_1 + \lambda^{-1}B_2)y = \omega^{-1}\lambda^{-1}(\lambda - 1 + \omega)y.$$

If $\lambda = \eta(\omega) > 0$, then $B_1 + \lambda^{-1}B_2 \geq 0$ and is irreducible, hence $y > 0$. Moreover,

(19) $$\phi(\lambda) = \omega^{-1}\lambda^{-1}(\lambda - 1 + \omega),$$

where ϕ is defined by (12). These conditions hold at least for $0 < \omega \leq 1$. Moreover, on this range $\eta(\omega)$ is itself a proper value, and actually exceeds in magnitude any other proper value of H. Since $\eta(\omega)$ is a continuous function of ω, as is any proper value, it remains a proper value for $\omega < \bar{\omega}$, with $\bar{\omega} > 1$, but not too large. It will be shown that $\eta(\omega)$ *does in fact continue to decrease when $1 < \omega < \bar{\omega}$ for some $\bar{\omega}$, when H is noncyclic.*

To show this, consider ω as a function of λ defined by (19):

(20) $$\omega - 1 = [\phi(\lambda) - 1]/[\lambda^{-1} - \phi(\lambda)].$$

It is known that $\eta(1) < 1$, $\phi(\gamma) = 1$; hence $\lambda = \gamma$, the denominator of the fraction is positive, and the numerator vanishes. If $\lambda > \gamma$ but not too large the denominator remains positive, but the numerator becomes negative since $\phi(\lambda)$ is strictly monotonically decreasing. Hence $\lambda > \gamma$ only for $\omega < 1$, but for $\omega > 1$ and not too large, $\lambda < \gamma$. The argument falls down when $\lambda \neq \eta(\omega)$.

If follows that $\eta(\omega)$ is minimized by some value of $\omega > 1$, but neither the minimal η nor the minimizing ω is known in general. However, in certain cases when A has property (A), both can be found. In fact, with the notation used above, the characteristic polynomial of $H(\omega)$ is expressible in terms of

$(\lambda - 1 + \omega)$, and

$$\delta[(\lambda - 1 + \omega)^2 I - \lambda \omega^2 M_1 M_2] = 0.$$

Hence if μ is a positive proper value of B, associated with it will be proper values of λ of H satisfying

$$(\lambda - 1 + \omega)^2 = \lambda \omega^2 \mu^2,$$

or

(21) $$\lambda - \omega \mu^{\frac{1}{2}} \lambda - 1 + \omega = 0.$$

The optimal ω will be that for which, as μ ranges over all proper values of B, the root λ of largest modulus is least. Such a condition is still not applicable in practice, in general, but it can be applied in case all proper values of $M_1 M_2$ are real and nonnegative. This is true, in particular, when $M_2 = M_1{}^T$, and A is positive definite. In that case it can be shown that the λ of largest magnitude is associated with the largest μ.

First it will be shown that for any fixed $\mu > 0$, the root λ of greatest modulus will be least when the two roots are equal. Equal roots occur when $\omega = \omega_0$ satisfies

$$\omega_0{}^2 \mu^2 - 4\omega_0 + 4 = 0,$$

hence

$$\omega_0 = 2/[1 \pm (1 - \mu^2)^{\frac{1}{2}}].$$

Corresponding to these values of ω_0, $\lambda_0{}^{\frac{1}{2}} = \omega_0/2$. Hence it will be shown that *when A is positive definite and has property (A) the optimal choice is given by*

(22)
$$\omega_0 = 2/[1 + (1 - \mu^2)^{\frac{1}{2}}],$$
$$\lambda_0{}^{\frac{1}{2}} = \mu/[1 + (1 - \mu^2)^{\frac{1}{2}}].$$

Since $\lambda_0{}^{\frac{1}{2}}$ is a monotonically increasing function of μ, if it can be established that the root λ of greatest modulus is least when the roots are equal, then the required result will follow.

For the roots to be complex, it is necessary that

$$\omega^2 \mu^2 - 4\omega + 4 < 0,$$

and in this event the modulus of each is $(\omega - 1)^{\frac{1}{2}}$. The inequality implies that $\omega > \omega_0$, hence that $\omega - 1 > \omega_0 - 1 = \lambda_0$. Therefore complex roots $\lambda^{\frac{1}{2}}$ can only exceed $\lambda_0{}^{\frac{1}{2}}$ in modulus.

If the roots are real and distinct, to show that the larger is greater than $\lambda_0{}^{\frac{1}{2}}$, it is sufficient to show that

$$\lambda_0 - \omega \mu \lambda_0{}^{\frac{1}{2}} + \omega - 1 < 0.$$

But since

$$\lambda_0 - \omega_0 \mu \lambda_0{}^{\frac{1}{2}} + \omega_0 - 1 = 0,$$

the above inequality reduces to

$$\mu \lambda_0{}^{\frac{1}{2}} < 1,$$

which is certainly true. This completes the proof and establishes that for the case considered,

$$\rho[H(\omega_0)] = \lambda_0.$$

There is an obvious extension of all foregoing results to a wider class of matrices, in that if P and Q are any nonsingular matrices, then

$$(PA_1Q)^{-1}(PA_2Q) = Q^{-1}A_1^{-1}A_2Q,$$

the matrix being similar to $A_1^{-1}A_2$. Hence all results obtained for any matrix A can be extended immediately to any matrix of the form PAQ where P and Q are arbitrary nonsingular matrices.

In case the matrix A is positive definite, Stein's theorem can be applied using A itself:

$$A - (I - AC^H)A(I - CA) = A(C^H + C - C^HAC)$$
$$= AC^H(C^{-1} + C^{H-1} - A)CA.$$

Hence a sufficient condition for convergence when $C_\nu = C$ for all ν is that

$$P = C^{-1} + C^{H-1} - A$$

be positive definite. In fact (Reich), *if A is symmetric and P is positive definite, then a necessary and sufficient condition for convergence is that A be positive definite.* Only the necessity needs to be proved.

Let

$$H = I - CA, \qquad R = AC^HPCA.$$

Then R is positive definite, and it is to be shown that $\rho(H) < 1$ implies that A is positive definite. But

$$A = R + H^HAH$$
$$= R + H^H(R + H^HAH)H$$
$$= \cdots\cdots\cdots\cdots\cdots\cdots$$
$$= R + H^HRH + (H^H)^2RH^2 + \cdots.$$

The infinite series converges if $\rho(H) < 1$, and each term is positive definite. Hence the sum is positive definite.

Assume A to be positive definite, and let

$$A = S - Q - Q^H,$$

where S is positive definite. Take

$$C = (S - Q)^{-1}.$$

Then

$$P = S$$

and the method converges.

In most applications, S is the diagonal of A, and Q the lower triangle. In "block relaxation," S is made up of blocks along the diagonal, each block being easily invertible. This amounts to forming the system

$$S^{-1}Ax = S^{-1}h.$$

More generally, let

(23) $$C(\omega) = \omega(S - \omega Q)^{-1}.$$

Then

$$P = (2\omega^{-1} - 1)S,$$

which leads to Ostrowski's result that the method converges if and only if

$$0 < \omega < 2.$$

In case A has property (A), the optimal value of ω is given by (22).

The matrices H, or $H(\omega)$, that arise in the use of the Gauss-Seidel method do not retain the symmetry of the original matrix A. Aitken proposed alternating,

(24) $$C_1 = (I - Q)^{-1}, \qquad C_2 = (I - Q^{\mathrm{H}})^{-1},$$

leading to the so-called "to-and-fro" methods. Then

(25) $$H_1 = (I - Q)^{-1}Q^{\mathrm{H}}, \qquad H_2 = (I - Q^{\mathrm{H}})^{-1}Q,$$
$$s_{2\nu+1} = H_1 s_{2\nu}, \qquad s_{2\nu+2} = H_2 s_{2\nu+1},$$

and therefore

(26) $$s_{2\nu+2} = H_2 H_1 s_{2\nu},$$

and the convergence properties are determined by the product

$$H_2 H_1 = (I - Q^{\mathrm{H}})^{-1}Q(I - Q)^{-1}Q^{\mathrm{H}}.$$

But Q and $(I - Q)^{-1}$ are commutative, and

(27) $$(I - Q^{\mathrm{H}})H_2 H_1 (I - Q^{\mathrm{H}})^{-1} = (I - Q)^{-1}QQ^{\mathrm{H}}(I - Q^{\mathrm{H}})^{-1},$$

which is symmetric and semidefinite. Hence $H_2 H_1$ is similar to a symmetric semidefinite matrix and all its roots are real and nonnegative. Indeed, if Q is triangular, it is nilpotent, and so is $(I - Q)^{-1}Q$, and hence the matrix (27) is a product of two nilpotent matrices. The spectral radius of (27) is the largest root ρ of

$$\delta[\rho I - (I - Q)^{-1}QQ^{\mathrm{H}}(I - Q^{\mathrm{H}})^{-1}] = 0,$$

and hence of

(28) $$\delta[\rho(A + QQ^{\mathrm{H}}) - QQ^{\mathrm{H}}] = 0.$$

If A is positive definite, then so is the matrix within brackets positive definite for any $\rho \geq 1$. On the other hand, the matrix

$$\rho I - (I - Q)^{-1}QQ^{\mathrm{H}}(I - Q^{\mathrm{H}})^{-1}$$

is certainly positive definite for ρ sufficiently large; hence so is

$$\rho(A + QQ^{\mathrm{H}}) - QQ^{\mathrm{H}}.$$

Therefore if A is not positive definite, (28) can be satisfied by some $\rho \geq 1$.

Consequently the *iteration* (25) *converges if and only if A is positive definite, hence if and only if the ordinary Gauss-Seidel iteration converges.*

In the "implicit" alternating direction methods, A can sometimes be expressed as a sum of positive definite matrices which also commute:

$$(29) \qquad \begin{aligned} A &= A_1 + A_2, \\ A_1 &= A_1{}^H, \qquad A_2 = A_2{}^H, \qquad A_1 A_2 = A_2 A_1. \end{aligned}$$

The commutativity implies that A_1 and A_2 have the same proper vectors:

$$A_1 = V\Lambda_1 V^H, \qquad A_2 = V\Lambda_2 V^H,$$

where Λ_1 and Λ_2 are diagonal, and V unitary or orthogonal. Then

$$(30) \qquad \begin{aligned} C_1 &= (\omega I + A_1)^{-1}, \qquad H_1 = (\omega I + A_1)^{-1}(\omega I - A_2), \\ C_2 &= (\omega I + A_2)^{-1}, \qquad H_2 = (\omega I + A_2)^{-1}(\omega I - A_1). \end{aligned}$$

Let $\alpha_i{}'$ and $\alpha_i{}''$ be the ith diagonal elements of Λ_1 and Λ_2. Then the spectral radius of $H_2 H_1$ is equal to

$$\rho_\omega(H_2 H_1) = \max \mid (\omega - \alpha_i{}')(\omega - \alpha_i{}'')(\omega + \alpha_i{}')^{-1}(\omega + \alpha_i{}'')^{-1} \mid ,$$

and the optimal choice of ω will be that one which minimizes ρ_ω. Similar schemes for particular types of problems are due to Peaceman and Rachford, Douglas and Rachford, and others, where, in fact, the parameter ω is allowed to vary as the iteration proceeds.

In the methods described so far the endeavor has been to make each individual remainder s_ν as small as possible. However, consider the following modification of the Jacobi method, applied to a positive definite matrix A with unit diagonal. Let

$$(31) \qquad C_\nu = \omega_\nu I, \qquad H_\nu = I - \omega_\nu A.$$

Hence

$$(32) \qquad s_{\nu+1} = \phi_\nu(A)s_1,$$

where

$$(33) \qquad \phi_\nu(\lambda) = (1 - \omega_\nu \lambda)(1 - \omega_{\nu-1}\lambda) \cdots (1 - \omega_1 \lambda).$$

If ν steps at least are contemplated, it is important only that the final remainder be small, and the magnitudes of the intermediate ones are not important. Hence a polynomial $\phi_\nu(\lambda)$ is required, satisfying

$$(34) \qquad \phi_\nu(0) = 1,$$

such that $\phi_\nu(\lambda)$ is as small as possible over a range that includes all the proper values of A, hence for

$$0 < \lambda_n \leq \lambda \leq \lambda_1,$$

where

$$\lambda_1(A) \geq \lambda_2(A) \geq \cdots \geq \lambda_n(A).$$

A natural choice is the Chebyshev polynomial defined over this range, and normalized by (34). But these polynomials are orthogonal with respect to a certain weight function, and are known to satisfy a certain three-term recursion relation. Hence it is not necessary to decide in advance on the value of ν, nor to know the zeros ω_i^{-1} explicitly.

The recursion for orthogonal polynomials is of the form

(35) $$\phi_{\nu+1}(\lambda) = (\beta_\nu \lambda + \delta_\nu)\phi_\nu(\lambda) + (1 - \delta_\nu)\phi_{\nu-1}(\lambda).$$

It is easy to verify that if $\phi_0(\lambda) = 1$ and $\phi_1(0) = 1$, then $\phi_\nu(0) = 1$ for every ν. It follows from (34) and (31) that

$$s_{\nu+1} = (\beta_\nu A + \delta_\nu I)s_\nu + (1 - \delta_\nu)s_{\nu-1},$$
$$s_{\nu+1} - s_\nu = \beta_\nu r_\nu + (\delta_\nu - 1)(s_\nu - s_{\nu-1}).$$

But
$$s_{\nu+1} - s_\nu = -(x_{\nu+1} - x_\nu).$$

Consequently (34) leads to

(36) $$x_{\nu+1} - x_\nu = (\delta_\nu - 1)(x_\nu - x_{\nu-1}) - \beta_\nu r_\nu.$$

For computation of the β_ν and δ_ν, reference is made to standard texts on orthogonal polynomials, and to Stiefel (1958) and Lanczos (1956) (see also Section 7.4 for a similar application to finding roots of a matrix). Stiefel also discusses the use of polynomials other than those of Chebyshev.

Naturally the method is equally effective for any matrix A that is similar to a positive definite matrix.

Somewhat more generally, given any iteration defining approximations $x_0, x_1, \cdots, x_\nu$, it is reasonable to consider an arithmetic mean of all of these:

$$x_\nu^* = \alpha_0^{(\nu)}x_0 + \alpha_1^{(\nu)}x_1 + \cdots + \alpha_\nu^{(\nu)}x_\nu,$$
$$1 = \alpha_0^{(\nu)} + \alpha_1^{(\nu)} + \cdots + \alpha_\nu^{(\nu)}.$$

Then
$$s_\nu^* = \alpha_0^{(\nu)}s_0 + \alpha_1^{(\nu)}s_1 + \cdots + \alpha_\nu^{(\nu)}s_\nu.$$

In case $H_\nu = H$ for all ν,

$$s_\nu^* = (\alpha_0^{(\nu)}I + \alpha_1^{(\nu)}H + \cdots + \alpha_\nu^{(\nu)}H^\nu)s_0$$
$$= \psi_\nu(H)s_0.$$

When $A = I - B$ and is positive definite, and $H = B$, ψ_ν is a polynomial in B and hence a polynomial in A, and a natural choice is again a Chebyshev polynomial. For more general choices of H no theory has been developed.

REFERENCES

Forsythe has remarked that the Gauss-Seidel method was not known to Gauss and not recommended by Seidel. Gauss did, however, use a method of relaxation, as the term is used here, annihilating at each step the largest residual (see Dedekind, 1901). For hand computation this is natural and more efficient than the straight cyclic process; for machine computation the search is time-consuming. When applied to the finite difference approximation to an elliptic partial differential equation, the method is sometimes called that of Liebmann (1918). An interesting exchange of correspondence relating to the method was carried on between Mehmke and Nekrasov (1892) where a special case of (4.01.13) occurs. The method there received a more systematic treatment than it had had previously, and Nekrasov (1884) had already discussed convergence. Hence, with some justice the method is called the method of Nekrasov in some of the Russian literature.

That the method converges for a positive definite matrix has been known for some time, but the converse was first proved by Reich (1949) in a form slightly less general than is stated here. The proof used here of the converse and of the direct theorem was suggested by the proof given by Weissinger (1953) of the direct theorem.

Most of the simpler methods have been discovered and rediscovered several times. The method of steepest descent is due to Cauchy (1847), who developed it for solving nonlinear systems of equations. Noteworthy among more general treatments are von Mises and Pollaczek-Geiringer (1929), Weissinger (1951, 1952), Forsythe (1953b), and a series of papers by Ostrowski. See also Householder (1953) and Durand (1961); special mention should be made of Faddeev and Faddeeva (1960) for the topics of this and later chapters.

The method of (4.1.8) is due to Schulz, but sometimes goes by the names of Hotelling and Bodewig. Ostrowski (1938a) discusses acceleration by factorization.

The proof of the theorem of Stein and Rosenberg (1948) given here is in part due to R. S. Varga (private communication).

The possibility of speeding convergence by introducing over-relaxation factors with the Jacobi iteration was considered by Richardson (1910), and Young (1954) showed how to obtain optimal coefficients from the Chebyshev polynomials. Lanczos (1952), Shortley (1953), Gavurin (1950), Abramov (1950), and others arrived in various ways at the use of these polynomials. Stiefel (1958) and Faddeev (1958) showed that a three-term recursion sufficed for the use of these and other polynomials for a like purpose.

There have been many papers on the determination and use of over-relaxation factors. These occur in the original Richardson method. Young (1954) who introduced property (A), Ostrowski (1953), Kahan (1958), Varga (1957), and Birkhoff and Varga (1959b) are only a few of the references that might be named. Frankel (1950) applied the method in a special case. For over-relaxation based upon a Gauss-Seidel iteration, most attention is paid to the matrices that arise from approximating a partial differential equation, where the algebraic system is large but the matrices sparse and of special form. These give rise in particular to the alternating-direction methods, on which there is a considerable amount of literature.

The method of (4.2.14) is a slightly generalized form of a method due to Kaczmarz (1937), and the method of (4.2.10) is similarly generalized from a method due to de la Garza (1951). The unified general treatment in Section 4.2 of the methods of projection follows Householder and Bauer (1961). Other sources are named in the text itself.

The error formula (4.01.13) was found by Nekrasov, and may well have been known to Gauss. Intermittently it is rediscovered and republished.

Upper bounds for lub (A^{-1}) and lub (A) involving determinants (hence of limited utility) are given by Wegner (1953), Richter (1954), and Kato (1960); see also Mirsky (1956c).

For a treatment of (4.3.21) in the complex case, see Kjellberg (1958).

For a much more extensive treatment of iterative methods with particular reference to applications to the solution of partial differential equations, Varga (1962) should be consulted.

PROBLEMS AND EXERCISES

1. The least-squares problem is that of solving for x in the system

$$\text{(i)} \quad Ax = h + d, \qquad A^H d = 0,$$

where the matrix A has n columns and $N > n$ rows. This is equivalent to the system

$$\text{(ii)} \quad A^H A x = A^H h.$$

Show that the method of (4.2.8) applied to the system (i) ignoring d (hence taking $r_v = h - Ax_v$) is equivalent to the application of the method of (4.2.24) to (ii).

2. Show that if each Y_v in (4.2.24) is a single vector y_v, and if these are the e_i in rotation, the resulting method is that of Gauss-Seidel.

3. Show likewise that the method of block relaxation described in Section 4.3 for positive definite matrices is equivalent to that described in Section 4.2, and amounts to taking the Y_v in (4.2.24) to be columns of the identity.

4. Show in detail that the inequality (4.2.27) applies also to Gastinel's method when A is positive definite.

5. Considering the equations $Ax = h$ to represent n hyperplanes, it is geometrically evident that by taking x_1 as the projection of x_0 upon the first plane; x_2 as the projection of x_1 upon the second; and so in rotation, the sequence must converge. Obtain the analytical expression for the iteration.

6. If A is Hermitian, $\| A \| = \alpha$, $\gamma^2\alpha^2 < 2$, then the iteration with $C_v = (-1)^v\gamma I$ converges [Bueckner, Bialy].

7. Suppose A is positive definite with roots

$$\alpha_1 \geq \alpha_2 \geq \cdots \geq \alpha_n.$$

The iteration with $H = I - \mu A$ converges if

$$2/\alpha_1 > \mu > 0,$$

and optimally if

$$\mu^{-1} = (\alpha_1 + \alpha_n)/2.$$

This differs from the method of steepest descent only in that the multiplier μ is fixed.

8. Let $A = 2I - (J + J^H)$. Then $\alpha_1 + \alpha_n = 4$, $\mu = \frac{1}{2}$. Hence the method of Exercise 7 coincides, in this case, with the method of Jacobi (simple iteration).

9. Let A have the form described in Exercise 1.51, where

$$B = 4I - (J + J^H).$$

Then $\alpha_1 + \alpha_n = 8$, $\mu = 4$, and again the method of Exercise 7 coincides with the method of Jacobi.

10. Let $A = A_1 + A_2$ and consider the system

$$(A_1 + \mu A_2)x(\mu) = h + (\mu - 1)A_2 v,$$

for arbitrary v. Evidently $x(1) = x$. Show that by expanding in powers of μ and equating like powers the result is equivalent to the iteration (4.3.4).

11. Let $\omega_n(\lambda, \mu) = \delta[\lambda(I - \mu J) - \mu J^H]$ and obtain the recursion for these polynomials. Referring to Exercise 1.50, show that for $A = I - \mu(J + J^H)$, both the Jacobi and the Gauss-Seidel iterations converge for arbitrary n if and only if $|\mu| \leq \frac{1}{2}$, and that the Gauss-Seidel iteration converges twice as fast:

$$\rho^2[\mu(J + J^H)] = \rho[\mu(I - \mu J)^{-1}J^H].$$

12. Multiply the matrix

$$A(\lambda, \mu) = \begin{pmatrix} \lambda I & Q_1 & 0 & 0 & \cdots \\ \mu^2 R_1 & \lambda I & Q_2 & 0 & \cdots \\ 0 & \mu^2 R_2 & \lambda I & Q_3 & \cdots \\ 0 & 0 & \mu^2 R_3 & \lambda I & \cdots \\ \cdots\cdots\cdots\cdots\cdots\cdots\cdots\cdots \end{pmatrix},$$

on the right by $M(\mu) = \mathrm{diag}\,(I, \mu I, \mu^2 I, \cdots)$, and on the left by $M^{-1} = M(\mu^{-1})$. Thence show that the determinants of $A(\lambda, \mu)$ and $A(\lambda/\mu, 1)$ differ only by a power of μ as a factor. Hence take $\mu^2 = \lambda$ and show that for $A(1, 1)$ the Gauss-Seidel iteration and the Jacobi iteration both diverge or both converge, and if they converge the former converges twice as fast. Similar conclusions can be drawn when block iteration is applied to

$$A = \begin{pmatrix} P_1 & -Q_1 & 0 & \cdots \\ -R_1 & P_2 & -Q_2 & \cdots \\ 0 & -R_2 & P_3 & \cdots \\ \cdots\cdots\cdots\cdots\cdots\cdots\cdots \end{pmatrix}.$$

13. If all C_ν are identical, then

$$r_\nu = A H^\nu A^{-1} r_0,$$

hence (Chapter 7) if H has a unique root λ satisfying

$$| \lambda | = \rho(H),$$

then for large ν, $r_{\nu+1} \doteq \lambda r_\nu$. Hence $(\lambda x_\nu - x_{\nu+1})/(\lambda - 1)$ should be an improved approximation to x (Lyusternik).

14. Let A be positive definite, and suppose numbers α and β are known such that $0 < \alpha \le \lambda_i(A) \le \beta$. In (4.3.36) let

$$x_1 - x_0 = 2r_0/(\beta - \alpha),$$

hence

$$s_1 = \phi_1(A)s_0,$$

where

$$\phi_1(\lambda) = [-2\lambda + (\beta + \alpha)]/(\beta + \alpha)$$
$$= \cos \phi/\cosh \omega,$$

and

$$\cos \phi = [-2\lambda + (\beta + \alpha)]/(\beta - \alpha),$$
$$\cosh \omega = (\beta + \alpha)/(\beta - \alpha).$$

Show that (4.3.36) with

$$\delta_\nu = 1 + \cosh (\nu - 1)\omega/\cosh (\nu + 1)\omega,$$
$$\beta_\nu = -4(\beta - \alpha)^{-1} \cosh \nu\omega/\cosh (\nu + 1)\omega$$

leads to

$$s_\nu = \phi_\nu(A)s_0,$$

where

$$\phi_\nu(\lambda) = \cos \nu\phi/\cos \nu\omega,$$

with ϕ and ω defined as above. These are the Chebyshev polynomials, and

$$\| s_\nu \|_s \le \| s_0 \|/\cosh \nu\omega.$$

15. In Exercise 12 consider $A(\lambda, \mu^{\frac{1}{2}})$, where

$$\lambda = \omega^{-1}(\mu - 1 + \omega).$$

Interpreting

$$A(1, 1) = I - B_1 - B_2,$$

where B_1 and B_2 are the lower and upper triangles, show that relation (4.3.21) holds also for this case.

16. For arbitrary nonsingular A and the system $Ax = k$, let u_1, $u_2, \cdots, u_n$ be linearly independent, and normalized by

$$u_i{}^H A A^H u_i = 1.$$

Then $I - 2A^H u_i u_i{}^H A$ is an elementary Hermitian. If $\mu_i > 0$, $\sum \mu_i = 1$, show that

$$x_{\nu+1} = x_\nu + 2 \sum \mu_i A^H u_i u_i{}^H r_i$$

is a centroid of the n reflections of x_ν into the hyperplanes $u_i{}^H(Ax - k) = 0$, and hence is interior to the sphere through x_ν with center at x. Analytically, show that the matrices

$$\sum \mu_i A^H u_i u_i{}^H A, \qquad I - 2 \sum \mu_i A^H u_i u_i{}^H A$$

are positive definite, and hence if

$$\sigma = \mathrm{glb} \left(\sum \mu_i A^H u_i u_i{}^H A \right),$$

then, with Euclidean norm,

$$\| s_{\nu+1} \| \leq \| s_\nu \| (1 - 2\sigma).$$

For each u_i a scalar multiple of e_i, the method is due to Cimmino.

17. (Sassenfeld's iteration). In general, if $A = A_1 + A_2$, then $A = A_1(I + \Gamma) + (A_2 - A_1\Gamma)$; hence if $I + \Gamma$ is nonsingular one may take

$$C = (I + \Gamma)^{-1} A_1{}^{-1}.$$

In particular, consider $\Gamma = \gamma I$. Then

$$H(\gamma) = (1 + \gamma)^{-1}(\gamma I - A_1{}^{-1}A_2).$$

Show that there is a real γ for which $\rho(H) < 1$ if and only if the convex hull of the roots $\lambda_i(-A_1{}^{-1}A_2)$ is entirely interior to one of the half-planes bounded by $\mathrm{Re}\ \lambda = 1$. Moreover, there exists a γ, possibly complex, for which $\rho(H) < 1$ if and only if the convex hull of the roots $\lambda_i(-A^{-1}A_2)$ does not contain the point 1 in or on the boundary.

18. Let $x_0, x_1, x_2, \cdots$ be formed with $C = -A_1{}^{-1}$, and $x_0{}^* = x_0, x_1{}^*$, $x_2{}^*, \cdots$ with $C = -(1 + \gamma)^{-1}A_1{}^{-1}$. Show that

$$x_\nu{}^* = (1 + \gamma)^{-\nu}(x_\nu + \nu\gamma x_{\nu-1} + \cdots + \gamma^\nu x_0).$$

19. In Exercise 17, show that if it is known only that $\rho(-A_1{}^{-1}A_2) < 1$, then the choice $\gamma = 0$ cannot be improved.

20. When A is ill-conditioned an iteration with

$$C^{-1} = A + \Gamma$$

is often suggested. If $\Gamma = \gamma I$, show that the iteration always converges for any γ interior to a certain polygon, not necessarily bounded, which contains the origin, and hence if A is positive definite it converges for any $\gamma > 0$ (and trivially for $\gamma = 0$). Convergence is best for small $| \gamma |$, but condition is best for $| \gamma |$ large.

21. Let S and $A = S - Q - Q^H$ be positive definite, let

$$\kappa(S - \omega Q)x = [(1 - \omega)S + \omega Q^H]x,$$

and show that

$$\kappa x^H[(2 - \omega)S - \omega A + \omega(B^H - B)]x$$
$$= x^H[(2 - \omega)S + \omega A + \omega(B^H - B)]x.$$

Hence obtain Ostrowski's result that $|\kappa| < 1$ when $0 < \omega < 2$ (J. Albrecht).

22. Apply (4.0.13) to show that in case of convergence

$$\| s_{\nu+1} \| \leq \| A^{-1}[(\omega^{-1} - 1)S + Q] \| \, \| x_{\nu+1} - x_\nu \|.$$

23. Let $A = I - B$ be positive definite and $s_\nu = B^\nu s_0$. It is sometimes suggested to take $x_\nu{}^*$ to be a mean of $x_0, x_1, \cdots, x_\nu$, and hence to form

$$x_\nu{}^* = \alpha_{\nu 0} x_0 + \alpha_{\nu 1} x_1 + \cdots + \alpha_{\nu\nu} x_\nu,$$
$$1 = \alpha_{\nu 0} + \alpha_{\nu 1} + \cdots + \alpha_{\nu\nu}.$$

Show that this is equivalent to the process indicated by (4.3.32), or, alternatively, that

$$s_\nu{}^* = \psi_\nu(B)s_0, \qquad \psi_\nu(1) = 1.$$

If $\rho(B) = \rho < 1$ the optimal choice is

$$\psi_\nu(\lambda) = \cos \nu\theta / \cosh \nu\omega_0$$

where

$$\lambda = \rho \cos \theta, \qquad 1 = \rho \cosh \omega_0.$$

Then

$$x_{\nu+1}^* = \beta_{\nu+1}(Bx_\nu{}^* + h - x_{\nu-1}^*) + x_{\nu-1}^*,$$
$$\beta_{\nu+1} = 2\rho^{-1} \cosh \nu\omega_0 / \cosh (\nu + 1)\omega_0.$$

Show also that

$$\beta_1 = 1, \qquad \beta_2 = 2/(2 - \rho^2), \qquad \beta_{\nu+1} = [1 - (\rho^2\beta_\nu/4)]^{-1}$$

(Golub and Varga).

24. Let $A = P + Q + \alpha I$ where $\alpha \geq 0$, P and Q are positive definite, $P^{-1} \geq 0$, $Q^{-1} \geq 0$, and $PQ = QP$ (cf. Exercise 1.12). If

$$C_0{}^{-1} = \beta I - Q, \qquad C_1{}^{-1} = \beta I - P,$$
$$2\beta + \alpha > 0,$$

then

$$\rho(H_1 H_0) < 1$$

[show that glb $[H + (\alpha + \beta)I] > \rho + \sigma/2$].

25. Let $A = I - B$,

$$B = \begin{pmatrix} 0 & 0 & M_1 \\ M_2 & 0 & 0 \\ 0 & M_3 & 0 \end{pmatrix}.$$

The Jacobi iteration and the Gauss-Seidel iteration both converge or both diverge; in the former case the Gauss-Seidel iteration converges three times as fast (Varga, who generalized to cyclic matrices of degree p). However, if

$$B = \begin{pmatrix} 0 & M_1 & 0 \\ 0 & 0 & M_2 \\ M_3 & 0 & 0 \end{pmatrix},$$

the Gauss-Seidel iteration converges only half again as fast.

26. For B as given at the outset in Exercise 25, let

$$C = (\omega^{-1}I - B_1)^{-1}$$

for over-relaxation. Show that for every root β of B there is a root λ of $H(\omega)$ satisfying

$$\lambda^2\omega^3\beta^3 = (\lambda + \omega - 1)^3,$$

and conversely.

27. For $x = Bx + h$, Abramov sets

$$x_{2\nu+1} = Bx_{2\nu} + h,$$
$$x_{2\nu+2} = 2(Bx_{2\nu+1} + h) - x_{2\nu}.$$

Show that the iteration converges if and only if the simple Jacobi iteration converges, and that in case of convergence it is about twice as fast (the double step is to be compared with two steps of the Jacobi iteration).

28. If A is an M-matrix and $Ax = h$, $Ay = k$, then $h \geq k \Rightarrow x \geq k$.

29. Let $C_\nu = C$ be fixed (Section 4.01), and let $x_0 = 0$, $x_1 = Ch$. Show inductively that (Bauer)

$$x_{2\nu} = (I + H^\nu)x_\nu.$$

Hence, by taking ν to be a power of 2, and forming explicitly the matrices $H, H^2, H^4, H^8, \cdots$ a quadratically convergent process is possible, but at the cost of a matrix multiplication at each step and the loss of the self-correcting feature.

CHAPTER 5

Direct Methods of Inversion

5.0. Uses of the Inverse. *A direct method* of finding X to satisfy $AX = M$ is one that defines X in terms of a finite number of operations on scalars. These operations are ordinarily the arithmetic ones, but may also include square-rooting. An example of a direct method of solution would be the use of Cramer's rule, which expresses the elements of X as quotients of determinants, since a determinant is defined as a sum of certain products of its elements. However, this particular method is completely impractical for general use.

Many methods are to be found in the literature, but all seem to be mere variations of only a few basic ones. Most methods, including all those in common use, are methods of factorization, expressing the matrix A, at least implicitly, as a product of easily invertible factors, or else forming a non-singular matrix C such that CA or AC is easily inverted. Some methods are analytic in form and proceed by progressive reduction of the matrix A; some are synthetic in form, starting with the inverse or the factors of a simple matrix and building up stepwise to the matrix itself. Still other methods, instead of transforming the system progressively, start with an arbitrary approximation x_0 and progressively transform the error until it vanishes after at most n steps. Such methods resemble those described in Section 4.2. The best known example is the method of conjugate gradients (Section 5.7), but there are others (Section 5.6).

In order to form $A^{-1}h$ it is not necessary to form A^{-1} explicitly, although advantages of having the inverse, at least approximately, have already been mentioned (Section 4.0). To form the determinant $\delta(A)$ requires only about the same number of operations as are required for forming $A^{-1}h$, many of them identical for the two problems. Since Cramer's rule requires the evaluation of $n + 1$ distinct determinants, this partly explains why Cramer's rule is not suitable for numerical work (although its importance as a theoretical tool is not to be minimized).

If the elements of a matrix A are perturbed slightly to give $A - F$, and if

$$(A - F)(x + d) = h, \qquad Ax = h,$$

then

$$d = (A - F)^{-1} Fx,$$

hence

$$\| d \| / \| x \| \leq \| (A - F)^{-1} \| \, \| F \|,$$

or

(1) $$(\| d \| / \| x \|)/(\| F \| / \| A \|) \leq \| (A - F)^{-1} \| \, \| A \|.$$

The vector d can be considered the error induced in x by an error F in A, and hence the left member of (1) represents the ratio of the relative error in x to the relative error in A. The right member is, for small F, approximately equal to

(2) $$\kappa(A) = \| A \| \, \| A^{-1} \|,$$

the "condition number" of A. This number has made its appearance before in Chapters 3 and 4, and in Chapter 4 in particular it limits the rates of convergence of some of the methods of projection. In the present context it, or rather its inverse, measures the stability of the problem, or the relative precision with which the matrix A defines the solution x of the system $Ax = h$. In other words, when κ is large, small relative errors in A can lead to large relative errors in the solution x.

Evidently if P and Q are any nonsingular matrices, the system $Ax = h$ is equivalent to

$$PAQy = k, \qquad k = Ph, \qquad x = Qy,$$

and when P and Q are diagonal matrices the transformation is easily carried out. It is reasonable to ask whether diagonal matrices P and Q can be found that would minimize $\kappa(PAQ)$. They certainly exist, but they are not known in general. However, there is empirical evidence that the minimum occurs when the elements of maximal modulus in the rows and columns of PAQ are comparable.

5.1. The Method of Modification. This is a synthetic method based upon a formula due to Sherman and Morrison, or a generalization due to Woodbury. The Sherman and Morrison formula may be considered a generalization of (1.1.3), and can be derived immediately from the observation that if B is nonsingular, then the inverse of $B - \sigma uv^H$ has the following form:

(1)

$$(B - \sigma uv^H)^{-1} = B^{-1} - \tau B^{-1} uv^H B^{-1},$$

$$\tau^{-1} + \sigma^{-1} = v^H B^{-1} u.$$

More generally (Woodbury),

(2)
$$(B - USV^H)^{-1} = B^{-1} - B^{-1}UTV^HB^{-1},$$
$$T^{-1} + S^{-1} = V^HB^{-1}U,$$

assuming the necessary inverses to exist. Each formula can be verified directly.

Ordinarily only the simpler form (1) is applied in practice. Let A be the matrix to be inverted, and let A_0 be any convenient matrix whose inverse is known. The choice $A_0 = I$ is not excluded. Let $u_1 = (A - A_0)e_1$, $A_1 = A_0 + u_1e_1^T$, and apply the formula to obtain A_1^{-1}. Let $u_2 = (A - A_1)e_2$, and apply the formula to obtain the inverse of $A_2 = A_1 + u_2e_2^T$. By proceeding in this way, $A_n = A$, and the inversion is completed after n steps. Obviously many variants are possible: the matrix could be built up a row at a time, either rows or columns could be taken in any order, and it is even possible to proceed element by element. Some have recommended proceeding element by element, selecting for any given step the element giving the largest, or at least a large, divisor. The search, however, has obvious disadvantages in machine computation.

In case the matrix is symmetric it might be desirable to retain the symmetry throughout, although it is not then possible to adjoin an entire row or an entire column in one step. If

(3)
$$\sigma = -\alpha_{ij}, \qquad u = v = e_i + e_j, \qquad i \neq j,$$

the result will be to adjoin the (i, j) and (j, i) elements, but at the same time elements of the diagonal will be modified by the same amount. By taking separately $u = v = e_i$ and $u = v = e_j$, the diagonal elements can be adjusted. An economical choice for A_0 is therefore the diagonal matrix obtained from A by subtracting from each diagonal element the sum of all other elements in the same row (or column). If this choice is made, only off-diagonal elements require adjustment.

An alternative to (3) is

(4)
$$\sigma = \alpha_{ij}, \qquad u = v = e_i - e_j.$$

Then A_0 should be formed by an initial augmentation of the diagonal elements.

For a matrix A that differs in only a few of its elements from one whose inverse is already known or easily obtained, and, in particular, for very sparse matrices, the method can clearly be very effective. Its effectiveness for machine computation requires more experience to judge.

Formally this method could be considered a method of factorization since in the first of (1) the matrix B^{-1} appears as a factor on the right. Nevertheless, it is perhaps not useful to think of it in this way, and the method seems to differ intrinsically from the others that will now be described.

5.2. Triangularization. The method of modification is entirely synthetic in that it starts with a matrix of the same order as A, but whose inverse is known, and makes a sequence of modifications until the inverse of A itself is obtained. Other methods are usually analytic in the sense that at each step the problem is reduced from that of inverting a matrix of order ν to that of inverting a matrix of order $\nu - 1$. However, in most of them, by an inconsequential reordering of certain steps, the method of factorization can be carried out synthetically in the sense that the inverse of a matrix of order $\nu + 1$ can be expressed in terms of the inverse of a submatrix of order ν.

In Section 1.3 it was shown that if A, along with every principal submatrix formed of the first ν rows and columns, $\nu = 1, 2, \cdots, n - 1$, is nonsingular, then there exists a unique factorization

$$(1) \qquad\qquad A = LDR,$$

where L is unit lower triangular, R unit upper triangular, and D diagonal. Also determinantal formulas for the elements of L, D, and R were given. If A itself is nonsingular, then there always exists a permutation matrix P such that PA (or AP) satisfies the above requirements. In any pivotal method (see below) the permutation is carried out as the computation proceeds. For the moment, however, let it be assumed that any required permutational transformation has been carried out in advance. There are several ways of making out the factorization.

Suppose

$$(2) \qquad\qquad A = \begin{pmatrix} A_{11} & A_{12} \\ A_{21} & A_{22} \end{pmatrix}$$

where A_{11} is a nonsingular square submatrix, and suppose that the factorization

$$(3) \qquad\qquad A_{11} = L_{11} D_1 R_{11}$$

has already been effected, with D_1 diagonal, and L_{11} and R_{11} unit lower and upper triangular matrices, respectively. Since A_{11} is nonsingular, by hypothesis, these matrices are also nonsingular. A simple algorithm for inverting a triangular matrix has already been given (Section 1.3), but it can be verified directly that

$$(4) \qquad
\begin{pmatrix} L_{11} & 0 \\ L_{21} & L_{22} \end{pmatrix}^{-1} = \begin{pmatrix} L_{11}^{-1} & 0 \\ -L_{22}^{-1} L_{21} L_{11}^{-1} & L_{22}^{-1} \end{pmatrix},$$

$$\begin{pmatrix} R_{11} & R_{12} \\ 0 & R_{22} \end{pmatrix}^{-1} = \begin{pmatrix} R_{11}^{-1} & -R_{11}^{-1} R_{12} R_{22}^{-1} \\ 0 & R_{22}^{-1} \end{pmatrix},$$

provided L_{11}, L_{22}, R_{11}, and R_{22} are all nonsingular. This exhibits such an algorithm, since either L_{11} or L_{22} can be interpreted to be a scalar, with L_{21} a column vector or row vector, as the case may be, so that the inverse can be formed by first inverting a scalar and then proceeding by successive bordering until the entire triangular matrix is inverted. Analogous remarks apply to the inversion of R, and they apply to arbitrary nonsingular triangular matrices, and not only unit triangular matrices.

Now, if possible, write

$$(5) \qquad \begin{pmatrix} A_{11} & A_{12} \\ A_{21} & A_{22} \end{pmatrix} = \begin{pmatrix} L_{11} & 0 \\ L_{21} & I \end{pmatrix} \begin{pmatrix} D_1 & 0 \\ 0 & \Delta_2 \end{pmatrix} \begin{pmatrix} R_{11} & R_{12} \\ 0 & I \end{pmatrix},$$

where nothing is said of the form of Δ_2. But clearly this is possible if

$$(6) \qquad \begin{aligned} L_{21} &= A_{21} R_{11}^{-1} D_1^{-1}, \qquad R_{12} = D_1^{-1} L_{11}^{-1} A_{12}, \\ \Delta_2 &= A_{22} - L_{21} D_1 R_{12} = A_{22} - A_{21} A_{11}^{-1} A_{12}, \end{aligned}$$

and, on the assumption that L_{11}, D_1, and R_{11} are known and can be inverted, it follows immediately that L_{21}, R_{12}, and Δ_2 are defined uniquely and can be obtained. Now if

$$(7) \qquad \Delta_2 = L_{22} D_2 R_{22},$$

where the factors have the indicated forms, then

$$(8) \qquad \begin{pmatrix} A_{11} & A_{12} \\ A_{21} & A_{22} \end{pmatrix} = \begin{pmatrix} L_{11} & 0 \\ L_{21} & L_{22} \end{pmatrix} \begin{pmatrix} D_1 & 0 \\ 0 & D_2 \end{pmatrix} \begin{pmatrix} R_{11} & R_{12} \\ 0 & R_{22} \end{pmatrix},$$

and the required factorization is complete.

There are many ways of applying these formulas, but first some obvious variants might be mentioned. Evidently (5) is equivalent to

$$(9) \qquad \begin{pmatrix} A_{11} & A_{12} \\ A_{21} & A_{22} \end{pmatrix} = \begin{pmatrix} L_{11} & 0 \\ L_{21} & I \end{pmatrix} \begin{pmatrix} (D_1 R_{11}) & (D_1 R_{12}) \\ 0 & \Delta_2 \end{pmatrix},$$

where D_1, R_{11}, and R_{12} are not obtained explicitly, but only in the indicated combinations. Alternatively,

$$(10) \qquad \begin{pmatrix} A_{11} & A_{12} \\ A_{21} & A_{22} \end{pmatrix} = \begin{pmatrix} (L_{11} D_1) & 0 \\ (L_{21} D_1) & \Delta_2 \end{pmatrix} \begin{pmatrix} R_{11} & R_{12} \\ 0 & I \end{pmatrix},$$

and, correspondingly,

$$(11) \qquad \begin{pmatrix} A_{11} & A_{12} \\ A_{21} & A_{22} \end{pmatrix} = \begin{pmatrix} L_{11} & 0 \\ L_{21} & L_{22} \end{pmatrix} \begin{pmatrix} (D_1 R_{11}) & (D_1 R_{12}) \\ 0 & (D_2 R_{22}) \end{pmatrix}$$

$$(12) \qquad = \begin{pmatrix} (L_{11} D_1) & 0 \\ (L_{21} D_1) & (L_{22} D_2) \end{pmatrix} \begin{pmatrix} R_{11} & R_{12} \\ 0 & R_{22} \end{pmatrix}.$$

More generally, if

(13) $$D = D'D'',$$

where D' and D'' are also diagonal, then

(14) $$\begin{pmatrix} A_{11} & A_{12} \\ A_{21} & A_{22} \end{pmatrix} = \begin{pmatrix} (L_{11}D_1') & 0 \\ (L_{21}D_1') & (L_{22}D_2') \end{pmatrix} \begin{pmatrix} (D_1''R_{11}) & (D_1''R_{12}) \\ 0 & (D_2''R_{22}) \end{pmatrix}.$$

The factorization (13) can be arbitrary, but in the interests of stability there are advantages in choosing $|D'|$ and $|D''|$ to be at least approximately equal.

In case A is Hermitian, then $L = R^H$ and Δ_2 is Hermitian. In case A is positive definite, then also Δ_2 is positive definite, and so is D, which is to say that $D \geq 0$. In that case $D' = D'' = D^{\frac{1}{2}}$ is a natural choice. This is the *method of Cholesky*, and is sometimes called the "square-rooting" method.

For definiteness, explicit attention will be confined mainly to (5) and (8), but it is understood that any of the forms (9) through (12) may be used instead. When $A_{11} = \alpha_{11}$, A_{12} and A_{21} are row and column vectors, the application of (5) reduces the factorization of a matrix A of order n to the factorization of a matrix Δ_2 of order $n - 1$. The same reduction applied to Δ_2 reduces the problem to the factorization of a matrix of order $n - 2$. At the final step only the factorization of a scalar is required. This is the method of *Gaussian elimination*. Ordinarily not D and R separately, but the product DR is obtained.

Naturally if $\alpha_{11} = 0$ the method cannot be applied as described. But if A is nonsingular, there must be a nonnull element in the first column. If $\alpha_{i1} \neq 0$, then replace A by $I_{1i}A$ (1.1.7) and proceed as before. Actually, unless A is positive definite, such an exchange should be made if $|\alpha_{i1}| > |\alpha_{11}|$. In other words, at the outset the element of largest modulus in the first column should be found and the row containing it should be put first. This is the search for the largest pivot. Naturally a similar search should be made at each stage of the elimination. In some cases it is even desirable to interchange columns so as to bring the largest element in the entire matrix to the pivotal position. But *for positive definite matrices the interchange is unnecessary*.

The *escalator method* starts also with $A_{11} = \alpha_{11}$, but applies (5) only to the leading second-order submatrix of the entire matrix, $A_{ij} = \alpha_{ij}, i, j = 1, 2$. When the factors of the second-order submatrix are obtained, this is identified with A_{11} and bordered, so that a further application of (5) gives the factors of the leading third-order submatrix. Thus inverses of leading submatrices of successively higher order are obtained, and finally the matrix itself is inverted.

The *method of Crout* is essentially Gaussian elimination, but applied in a manner to conserve storage requirements. The same device is also a part of the Gauss-Jordan method (see below). In fact, the observation upon which it is based applies equally to any method of triangularization, including the escalator method. This observation is simply that at any stage (5) of the factorization process, no further reference to the original matrix A is required, and that for each position (i, j), it is necessary to store an element of one and only one of the three factors. Thus a diagonal element is necessarily 1 in L_{11} and in R_{11} and does not need to be stored. Hence along the diagonal only the diagonal elements of D_1 and Δ_2 require storage. In any off-diagonal position, only one of the three factors has a nonzero element and only this requires storage. As the computation proceeds, and a column is adjoined to L_{21} and a row to R_{12}, the order of Δ_2 is reduced, so that the positions previously assigned to Δ_2 are written over by elements of the new Δ_2, the new column in L_{21}, the new row in R_{12}, and the new diagonal element in D_1. Analogous remarks apply to the escalator method.

In practice it is more common to form explicitly

$$(15) \qquad \begin{pmatrix} \Lambda_{11} & 0 \\ \Lambda_{21} & I \end{pmatrix} \begin{pmatrix} A_{11} & A_{12} \\ A_{21} & A_{22} \end{pmatrix} = \begin{pmatrix} D_1 & 0 \\ 0 & \Delta_2 \end{pmatrix} \begin{pmatrix} R_{11} & R_{12} \\ 0 & I \end{pmatrix},$$

where

$$(16) \qquad \Lambda_{11} = L_{11}^{-1}, \qquad \Lambda_{21} = -L_{21}\Lambda_{11}.$$

In fact, in Gaussian elimination the right member is formed by multiplying by matrices $L_i^{-1}(-l_i)$ (Section 1.1) in sequence, and the left-hand multiplier in (15) is a product of these. Moreover,

$$L = L_1(-l_1)L_2(-l_2) \cdots L_{n-1}(-l_{n-1}).$$

For solving the system of equations $Ax = h$, this left-hand multiplier is not required explicitly. Instead the system is progressively transformed by

$$L_1^{-1}Ax = L_1^{-1}h,$$
$$(17) \qquad L_2^{-1}L_1^{-1}Ax = L_2^{-1}L_1^{-1}h,$$
$$\cdots\cdots\cdots\cdots$$

and in the general step the matrix multiplying x on the left has the form (15). But L^{-1} is required explicitly if A^{-1} is required, and the Crout storage technique is equally applicable here.

It is sometimes argued in the literature that, since it is L^{-1} that is required for A^{-1} and not L, fewer operations are required when L^{-1} is formed directly by the stages (15). At first sight this appears plausible, but it is actually not the case. In fact, each l_i must be formed explicitly, and as (1.2.1) shows, this is the ith column of L itself apart from sign. Hence, whether L is first formed completely, by successively storing the l_i, and then inverted, or whether the products $L_i^{-1} \cdots L_1^{-1}$ are formed explicitly as soon as L_i is formed, is immaterial so far as the arithmetic operations are concerned. In fact, there is some added flexibility available when L is known completely, since then the inversion can be made either by proceeding to the left and up or to the right and down.

The reduction indicated in (15) can be carried a stage further, since (15) implies

(18)
$$\begin{pmatrix} (D_1 R_{11})^{-1} & 0 \\ 0 & I \end{pmatrix} \begin{pmatrix} \Lambda_{11} & 0 \\ \Lambda_{12} & I \end{pmatrix} \begin{pmatrix} A_{11} & A_{12} \\ A_{21} & A_{22} \end{pmatrix} = \begin{pmatrix} I & R_{11}^{-1} R_{12} \\ 0 & \Delta_2 \end{pmatrix}.$$

Note that

(19)
$$R_{11}^{-1} R_{12} = A_{11}^{-1} A_{12}.$$

Evidently

$$\begin{pmatrix} I & 0 \\ 0 & L_{22} \end{pmatrix} \begin{pmatrix} I & 0 \\ 0 & D_2 \end{pmatrix} \begin{pmatrix} I & R_{11}^{-1} R_{12} \\ 0 & R_{22} \end{pmatrix} = \begin{pmatrix} I & R_{11}^{-1} R_{12} \\ 0 & \Delta_2 \end{pmatrix}$$

because of (7), and

$$\begin{pmatrix} I & 0 \\ 0 & D_2^{-1} \end{pmatrix} \begin{pmatrix} I & 0 \\ 0 & L_{22}^{-1} \end{pmatrix} \begin{pmatrix} R_{11}^{-1} & 0 \\ 0 & I \end{pmatrix} \begin{pmatrix} D_1^{-1} & 0 \\ 0 & I \end{pmatrix} \begin{pmatrix} L_{11}^{-1} & 0 \\ -L_{21} L_{11}^{-1} & I \end{pmatrix}$$

$$= \begin{pmatrix} R_{11}^{-1} & 0 \\ 0 & I \end{pmatrix} \begin{pmatrix} D_1^{-1} & 0 \\ 0 & D_2^{-1} \end{pmatrix} \begin{pmatrix} L_{11} & 0 \\ L_{21} & L_{22} \end{pmatrix}^{-1}$$

Consequently all arithmetic operations are the same, and the reduction (18), which progressively inverts both L and R as they are formed, merely has the effect of inverting them in a particular way (down and to the right for both).

When this scheme is applied and combined with the device for economizing storage, the result is the *Gauss-Jordan method*. Since

(20)
$$DR = PD$$

where P is also unit upper triangular, a possible and common variant is

(21)
$$\begin{pmatrix} P_{11}^{-1} & 0 \\ 0 & I \end{pmatrix} \begin{pmatrix} \Lambda_{11} & 0 \\ \Lambda_{21} & I \end{pmatrix} \begin{pmatrix} A_{11} & A_{12} \\ A_{21} & A_{22} \end{pmatrix} = \begin{pmatrix} D_1 & P_{11}^{-1} P_{12} D_1 \\ 0 & \Delta_2 \end{pmatrix}$$

which leads finally to

(22)
$$P^{-1} L^{-1} A = D.$$

In solving the system $Ax = h$, neither P^{-1} nor L^{-1} nor the product is required explicitly, but the final form of the system is

$$Dx = P^{-1} L^{-1} h.$$

It is always possible to carry out any of these reductions provided all of the leading submatrices of A are nonsingular. In practice one of these may turn out to be at least nearly singular, and a permutation of rows should be made. If A itself is nonsingular, it is always possible at each stage to permute the rows of Δ_2, and the corresponding rows throughout, so that for each i,

$$| l_i | \le e.$$

This means bringing to the top position the numerically largest element in the first column of Δ_2. In some instances it is even advantageous to permute columns so that the largest element in Δ_2 is brought to the leading (pivotal) position.

Other variants of the same general principle can be devised, and some have appeared in the literature as "new." For example, if C is any nonsingular matrix, then the matrix CA (or AC) can be factored in place of A. Evidently, if

$$CA = LDR,$$

then

$$A^{-1} = R^{-1}D^{-1}L^{-1}C,$$

and if CA is better conditioned than A itself, or if C is an approximation to A^{-1}, but not an adequate approximation, then factorization of CA is worthwhile. Also the factorization

(23) $$A = RDL$$

is possible, with the lower triangle on the right and upper triangle on the left. These are not, of course, the same factors as those in (1).

The identity

(24) $$\begin{pmatrix} A_{11} & A_{12} \\ A_{21} & A_{22} \end{pmatrix} = \begin{pmatrix} A_{11} & 0 \\ A_{21} & I \end{pmatrix} \begin{pmatrix} I & A_{11}^{-1}A_{12} \\ 0 & \Delta_2 \end{pmatrix},$$

$$\Delta_2 = A_{22} - A_{21}A_{11}^{-1}A_{12}$$

has already appeared in Chapter 1. An equivalent form is

$$\begin{pmatrix} A_{11}^{-1} & 0 \\ -A_{21}A_{11}^{-1} & I \end{pmatrix} \begin{pmatrix} A_{11} & A_{12} \\ A_{21} & A_{22} \end{pmatrix} = \begin{pmatrix} I & A_{11}^{-1}A_{12} \\ 0 & \Delta_2 \end{pmatrix},$$

for the same Δ_2. When the inverse of a particular submatrix is known at the outset, or easily obtainable, then (24) reduces the problem immediately to that of inverting Δ_2. The so-called *method of partitioning* amounts to a partitioning of the matrix A arbitrarily and inverting first the submatrix A_{11} and then the resulting matrix Δ_2. This is clearly possible, and may be required for efficient transfers among hierarchies of storage. But it is clear from the foregoing discussion that the arithmetic operations are identical with those of simple Gaussian elimination, and it introduces artificial restrictions in the search for an optimal pivot. Hence it should be used only when clearly called for, and then with due recognition of the possible hazards.

5.3. A More General Formulation. The Gauss and the Gauss-Jordan methods can be regarded as particular applications of the following lemma: *If P is nonsingular of order n, Q is of rank n − r at most, and P − Q of rank r + 1 at least, then there exists a nonsingular elementary matrix I − $\sigma u v^H$ such that*

$$(1) \qquad (I - \sigma u v^H)(P - Q) = P - R,$$

where R is of rank n − r − 1 at most. If P is diagonal, possibly the identity, or triangular, or otherwise easily invertible, then by repeated application of this lemma A^{-1} can be expressed as a product of P^{-1} and at most n elementary matrices.

The assertion of the lemma is trivial unless Q is of rank n − r at least; hence its rank will be assumed to be exactly n − r. Then there exists a matrix W of r linearly independent columns, such that

$$QW = 0.$$

The proof will be made by showing that it is possible to choose σ, u, v, and w, so that the matrix (W, w) is of rank r + 1, and both $RW = 0$ and $Rw = 0$.

Since

$$R = Q + \sigma u v^H(P - Q),$$

and since necessarily $\sigma u \neq 0$, it follows that $RW = 0$ if and only if

$$v^H P W = 0.$$

Hence v must lie in the space orthogonal to PW. Let V be any matrix of n − r columns such that (W, V) is nonsingular, and let

$$(X, Y)^H = (W, V)^{-1} P^{-1}.$$

Then

$$(2) \qquad \begin{array}{ll} X^H P W = I, & X^H P V = 0, \\ Y^H P W = 0, & Y^H P V = I. \end{array}$$

Hence v is to be a linear combination of columns of Y:

$$(3) \qquad v = Yf.$$

The requirement that $Rw = 0$ will be fulfilled if

$$(4) \qquad u = Qw, \qquad 1 + \sigma v^H(P - Q)w = 0.$$

Hence it is necessary that

$$(5) \qquad v^H(P - Q)w \neq 0,$$

and the nonsingularity of $I - \sigma u v^H$, which is equivalent to

$$\delta(I - \sigma u v^H) = 1 - \sigma v^H u = 1 - \sigma v^H Q w \neq 0,$$

implies that

$$(6) \qquad v^H P w \neq 0.$$

But conversely, if v satisfies (3) for some f, w satisfies (5) and (6), and u and σ are given by (4), then the required elementary matrix is obtained and the lemma established. It remains only to show that conditions (3), (5), and (6) are always consistent.

For this, choose w in the form

$$w = Vg,$$

and the problem is that of finding f and g so that the inequalities

$$f^H Y^H P V g \neq 0, \qquad f^H Y^H (P - Q) V g \neq 0,$$

are both satisfied. Since

$$f^H Y^H P V g = f^H g,$$

these can certainly be satisfied if the matrix

$$M = Y^H (P - Q) V \neq 0.$$

Suppose $M = 0$. Then the columns of $(P - Q)^H Y$, being orthogonal to those of V, must be linear combinations of those of $P^H X$, which is to say that

$$(P - Q)^H Y = P^H X S,$$

for some S. But by (2), $W^H P^H Y = 0$, and by the initial choice of W, $W^H Q^H = 0$, whence (2) implies that

$$W^H P^H X S = S = 0.$$

Hence $M = 0$ implies that $Y^H (P - Q) = 0$. But Y is of rank $n - r$, and $P - Q$ of rank $r + 1$ by hypothesis, and hence $Y^H (P - Q)$ of rank 1 at least. Hence $M \neq 0$, and the proof is complete.

In applying this lemma, it is not necessary that P be fixed throughout. It is possible, instead, to let

$$A = P_0 - Q_0,$$
$$(I - \sigma_1 u_1 v_1^H)(P_0 - Q_0) = P_0 - R_1 = P_1 - Q_1,$$
$$(I - \sigma_2 u_2 v_2^H)(P_1 - Q_1) = P_1 - R_2 = P_2 - Q_2,$$
$$\cdots \cdots \cdots \cdots \cdots$$

where each P_i is nonsingular, and each Q_i has at most the rank of R_i. In Gaussian elimination, P_i is the upper triangle, including the diagonal, of $P_{i-1} - R_i$, and $v_i = e_i$. In one form of the Gauss-Jordan method, P_i is the diagonal of $P_{i-1} - R_i$, and, again, $v_i = e_i$. In either case $v_i^H u_i = 0$,

$$\delta(I - \sigma_i u_i v_i^H) = 1.$$

The other most natural form of the Gauss-Jordan method is to take $P = I$ throughout.

5.4. Orthogonal Triangularization. In Section 1.1 it was shown that any nonsingular matrix A can be expressed as a product of a unitary matrix and an upper triangle. The same resolution applied to A^H expresses A as a product of a lower triangular matrix and a unitary matrix. It is sufficient to consider the first form. Since any triangular matrix is expressible as a product of a diagonal matrix and a unit triangular matrix the factorization can be written

$$(1) \qquad\qquad A = WDR, \qquad W^H W = I.$$

If A is rectangular, with linearly independent columns, the reduction is still possible with W a matrix of orthonormal columns, satisfying (1). If Δ is any diagonal matrix whose diagonal elements are of unit modulus,

$$|\, \Delta \,| = I,$$

then an equivalent factorization is

$$(2) \qquad\qquad A = (W\Delta)(\Delta^{-1}D)R,$$

since the columns of $W\Delta$ are also orthonormal. However, the proof given in Section 1.1 shows that R is unique, and that W and D are unique up to a factor Δ. Equivalent factorizations are evidently $(WD)R$, where the columns of WD are left unnormalized, $W(DR)$ where DR is merely upper triangular but not unit upper triangular, and $(WD')(D''R)$, where $D'D'' = D$, and D' and D'' are selected in any way, presumably to be equal or nearly equal in modulus.

The method used in Section 1.1 was to form W as a product of elementary Hermitian matrices. After the νth step the reduction appears in the form

$$(3) \qquad \begin{pmatrix} H_{11} & H_{12} \\ H_{21} & H_{22} \end{pmatrix} \begin{pmatrix} A_{11} & A_{12} \\ A_{21} & A_{22} \end{pmatrix} = \begin{pmatrix} D_1 & 0 \\ 0 & \Delta_2 \end{pmatrix} \begin{pmatrix} R_{11} & R_{12} \\ 0 & I \end{pmatrix},$$

and only the matrix Δ_2 of order $n - \nu$ remains to be further reduced. At the νth step the vector w_ν in the reducing matrix $I - 2w_\nu w_\nu^H$ is null above the νth element.

Consider the first step in the reduction. Omitting subscripts, and taking a to be the first column of A, it is required to find w so that

$$(I - 2ww^H)a = \alpha e_1, \qquad w^H w = 1.$$

Since the unitary transformation cannot change the Euclidean norm,

$$|\, \alpha \,|^2 = \|\, a \,\|^2.$$

Hence

$$a - \alpha e_1 = 2\mu w, \qquad \mu = w^H a$$
$$|\, \alpha \,|^2 - \alpha a^H e_1 = 2 \,|\, \mu \,|^2.$$

Hence $\alpha a^H e_1$ must be real, but either sign is permitted. However, a singularity would arise with $\alpha a^H e_1 > 0$ and nearly equal to $|\alpha|^2$. Hence choose α so that

$$\alpha a^H e_1 \le 0.$$

If $a^H e_1$ is complex, a square root is required to compute its modulus, and one other for computing α.

Some economy in the arithmetic operations required in forming A^{-1} can be had by not forming W^H explicitly, but, instead, multiplying

$$(DR)^{-1} W_{n-1} W_{n-2} \cdots W_1$$

from left to right.

For reducing real matrices, Givens proposed forming the orthogonal matrix W as a product of plane rotations. The elements below the diagonal in A are eliminated in the order

$$(2,1), \qquad (3,1), \qquad \cdots, \qquad (n,1)$$
$$(3,2), \qquad \cdots, \qquad (n,2)$$
$$\cdots$$
$$(n, n-1),$$

where the element in the (i, j) position is eliminated by a rotation in the (i, j)-plane. Since each rotation requires one square root, a total of $n(n-1)/2$ square roots are required for the complete reduction, as well as a larger number of arithmetic operations than for the method just described. Since each rotation matrix has a determinant of $+1$, whereas $\delta(W_i) = -1$ for every i, it is not easy to relate the arithmetic in the two methods, but it was shown above that the methods lead to final matrices W that can differ at most by a factor Δ with $|\Delta| = I$.

5.5. Orthogonalization. In (5.4.1), since R is a unit upper triangular matrix, the first column of W is simply the normalized first column of A, and

$$|\delta_1| = \| a_1 \|,$$

where δ_1 is the first element of D, and a_1 the first column of A. The second column of W is the normalized component of a_2, the second column of A, that is orthogonal to a_1, and $|\delta_2|$ is the norm of the component itself. In general, the νth column of W is the normalized component of a_ν that is orthogonal to all previous columns, and $|\delta_\nu|$ is the norm of that component.

The Schmidt orthogonalization proceeds by building up W and R, column by column, in this way: $a_{\nu+1}$ is projected upon the space of $a_1, \cdots, a_\nu$; the difference between $a_{\nu+1}$ and its projection is normalized and introduced as a column of W.

To exhibit the algorithm, let A_ν and W_ν represent the matrices of the first ν columns of A and W, respectively; let D_ν and R_ν represent the νth-order principal submatrices of D and R. Then

(1) $$A_\nu = W_\nu D_\nu R_\nu, \qquad W_\nu^{\mathrm{H}} W_\nu = I.$$

It is required to form $r_{\nu+1}$, $\delta_{\nu+1}$, and $w_{\nu+1}$ so that

(2) $$a_{\nu+1} = W_\nu D_\nu r_{\nu+1} + \delta_{\nu+1} w_{\nu+1},$$
$$w_{\nu+1}^{\mathrm{H}} w_{\nu+1} = 1, \qquad w_{\nu+1}^{\mathrm{H}} W_\nu = 0.$$

Then

(3) $$r_{\nu+1} = D_\nu^{-1} W_\nu^{\mathrm{H}} a_{\nu+1},$$

hence

(4) $$\delta_{\nu+1} w_{\nu+1} = (I - W_\nu W_{\nu+1}^{\mathrm{H}}) a_{\nu+1},$$

(5) $$| \delta_{\nu+1} |^2 = a_{\nu+1}^{\mathrm{H}} (I - W_\nu W_{\nu+1}^{\mathrm{H}}) a_{\nu+1}.$$

Therefore all required quantities are completely determined, and

$$R_{\nu+1} = \begin{pmatrix} R_\nu & r_{\nu+1} \\ 0 & 1 \end{pmatrix}.$$

The projection of $a_{\nu+1}$ upon the space of A_ν is evidently

$$W_\nu W_\nu^{\mathrm{H}} a_{\nu+1}.$$

The least-squares problem is that of finding a vector x_ν such that

(6) $$A_\nu x_\nu + a_{\nu+1} = d_{\nu+1},$$

where $d_{\nu+1}^{\mathrm{H}} d_{\nu+1}$ is minimized. Evidently $d_{\nu+1}$ must be orthogonal to the columns of A_ν, and hence

$$d_{\nu+1} = \delta_{\nu+1} w_{\nu+1}$$

in the above formulation. In least-square curve-fitting, functions ϕ_i of an independent variable or of independent variables are given, along with their values at specified points. The elements of a_i, $i = 1, 2, \cdots, \nu$, are the values of ϕ_i at these points, while the elements of $-a_{\nu+1}$ are possibly measured values of a function ϕ at the same points. The elements of x_ν, the solution of (6), are then the coefficients of the linear combination of the ϕ_i that gives the best representation of the function ϕ in the sense of least squares.

An obvious possible variant of the method in general is to premultiply (or postmultiply) A by a nonsingular matrix C, and to apply the method to CA (or to AC). Almost equally obvious is the possibility of minimizing $\delta_{\nu+1} w_{\nu+1}$ in another ellipsoidal norm. In the least-squares problem such a norm is commonly given by a diagonal matrix.

The factorization is sometimes expressed in the form

(7) $$AP = WD,$$

where

(8) $$P = R^{-1},$$

but P is obtained directly. The first column of WD is just the first column of A:

$$\delta_1 w_1 = a_1.$$

The method can be described geometrically as follows: From each column of A except a_1 is subtracted the orthogonal projection of that column upon a_1. Evidently the residual is orthogonal to a_1, and the residual in the second column is precisely $\delta_2 w_2$. Next, every column except the first two is replaced by the result of subtracting its projection upon w_2. Hence each residual is orthogonal to both w_1 and w_2 (hence to a_1 and a_2), and the residual in the third column is $\delta_3 w_3$. A continuation of this process yields the complete reduction.

The first step is typical of all others, and requires the formation of an elementary matrix $I - e_1 p_1{}^H$ such that

$$a_1{}^H A(I - e_1 p_1{}^H) = |\,\delta_1\,|^2\, e_1{}^T,$$
$$|\,\delta_1\,|^2 = a_1{}^H a_1, \qquad e_1{}^T p_1 = 0.$$

The solution is

$$p_1{}^H = |\,\delta_1\,|^{-2}\, a_1{}^H A - e_1{}^T.$$

To put the matter otherwise, let

$$A_1 = A, \qquad A_{i+1} = A_i(I - e_i p_i{}^H), \qquad A_n = WD,$$
$$e_j{}^T p_i = 0, \qquad j \le i.$$

Then A_i and A_{i+1} agree in the first i columns, and, in particular, $A = A_1$ and WD agree in the first column, δ_1 being a normalizing factor. Hence in

$$R = (I + e_{n-1} p_{n-1}^H) \cdots (I + e_1 p_1{}^H) = D^{-1} W^H A,$$

the first row of $D^{-1} W^H$ is known, therefore the first row of $D^{-1} W^H A$ can be formed, and this is $e_1{}^T + p_1{}^H$ which is the first row of R. Thus p_1 and thence A_2 can be formed. But the second column of A_2 is the second column of WD, therefore the second row of

$$(I + e_{n-1} p_{n-1}^H) \cdots (I + e_2 p_2{}^H) = D^{-1} W^H A_2$$

can be formed, and hence p_2 obtained. By continuing this process, each p_i is obtained in sequence, and since

$$P^{-1} = R = I + e_1 p_1{}^H + \cdots + e_{n-1} p_{n-1}^H,$$

it follows that the matrix R is a byproduct, whether or not it is recorded explicitly. Hence the "direct" formation of P does not save arithmetic operations, although storage and redtape operations may be reduced.

Note that in the first method here described, the matrix R is formed a column at a time, and these become progressively longer. In the second method the matrix R is formed a row at a time and these become progressively shorter. In either method of orthogonal triangularization (Section 5.4) it is also true that each step forms explicitly a new row of R, beginning with the first. Either method of orthogonalization requires only $n - 1$ square roots, as does the first method of orthogonal triangularization. But the method of orthogonal triangularization has the theoretical advantage that each transformation, being unitary, has the spherical condition number $\kappa = 1$, and hence each transformed matrix has the condition number of A itself, which is to say that the condition cannot deteriorate. However, it is not clear when, if ever, this theoretical advantage is worth the additional computational labor, and neither method seems to compete, in practice, with simple Gaussian elimination, even when the problem is sufficiently unstable to require double precision.

5.6. Orthogonalization and Projection. Since orthogonalization involves taking a projection, it is not surprising that the method of orthogonalization should be closely related to the methods of successive approximation that are based on projection. Consider in particular the method given by (4.2.10). With slight modification of the indices to accord with present notation, this method starts with an arbitrary x_1 and y_1 and forms

(1)
$$x_2 = x_1 + \mu_1 y_1, \qquad r_2 = r_1 - \mu_1 A y_1,$$
$$\mu_1 = y_1^H A^H r_1 / y_1^H A^H A y_1.$$

The result of this is that

$$r_2^H A y_1 = 0,$$

which is to say that r_2 is made orthogonal to Ay_1. If it were possible to choose successive y_i so that, in general, $r_{\nu+1}$, besides being orthogonal to Ay_ν, is also orthogonal to $Ay_1, \cdots, Ay_{\nu-1}$, and if all y_i are linearly independent, then $r_{n+1} = 0$, $x_{n+1} = x$, since r_{n+1} would then be orthogonal to all vectors in the space. The method of successive approximation will then have become a direct method. But $r_{\nu+1}$ is a linear combination of r_ν and of Ay_ν. Hence it is required only that Ay_ν be chosen orthogonal to all vectors $Ay_1, \cdots, Ay_{\nu-1}$.

For carrying this out, let P be an arbitrary nonsingular matrix, and let

$$P_\nu = (p_1, \cdots, p_\nu), \qquad P_n = P.$$

Then y_ν will be selected as a suitable linear combination of the columns of

P_ν; and hence
$$Y_\nu = (y_1, \cdots, y_\nu) = P_\nu T_\nu^{-1},$$
where T_ν is an upper triangular matrix. It is required that

(2) $$Y_\nu^H A^H A Y_\nu = D_\nu^2$$

be diagonal. If this can be accomplished for $\nu = n$, then, dropping the subscript on Y_n,
$$A Y D^{-1} = W$$
is a unitary matrix and
$$A^{-1} = Y D^{-1} W^H.$$

Since Y_ν has appeared as a postmultiplier of A, its inverse is not required, and it is not necessary that it be triangular or of any other particular form.

Assuming T_ν to be unit upper triangular, it will be required that

(3) $$p_{\nu+1} = Y_\nu t_{\nu+1} + y_{\nu+1}, \qquad y_{\nu+1}^H A^H A Y_\nu = 0.$$

Consequently, application of (2) leads to
$$t_{\nu+1} = D_\nu^{-2} Y_\nu^H A^H A p_{\nu+1};$$
with $t_{\nu+1}$ it is possible to form $y_{\nu+1}$, and thence
$$\delta_{\nu+1}^2 = y_{\nu+1}^H A^H A y_{\nu+1}.$$

The choice $P = I$ makes Y an upper triangular matrix, and it in fact is the matrix $P = R^{-1}$ of (5.5.7). It was shown there that the direct formation of R^{-1} when R is triangular does not save arithmetic operations. However, if the primary interest is in solving the system
$$Ax = h,$$
there is an incidental advantage in that given any initial approximation x_1 to x, each step provides an improved approximation:

(4)
$$x_{\nu+1} = x_\nu + \mu_\nu y_\nu, \qquad x_{n+1} = x,$$
$$\mu_\nu = y_\nu^H A^H r_\nu / y_\nu^H A^H A y_\nu$$
$$= y_\nu^H A^H r_\nu / \delta_\nu^2.$$

If the first of (4) is multiplied by A, it leads to the sequence
$$\mu_1 A y_1 = r_1 - r_2,$$
$$\cdots \cdots$$
$$\mu_{n-1} A y_{n-1} = r_{n-1} - r_n,$$
$$\mu_n A y_n = r_n,$$
and from these equations it follows that if

(5) $$R = (r_1, \cdots, r_n), \qquad M = \operatorname{diag}(\mu_1, \cdots, \mu_n),$$

then (this R is not to be confused with that of previous sections),

(6) $$A Y M = R(I - J).$$

Hence

$$D^2M = Y^H A^H R(I - J),$$
(7)
$$Y^H A^H R = D^2M(I + J + J^2 + \cdots + J^{n-1}).$$

The matrix on the right is lower triangular, hence so is that on the left, as had been required. If Y has been made upper triangular, then Y^H is lower triangular and it follows that $A^H R$ must be lower triangular. Consequently in an orthogonalization by means of an upper triangular matrix, each remainder $r_{\nu+1}$ is orthogonal to the first ν columns of A.

5.7. The Method of Conjugate Gradients. If A is positive definite, the method of steepest descent takes $y_1 = r_1$, and chooses μ_1 to minimize $r_2{}^H A^{-1} r_2 = s_2{}^H A s_2$; hence

$$\mu_1 = r_1{}^H r_1 / r_1{}^H A r_1,$$

and thereby r_2 is made orthogonal to r_1. Also in the method of steepest descent, y_2 would be taken equal to r_2, but if y_2 is, instead, taken to be a suitable linear combination of r_1 and r_2, then r_3 can be made orthogonal to both r_1 and r_2. This corresponds to taking the vectors p_ν of Section 5.6 to be the r_ν. Not only that, but since each $r_{\nu+1}$ is then a linear combination of $r_1, r_2, \cdots, r_\nu$ and Ar_ν, this becomes a special instance of the Lanczos bi-orthogonalization with $b_\nu = c_\nu = r_\nu$.

As in (5.6.6),

(1) $$AYM = R(I - J),$$

where M is diagonal. Moreover, it is required that

(2) $$R = YT, \qquad R^H R = P,$$

where T is an upper triangular and P a diagonal matrix. Hence

$$AR = R(I - J)M^{-1}T,$$

and therefore

$$R^H AR = P(I - J)M^{-1}T.$$

But the matrix on the right is null below the first subdiagonal, and, being equal to the Hermitian matrix on the left, must be null above the first superdiagonal, and has the form

$$T = I - \Gamma,$$

where Γ is null everywhere except along the first superdiagonal. Therefore

(3) $$R = Y(1 - \Gamma),$$

(4) $$AR = R(I - J)M^{-1}(I - \Gamma),$$

(5) $$R^H AR = P(I - J)M^{-1}(I - \Gamma)$$
$$= (I - \Gamma^H)M^{-1}(I - J^T)P,$$

since M and P are real diagonals. Moreover, from (3) and (5) it follows that

$$(6) \qquad Y^H A Y = M^{-1}(I - J^T)P(I - \Gamma)^{-1} = \Omega,$$

and this is both upper triangular and Hermitian, hence diagonal. Consequently

$$(I - J^T)P = M\Omega(I - \Gamma),$$

which implies the two relations

$$(7) \qquad P = M\Omega, \qquad P\Gamma = J^T P.$$

The algorithm for obtaining the r_i and y_i in sequence is given by the vector relations, equivalent to (3) and (1),

$$(8) \qquad \begin{aligned} r_1 &= y_1, & \mu_1 A y_1 &= r_1 - r_2, \\ r_2 &= y_2 - \gamma_1 y_1, & \mu_2 A y_2 &= r_2 - r_3, \\ r_3 &= y_3 - \gamma_2 y_2, & \mu_3 A y_3 &= r_3 - r_4, \\ &\cdots\cdots\cdots \end{aligned}$$

and the scalar relations, equivalent to the second of (2) and (6) and (7),

$$(9) \qquad \begin{aligned} r_i^H r_i &= \rho_i, & y_i^H A y_i &= \omega_i, \\ \mu_i &= \rho_i/\omega_i, & \gamma_i &= \rho_{i+1}/\rho_i. \end{aligned}$$

The vectors x_i satisfy, by definition,

$$(10) \qquad \begin{aligned} x_2 &= x_1 + \mu_1 y_1, \\ x_3 &= x_2 + \mu_2 y_2. \\ &\cdots\cdots\cdots \\ x &= x_n + \mu_n y_n. \end{aligned}$$

From (6) it follows that

$$(11) \qquad A^{-1} = Y\Omega^{-1}Y^H.$$

Equation (4) is to be compared with (1.5.8) in the Lanczos reduction. The matrix R in this development corresponds to the two matrices B and C which appear there (since A is Hermitian a single sequence of vectors is sufficient), but the normalization is different. Also the auxiliary sequence of vectors y_i is introduced here but not in Section 1.5.

The following relations can be derived from the foregoing development:

$$(12) \qquad Y^H R = P(I - J)^{-1},$$

$$(13) \qquad R^H A Y = \Omega M(I - J)M^{-1} = (I - \Gamma^H)\Omega,$$

$$(14) \qquad Y^H Y = P(I - J)^{-1}(I - \Gamma)^{-1},$$

$$(15) \qquad R^H A R = (I - J^H)\Omega(I - \Gamma),$$

$$(16) \qquad R^H A^{-1} R = (I - J^T)^{-1}PM(I - J)^{-1},$$

$$(17) \qquad A Y = Y(I - \Gamma)(I - J)M^{-1}.$$

From these it can be shown that

$$(18) \qquad \| r_i \|^2 \kappa(A) \geq \| y_i \|^2 \geq \| r_i \|^2 = \rho_i,$$

$$(19) \qquad r_i^{\mathrm{H}} A r_i \geq y_i^{\mathrm{H}} A y_i,$$

$$(20) \qquad s_i^{\mathrm{H}} A s_i = \mu_i \rho_i + \mu_{i+1} \rho_{i+1} + \cdots + \mu_n \rho_n,$$

$$(21) \qquad \lambda_n(A) \leq \mu_i^{-1} \leq \lambda_1(A),$$

$$(22) \qquad \rho_i \kappa(A) \geq \rho_i + \rho_{i+1} + \cdots + \rho_n,$$

$$(23) \qquad 1 + \mu_i + \mu_i \mu_{i-1} + \cdots + \mu_i \cdots \mu_1 \leq \kappa(A).$$

Here the norms, and the condition number κ, are spherical. Evidently if

$$S = A^{-1} R$$

is the matrix of the error vectors s_i, (16) is equivalent to

$$S^{\mathrm{H}} A S = (I - J^{\mathrm{T}})^{-1} P M (I - J)^{-1}.$$

Since all scalars are nonnegative, relations (18) through (23) provide bounds in terms of $\lambda_1(A)$ and $\lambda_n(A)$.

REFERENCES

For practical techniques and the establishment of error bounds, papers and two forthcoming books by Wilkinson are unsurpassed. See also, *Modern Computing Methods*, published by the National Physical Laboratory (1961). Attention may be called again to Faddeev and Faddeeva (1960), and to Durand (1961).

Among the many papers in the literature that allegedly give new methods, whether direct or otherwise, the great majority are rediscoveries, or, as in the method of Crout (1941), present a more efficient organization of the data or ordering of the steps. Notable exceptions in recent years are the method of conjugate gradients, due to Hestenes and Stiefel (1952), based upon the Lanczos algorithm, and the method of modification. The latter, as a method of inversion, is due to Sherman and Morrison (1949; see also Sherman, 1953), but the method is based upon (5.1.1) which is due to Bartlett (1951). The method of Givens (1958) seems also to be an exception. The Gauss-Jordan method, so-called, seems to have been described first by Clasen (1888; see also Mehmke, 1930, and Mansion, 1888). Since it can be regarded as a modification of Gaussian elimination, the name of Gauss is properly applied, but that of Jordan seems to be due to an error, since the method was described only in the third edition of his *Handbuch der Vermessungskunde*, prepared after his death.

The essential identity of Gaussian elimination and the method of Chió for evaluating a determinant will be recognized by anyone acquainted with both.

A curiosity in the history of the subject is the limited vogue of Cracovians, as developed by Banachiewicz. Cracovians are matrices with a row-by-row rule of composition, and clearly any formula in Cracovians has a corresponding formula in matrices, and conversely. Some consider Cracovians more convenient for purposes of equation-solving, and the method of Banachiewicz is the method of Cholesky expressed by Cracovians.

The classical paper by von Neumann and Goldstine (1947) established a technique and a standard that far exceeds in importance the detailed results. Later work, especially by Wilkinson, has shown that the error estimates themselves were much too pessimistic. Another important early paper on the analysis of errors is Turing (1948).

For a further discussion of scaling and minimal condition, see the paper presented by F. L. Bauer at the Matrix Symposium, IFIP Conference, 1962.

PROBLEMS AND EXERCISES

1. Obtain the conditions for the factorization(5.2.23), and show that the factorization, when possible, is unique.

2. Apply the two factorizations (5.2.1) and (5.2.23) to

$$
A = \begin{pmatrix}
n & n-1 & n-2 & \cdots & 2 & 1 \\
n-1 & n-1 & n-2 & \cdots & 2 & 1 \\
0 & n-2 & n-2 & \cdots & 2 & 1 \\
& & \cdots\cdots\cdots\cdots & & & \\
0 & 0 & 0 & \cdots & 2 & 1 \\
0 & 0 & 0 & \cdots & 1 & 1
\end{pmatrix}.
$$

3. Let A be of rank ρ, and

$$A_{11} = L_{11}D_1R_{11}$$

nonsingular and of order ρ. Then the factorization

$$
\begin{pmatrix} A_{11} & A_{12} \\ A_{21} & A_{22} \end{pmatrix} = \begin{pmatrix} L_{11} & 0 \\ L_{21} & I \end{pmatrix} \begin{pmatrix} D_1 & 0 \\ 0 & 0 \end{pmatrix} \begin{pmatrix} R_{11} & R_{12} \\ 0 & I \end{pmatrix}
$$

is possible and unique, and the last $n - \rho$ columns of R^{-1} satisfy the homogeneous equations $Ax = 0$.

4. For solving the system $Ax = h$, let $p_0^{(0)}, p_1^{(0)}, \cdots, p_n^{(0)}$ represent $n + 1$ points not on the same hyperplane. From $p_0^{(0)}$, project each of the other points $p_i^{(0)}$ upon the first hyperplane of the system $Ax = h$ to form the points $p_1^{(1)}, p_2^{(1)}, \cdots, p_n^{(1)}$. From $p_1^{(1)}$ project each of the others upon the second hyperplane. At the last step project $p_n^{(n-1)}$ from $p_{n-1}^{(n-1)}$ upon the last hyperplane to form $x = p_n^{(n)}$ (Purcell, Pietrzykowski). Show that this is done analytically by forming the augmented matrix

$$B = (A, h),$$

and the nonsingular matrix

$$P = \begin{pmatrix} p_0^{(0)} & p_1^{(0)} & \cdots & p_n^{(0)} \\ -1 & -1 & \cdots & -1 \end{pmatrix}$$

of order $n + 1$, then multiplying the matrix BP successively on the right by elementary upper triangular matrices (Section 1.1) R_i to reduce the product to lower trapezoidal form. The last column of $PR_1R_2 \cdots R_n$ provides the solution (if B is made formally into a square matrix by adjoining a null row below, this corresponds to Exercise 3 with BP in place of A, and $BPR^{-1} = LD$, where the last element in the diagonal of D is zero).

5. Write out in detail in terms of the elements a step in the Gauss-Jordan reduction. Show that if the same scheme is applied to the entire matrix (A, B) the result will be the matrix $(A^{-1}, A^{-1}B)$.

6. Write out in terms of the elements a step in the Gauss-Jordan reduction

$$\begin{pmatrix} A \\ B \end{pmatrix} \rightarrow \begin{pmatrix} A^{-1} \\ BA^{-1} \end{pmatrix}.$$

7. Define $D(a)$ to be the diagonal matrix for which

$$D(a)e = a, \qquad e^T D(a) = a^T.$$

Since $e^H J - e^H = e_n^H$, if $e_n^H y = 0$, then

$$e^H(J - I)D(a)y = 0$$

for any a. Hence the substitution

$$D(A^H e_1)x = (J - I)D(A^H e_1)y + e_1^T h e_1$$

reduces the system to one of order $n - 1$, provided $D(A^H e_1)$ is nonsingular (Peres).

8. Expand $(B - \sigma uv^H)^{-1}$ formally in powers of σ and hence obtain (5.1.1). Do likewise with $(B - USV^H)^{-1}$.

9. Solving the complex system

$$(A + iB)(x + iy) = h + ik,$$

where A, B, x, y, h, and k are real, is equivalent to solving

$$\begin{pmatrix} A & -B \\ B & A \end{pmatrix} \begin{pmatrix} x \\ y \end{pmatrix} = \begin{pmatrix} h \\ k \end{pmatrix}.$$

Hence show that if the necessary inverses exist,

$$(A + iB)^{-1} = P + iQ,$$

where

$$P = \Delta_2^{-1}, \qquad Q = -\Delta_2^{-1} BA^{-1}.$$

10. Write out in detail the steps required in applying the Givens method of orthogonal triangularization, Section 5.4.

11. If the diagonal blocks are square, show that each of the following factorizations is possible given only suitable nonsingularity conditions:

$$
\begin{pmatrix} A_1 & B_1 & 0 & \cdots \\ C_1 & A_2 & B_2 & \cdots \\ 0 & C_2 & A_3 & \cdots \\ & \cdots \end{pmatrix} = \begin{pmatrix} I & 0 & 0 & \cdots \\ L_1 & I & 0 & \cdots \\ 0 & L_2 & I & \cdots \\ & \cdots \end{pmatrix} \begin{pmatrix} R_1 & B_1 & 0 & \cdots \\ 0 & R_2 & B_2 & \cdots \\ 0 & 0 & R_3 & \cdots \\ & \cdots \end{pmatrix}
$$

$$
= \begin{pmatrix} M_1 & 0 & 0 & \cdots \\ C_1 & M_2 & 0 & \cdots \\ 0 & C_2 & M_2 & \cdots \\ & \cdots \end{pmatrix} \begin{pmatrix} I & P_1 & 0 & \cdots \\ 0 & I & P_2 & \cdots \\ 0 & 0 & I & \cdots \\ & \cdots \end{pmatrix}
$$

$$
= \begin{pmatrix} I & P_1' & 0 & \cdots \\ 0 & I & P_2' & \cdots \\ 0 & 0 & I & \cdots \\ & \cdots \end{pmatrix} \begin{pmatrix} M_1' & 0 & 0 & \cdots \\ C_1 & M_2' & 0 & \cdots \\ 0 & C_2 & M_3' & \cdots \\ & \cdots \end{pmatrix}
$$

$$
= \begin{pmatrix} R_1' & B_1 & 0 & \cdots \\ 0 & R_2' & B_2 & \cdots \\ 0 & 0 & R_3' & \cdots \\ & \cdots \end{pmatrix} \begin{pmatrix} I & 0 & 0 & \cdots \\ L_1' & I & 0 & \cdots \\ 0 & L_2' & I & \cdots \\ & \cdots \end{pmatrix}.
$$

In each case obtain the recursions defining L_i, R_i, $\cdots$.

12. Let $\omega(\nu) = [2 \cdot 3^{\frac{1}{2}} \cdot 4^{\frac{1}{3}} \cdots \nu^{1/(\nu-1)}]^{\frac{1}{2}}$. Suppose the matrix A is scaled so that $|A| \leq ee^{\mathrm{T}}$. In Gaussian factorization

$$A = LDR,$$

if each pivotal element is the largest possible, then [Wilkinson]

$$|\delta_n| \leq |\delta_{n-\nu+1}| \, \nu^{\frac{1}{2}} \omega(\nu).$$

(Make use of Hadamard's theorem in the form, if $|A| \leq ee^{\mathrm{T}}$, then $\delta(A) \leq n^{n/2}$.)

13. The largest element of a positive definite matrix lies on the diagonal.

14. Let $A = LL^{\mathrm{H}}$ where L is lower triangular, and let $|A| \leq ee^{\mathrm{T}}$. By the partitioning

$$
L = \begin{pmatrix} M & 0 \\ m^{\mathrm{H}} & \mu \end{pmatrix},
$$

show that $|L| \leq ee^{\mathrm{T}}$.

15. Let

$$A = \begin{pmatrix} \beta & b^H \\ b & B \end{pmatrix} = \begin{pmatrix} 1 & 0 \\ l & I \end{pmatrix}\begin{pmatrix} \beta & 0 \\ 0 & \Delta \end{pmatrix}\begin{pmatrix} 1 & l^H \\ 0 & I \end{pmatrix}$$

be positive definite. Show that $|B| \le \delta ee^T \Rightarrow |\Delta| \le \delta ee^T$. Combine this with the previous result to show that, in successive Gaussian transforms of a positive definite matrix, the largest element cannot exceed that of A itself.

16. Prove in detail the identities (5.7.12) and following.

17. Use (5.7.4) and the second of (5.7.2) to obtain a recursion for the r_i alone.

18. Use (5.7.17) and (5.7.6) to obtain a recursion for the y_i alone.

19. In the Lanczos reduction, Section 1.5, let

$$A = \text{diag}\,(\alpha_1, \alpha_2, \cdots, \alpha_n),$$
$$b_1 = c_1 = e.$$

Then $B = C$ is orthogonal by columns. Moreover, the value of the ith element in the νth column of B is a polynomial $\beta_\nu(\alpha)$ of degree $\nu - 1$ in α evaluated for $\alpha = \alpha_i$:

$$b_\nu^T = [\beta_\nu(\alpha_1), \quad \beta_\nu(\alpha_2), \cdots, \beta_\nu(\alpha_n)],$$

assuming all quantities real. Let

$$B_\nu = (b_1, b_2, \cdots, b_\nu), \qquad P_\nu = B_\nu^T B_\nu.$$

Let the ith element of the vector h represent the result of a measurement at $\alpha = \alpha_i$ of a quantity whose value is a function of α. Then the normal equations for a least-squares polynomial fit can be written

$$P_\nu y = B_\nu^T h,$$

since $\eta_1\beta_1(\alpha) + \eta_2\beta_2(\alpha) + \cdots + \eta_\nu\beta_\nu(\alpha)$ is that polynomial of degree $\nu - 1$ in α that best fits the data in the least-squares sense. The polynomials $\beta_\nu(\alpha)$ are orthogonal over the points $\alpha_1, \cdots, \alpha_n$, which is to say that

$$\sum_i \beta_\nu(\alpha_i)\beta_\mu(\alpha_i) = 0, \nu \ne \mu.$$

Obtain the algorithm in scalar form. [The algorithm has at least two advantages over direct determination of the coefficients of the powers of α. First, it appears to be more stable. Second, the determination of the polynomial of degree ν is based upon that of degree $\nu - 1$ and does not require separate computation.]

20. Apply the obvious identity

$$(u_1, u_2, \cdots, u_n)(v_1, v_2, \cdots, v_n)^H = u_1 v_1^H + \cdots + u_n v_n^H$$

along with Wedderburn's theorem (Exercise 1.34) to obtain Egervary's "rank-reducing" transformations, and thereby derive the methods of triangularization and of orthogonal triangularization.

21. If A is Hermitian show inductively that given any nonsingular Y, a unit upper triangular matrix T can be constructed such that $Y = XT$ and $X^H A X = D$ is diagonal. Hence $x = XD^{-1}X^H h$ satisfies $Ax = h$ (Fox, Huskey, and Wilkinson). This is Morris's escalator method.

CHAPTER 6

Proper Values and Vectors: Normalization
and Reduction of the Matrix

6.0. Purpose of Normalization. For inverting a matrix and for solving
a system of linear equations it is meaningful and convenient to distinguish
direct methods from methods of successive approximation. For finding the
characteristic roots of a matrix it is common to say that a method is direct
in case it leads explicitly to the characteristic polynomial in a finite number
of steps, and otherwise to call it an iterative method, or a method of successive approximation. This distinction, however, is somewhat misleading,
since even when the characteristic polynomial is known explicitly it is
ordinarily necessary to use a method of successive approximation to obtain
its zeros. Moreover, some of these methods, including in particular the
methods of the Bernoulli type, when interpreted as operations applied to
the companion matrix, are only special cases of methods that apply to any
matrix, hence could have been applied to the original, unreduced, matrix.

It seems appropriate, barring very special cases, therefore, to assume
always at the outset that at some stage a method of successive approximation must be used, and to consider whether or not it is best to first reduce
the matrix to some simpler form, possibly that of the companion matrix.
The purpose of this chapter is to describe certain methods for making such
a reduction. Also described in this chapter are methods of "deflation" to
be applied when some roots are already known. These methods correspond
to that of "dividing out" known roots of an algebraic equation, and they
permit the replacement of the original matrix by one of lower order whose
roots are the yet unknown roots of the original.

There are two important forms intermediate between a completely
unreduced "full" matrix, and the *Frobenius form* of the companion matrix
to the characteristic polynomial, i.e., the form (6.1.8). These are the
Hessenberg (sometimes called almost triangular), and the Jacobi (tridiagonal)
forms. The *Hessenberg form* is null below the first subdiagonal (or above the
first superdiagonal); the *Jacobi form* is null both below the first subdiagonal
and above the first superdiagonal. Any matrix can be reduced by purely
rational transformations to any of the three forms, Hessenberg, Jacobi, and
Frobenius, although some methods that involve square roots will also be
described. The Lanczos reduction to Jacobi form is rational and has been
described already in Section 1.4; the Jacobi form is, moreover, a special
Hessenberg form.

In a general matrix of order n there are n^2 independent elements and the roots are functions of these n^2 parameters; in the Hessenberg form there are only $(n^2 + 3n - 2)/2$ independent elements, the others being zeros; in the Jacobi form there are $3n - 2$; and in the Frobenius form only n. The n independent elements in the Frobenius form are the coefficients of the characteristic polynomial. According to the form used, the roots may be expressed (implicitly) as functions of n^2 independent variables, of $(n^2 + 3n - 2)/2$, of $3n - 2$, or of only n. Clearly it is easier to work with n independent parameters than with n^2, which is the reason for making the reduction. On the other hand, in compressing the data the stability may be decreased, so that a small change in an element of the original matrix may have little effect on a root, whereas a small change in an element of the Frobenius form, or other reduced form, may be catastrophic (Exercise 18). For this reason the complete reduction to Frobenius form is to be undertaken only with caution. It is possible, and perhaps advisable if the complete reduction is to be made, to make it stagewise: first reducing to Hessenberg form; then from Hessenberg form to Jacobi form; and finally from Jacobi form to Frobenius form. In the last two reductions the use of multiple precision is often essential.

The problem of finding the coefficients of the characteristic polynomial $\delta(\lambda I - A)$ can be considered a special case of that of finding the coefficients of $\delta[P(\lambda)]$, where $P(\lambda)$ is a matrix whose elements are polynomials in λ. One way of doing this is to evaluate $\delta[P(\lambda_i)]$ for a sufficient number of distinct values of λ_i, then to pass an interpolation polynomial through the points so obtained. Such a method is available in particular when $P(\lambda) = \lambda I - A$, and is sometimes used in practice. There are required $n + 1$ distinct values λ_i, in this case. The method is not to be recommended in this special case unless, perhaps, the matrix is small. When

$$P(\lambda) = P_0 \lambda + P_1,$$

where P_0 and P_1 are constant matrices, if either P_0 or P_1 is nonsingular the problem reduces immediately to the characteristic value problem. Some aspects of the more general problem will be indicated in the exercises.

For the characteristic value problem itself, most methods used in practice can be derived from a basic idea due to Krylov and already utilized in the development of the Lanczos reduction. These methods make use of the fact that the coefficients expressing the linear dependence of consecutive vectors of a Krylov sequence are coefficients of a divisor of the characteristic polynomial. These methods will be shown to involve the formation, at least implicitly, of a Krylov sequence. Some of these methods yield a similarity transformation of the original matrix to a simpler form (Hessenberg, Jacobi, or Frobenius), and some do not. When a similarity reduction has been made and a root λ found, a proper vector of the reduced matrix can be found and then transformed to a proper vector for the original matrix. Thus, if

$$AP = PB,$$

and if

$$By = \lambda y,$$

then

$$APy = \lambda Py,$$

so that y is a proper vector for B and Py is a proper vector for A. However, it is equally possible to solve directly the homogeneous system

$$(\lambda I - A)x = 0,$$

and when no similarity transformation has been made this may be necessary. When there is likelihood of instability it may be advantageous to do this in any event. It may turn out, then, that the computed λ and x require improvement by applying one of the methods of the next chapter.

In the next section the original method of Krylov will be discussed, and in succeeding sections some other standard methods will be described and related to the method of Krylov. There remain two other methods which seem to be not reducible to the method of Krylov. One is historically much older, and is due to Leverrier. It is of interest for theoretical and for historical reasons, but is not to be recommended, at least for large matrices, because it requires too many operations and it tends to be unstable. The other is an escalator method and has been developed independently by Samuelson and by Bryan. Both of these methods provide the vectors along with the roots. The final section will deal with methods of deflation.

6.1. The Method of Krylov. Let $v = v_1 \neq 0$ be an arbitrary nonnull vector, and let

(1) $$v_{i+1} = Av_i.$$

In the sequence of vectors v_i, there will be a first, say v_{m+1}, that is expressible as a linear combination of those that precede it. Hence if

(2) $$V = (v_1, v_2, \cdots, v_m),$$

then V is of rank m and

(3) $$Vf + v_{m+1} = 0$$

for some vector f. If the vector $l = l(\lambda)$ is defined by

(4) $$l^T = (1, \lambda, \cdots, \lambda^{m-1}),$$

and the polynomial

(5) $$\psi(\lambda) = l^T f + \lambda^m,$$

then (3) is equivalent to

$$\psi(A)v = 0,$$

which states that $\psi(\lambda)$ annihilates v. But since V is of rank m, there is no polynomial of lower degree that annihilates v, whence $\psi(\lambda)$ is the minimal polynomial for v. Hence $\psi(\lambda)$ divides the minimal polynomial for A, and if $m = n$, then $\psi(\lambda)$ is the characteristic polynomial for A.

Consistency of the relations (3) with $\psi(\lambda) = 0$ requires that for any proper value λ the matrix

$$(6) \qquad \begin{pmatrix} 1 & \lambda & \cdots & \lambda^m \\ v_1 & v_2 & \cdots & v_{m+1} \end{pmatrix}$$

be of rank m. Hence if

$$(v_1', v_2', \cdots, v_{m+1}')$$

is any matrix of m linearly independent rows of V, then

$$(7) \qquad \psi(\lambda) \equiv \delta \begin{pmatrix} 1 & \lambda & \cdots & \lambda^m \\ v_1' & v_2' & \cdots & v_{m+1}' \end{pmatrix} / \delta(v_1', v_2', \cdots, v_m').$$

In particular, if $m = n$, this is a determinantal expression for the characteristic polynomial $\phi(\lambda)$ in which λ occurs along a single row instead of along the diagonal. Moreover, if $v_1 = e_1$, of dimension n, then $v_1' = e_1'$ of dimension m, and the determinant in the numerator can be reduced immediately to one of order m. This Krylov does.

Supposing f has been found, by whatever method, let

$$(8) \qquad F = J - f e_m^{\mathrm{T}},$$

the two vectors f and e_m being of m dimensions. Then

$$(9) \qquad AV = VF,$$

as can be verified directly. Then F is by definition *the companion matrix* to $\psi(\lambda)$. Conversely, suppose W is any matrix of rank m such that

$$(10) \qquad AW = WF.$$

By comparing the two sides of this equation, column by column, it appears that the columns of W form a Krylov sequence, and that

$$Wf + w_{m+1} = 0.$$

Hence w_1, the first column of W, has $\psi(\lambda)$ as its minimal polynomial.

Suppose $m < n$. Let $V = V_1$, $F = F_1$:

$$AV_1 = V_1 F_1.$$

Let V_2 be a matrix of rank $n - m$ such that (V_1, V_2) is nonsingular. Then

$$A(V_1, V_2) = (V_1, V_2) \begin{pmatrix} F_1 & F_{12} \\ 0 & F_{22} \end{pmatrix},$$

where

$$(V_1, V_2)^{-1} A V_2 = \begin{pmatrix} F_{12} \\ F_{22} \end{pmatrix}.$$

The characteristic equation of A is

$$\delta(\lambda I - A) = \delta(\lambda I - F_1)\, \delta(\lambda I - F_{22}).$$

While F_{22} will not, in general, have the Frobenius form, the same transformation can be applied to it. By continuing, the complete characteristic polynomial A can be obtained as a product

$$\delta(\lambda I - F_1)\, \delta(\lambda I - F_2)\cdots$$

(cf. Section 1.5).

The Krylov method is therefore quite general and is based upon the simple observation that in any sequence of iterates (1) there will be a smallest set of consecutive iterates which are linearly dependent, and that the coefficients of a vanishing combination are coefficients of a divisor of the characteristic polynomial. The methods to be described next will be shown to be based upon such a sequence and to represent particular methods of evaluating these coefficients.

6.2. The Weber-Voetter Method. The Weber-Voetter method does not make a similarity transformation of the matrix A, but operates upon the characteristic determinant. Hence the relation to the Krylov method is not immediately apparent. First the method will be described, and then its relation to the Krylov method will be established. In order for the method to go through to completion it is necessary that certain quantities that occur in the course of the computation be nonvanishing. At the outset these conditions will be assumed. Later, however, it will be shown that they can be expected to hold "in general," and possible remedial steps will be indicated in case they fail.

The purpose of the method is to form a matrix $W(\lambda)$ such that

$$(1) \qquad (A - \lambda I)W(\lambda) = Q(\lambda) = \begin{pmatrix} * & * & * & * & \cdots \\ * & 0 & 0 & 0 & \cdots \\ * & * & 0 & 0 & \cdots \\ * & * & * & 0 & \cdots \\ & & \cdots\cdots\cdots & & \end{pmatrix},$$

where asterisks indicate elements not required to vanish, and where λ occurs only in the first row of $Q(\lambda)$. In fact,

$$(2) \qquad Q(\lambda) = Q_0 - \lambda e_1 q^{H}(\lambda),$$

and the ith element of q^{H} is of degree $i - 1$ in λ. The determinant of the matrix $W(\lambda)$ in (1) is to be a constant, and hence if the reduction (1) is possible, the element $(1, n)$ in $Q(\lambda)$ is the characteristic polynomial, apart, perhaps, from a constant multiplier.

The matrix $W(\lambda)$ is formed as a product of matrices

(3) $$W(\lambda) = M_2(\lambda)M_3(\lambda) \cdots M_n(\lambda),$$

where

(4) $$M_i(\lambda) = M_i + \lambda e_{i-1}e_i^{\mathrm{T}}, \qquad i = 2, 3, \cdots, n,$$

and where M_i differs from I only in the ith column. The matrices $M_i(\lambda)$ will be formed in sequence as follows: Let

(5) $$P_1(\lambda) = A - \lambda I,$$
$$P_{i+1}(\lambda) = P_i(\lambda)M_{i+1}(\lambda).$$

Clearly the first i columns of $P_i(\lambda)$ and $P_{i+1}(\lambda)$ will be alike, and these must be the first i columns of $Q(\lambda)$. Moreover the last $n - i$ columns of $P_i(\lambda)$ are the same as those of the original matrix $A - \lambda I$. With an obvious partitioning, it is to be shown that sub-vectors $m_{1,i+1}$ and $m_{2,i+1}$, with constant elements are, "in general," determined uniquely by the requirement that the vector

$$\begin{pmatrix} Q_{11}(\lambda) & A_{12} \\ Q_{21} & A_{22} - \lambda I \end{pmatrix} \begin{pmatrix} m_{1,i+1} + \lambda e_i \\ m_{2,i+1} \end{pmatrix}$$

shall agree in form with column $i + 1$ of $Q(\lambda)$.

Evidently the choice

(6) $$m_{2,i+1} = Q_{21}e_i$$

frees the last $n - i$ elements of λ. The first i elements are those of

(7) $$\lambda Q_{11}(\lambda)e_i + Q_{11}(\lambda)m_{1,i+1} + A_{12}m_{2,i+1}.$$

But only the first element contains λ, and only the first element of $\lambda Q_{11}(\lambda)e_i$ is nonvanishing. Because of the form of $Q_{11}(\lambda)$, and because $m_{2,i+1}$ has already been determined, it follows that each of the first $i - 1$ elements of $m_{1,i+1}$ can be selected in sequence, beginning with the first, to make all elements in (7) vanish except the first, provided only

$$(e_2^{\mathrm{T}}Qe_1)(e_3^{\mathrm{T}}Qe_2) \cdots (e_i^{\mathrm{T}}Qe_{i-1}) \neq 0.$$

This will be assumed. Finally, the ith element in $m_{1,i+1}$ can be selected to annihilate the first element in

$$Q_{21}m_{1,i+1} + A_{22}m_{2,i-1},$$

and this is possible provided

(8) $$e_{i+1}^{\mathrm{T}}Qe_i \neq 0.$$

Thus the algorithm can be carried through to completion, and the matrix $Q(\lambda)$ can be formed a column at a time unless one of the conditions (8) fails.

Note, however, that in (8), Q can be replaced by $P_i(\lambda)$, since the first i columns in $P_i(\lambda)$ are those of $Q(\lambda)$. Assuming it has been possible to form $P_i(\lambda)$, it will be possible to form $P_{i+1}(\lambda)$ unless

$$(9) \qquad e_{i+1}^T P_i(\lambda) e_i = 0.$$

But if this should happen, it may be that for some $j > i + 1$,

$$e_j^T P_i(\lambda) e_i \neq 0.$$

In that event a permutational transformation can be applied which, in A, interchanges rows $i + 1$ and j, and also columns $i + 1$ and j, and in $P_i(\lambda)$ interchanges rows $i + 1$ and j. The algorithm can then proceed as before. Even when (9) does not hold, a permutation might be desirable for purposes of stability.

To relate this method to that of Krylov, consider the matrix (6.1.6) with $m = n$. By an obvious transformation, the determinant of this matrix is seen to be equal to that of

$$(10) \qquad K(\lambda) = [(A - \lambda I)v, (A^2 - \lambda^2 I)v, \cdots, (A^n - \lambda^n I)v].$$

Moreover

$$(11) \qquad K(\lambda) = (A - \lambda I)H(\lambda),$$

where

$$(12) \qquad H(\lambda) = V(I - \lambda J^T)^{-1},$$

the matrix V being that of (6.1.2) with $m = n$. This factorization can be verified directly.

Two lemmas are of some interest in themselves, and will be of assistance in completing the discussion. The first is: If V is defined by (6.1.2) and (6.1.1) with $m = n$ and $v = v_1 = e_1$, and if every leading principal minor determinant of V, including the determinant of V itself, is nonvanishing, then there exists a unique unit upper triangular matrix R such that AVR has the form of $Q(0)$ in (1).

To prove this, observe that, when $v_1 = e_1$, then any leading principal minor determinant of V except the first is a leading principal minor determinant of the submatrix of V which remains after the deletion of the first row and first column. Hence let $P = (e_n, e_1, e_2, \cdots, e_{n-1})$ and consider the triangular factorization of PAV. This is unique when the hypothesis of the lemma is fulfilled; the unit upper triangular factor is the inverse R^{-1} of the required R, and the lower triangular factor is $PQ(0)$.

Now consider the product

$$K(\lambda)R = (A - \lambda I)V(I - \lambda J^T)^{-1}R.$$

It is readily verified that, considered as a polynomial in λ, this matrix has the form of the matrix $Q(\lambda)$ obtained by the Weber-Voetter method, and that the product $V(I - \lambda J^T)^{-1}R$ has the form of $W(\lambda)$. It will be shown next that, in fact,

(13) $$H(\lambda)R = W(\lambda), \qquad K(\lambda)R = Q(\lambda),$$

where the $W(\lambda)$ and $Q(\lambda)$ are the matrices defined by the Weber-Voetter method, where R is defined in Lemma 1, and where $H(\lambda)$ and $K(\lambda)$ are given by (10) and (12). This will follow from the fact that, given the above hypothesis, matrices $W(\lambda)$ and $Q(\lambda)$ satisfying (1) are uniquely defined by the requirement that as functions of λ they have the forms obtained in the Weber-Voetter method.

For $W(\lambda)$ this form is

(14) $$W(\lambda) = e_1(1, \lambda, \cdots, \lambda^{n-1}) + W_1 + \lambda W_2 + \cdots + \lambda^{n-2}W_{n-1},$$

where each W_i is constant and, moreover, vanishes in its first i columns:

(15) $$W_i e_j = 0, \qquad j \le i.$$

Let

(16) $$Q(\lambda) = Q_0 + \lambda Q_1 + \cdots + \lambda^n Q_n.$$

From (1) it follows that

(17)
$$
\begin{aligned}
Q_0 &= A(e_1 e_1^{\mathrm{T}} + W_1) \\
Q_1 &= A(e_1 e_2^{\mathrm{T}} + W_2) - (e_1 e_1^{\mathrm{T}} + W_1), \\
Q_2 &= A(e_1 e_3^{\mathrm{T}} + W_3) - (e_1 e_2^{\mathrm{T}} + W_2), \\
&\cdots\cdots\cdots\cdots\cdots\cdots\cdots\cdots\cdots\cdots\cdots\cdots \\
Q_n &= -e_1 e_n^{\mathrm{T}}.
\end{aligned}
$$

First

$$Q_0 e_1 = A e_1 = v_2$$

and the first column of Q_0 is determined. Next,

$$
\begin{aligned}
Q_1 e_1 &= -e_1, \\
Q_1 e_2 &= A e_1 - W_1 e_2.
\end{aligned}
$$

Since Q_1 is null below the first row, the second relation implies that, for some scalar κ_{12},

$$W_1 e_2 = A e_1 + \kappa_{12} e_1.$$

Hence

$$
\begin{aligned}
Q_0 e_2 &= A W_1 e_2 \\
&= A^2 e_1 + \kappa_{12} A e_1 \\
&= v_3 + \kappa_{12} v_2.
\end{aligned}
$$

But

$$e_2^{\mathrm{T}} Q_0 e_2 = 0 = e_2^{\mathrm{T}} v_3 + \kappa_{12} e_2^{\mathrm{T}} v_2.$$

Therefore κ_{12} can be determined provided $e_2^{\mathrm{T}} v_2 \ne 0$, hence provided the second leading principal minor determinant of V is nonvanishing. But given κ_{12}, the second columns of Q_0, Q_1, and W_1 are known. The next step is sufficiently illustrative of all the others. Evidently

$$Q_2 e_1 = 0$$

identically, and

$$Q_2 e_2 = -e_1,$$
$$Q_2 e_3 = A e_1 - W_2 e_3,$$
$$Q_1 e_3 = A W_2 e_3 - W_1 e_3,$$
$$Q_0 e_3 = A W_1 e_3.$$

The second and third of these imply that for some scalars κ_{13} and κ_{23},

$$-\kappa_{23} e_1 = A e_1 - W_2 e_3,$$
$$-\kappa_{13} e_1 = A W_2 e_3 - W_1 e_3,$$

hence that

$$W_2 e_3 = A e_1 + \kappa_{23} e_1,$$
$$W_1 e_3 = A^2 e_1 + \kappa_{23} A e_1 + \kappa_{13} e_1,$$

and this, with the fourth of the above relations gives

$$Q_0 e_3 = A^3 e_1 + \kappa_{23} A^2 e_1 + \kappa_{13} A e_1,$$
$$- v_4 + \kappa_{23} v_3 + \kappa_{13} v_2.$$

But

$$e_2^T Q_0 e_3 = e_3^T Q_0 e_3 = 0.$$

Hence there are two equations for determining κ_{13} and κ_{23}, and they have a unique solution when the third leading principal minor determinant of V is nonvanishing. When κ_{13} and κ_{23} are known, the third column of W_1, W_2, Q_0, Q_1, and Q_2 are given.

This shows that the matrix $W(\lambda)$, which is the product of all the matrices $M_i(\lambda)$ formed in the Weber-Voetter method, is in fact based upon the matrix V of Krylov vectors beginning with $v_1 = e_1$. To interpret the conditions (8) for carrying through algorithms, observe that, since

$$A V R = Q_0,$$

and R is unit upper triangular, the product of the first ν elements of the form (8) is equal to the determinant of the submatrix of AV that remains after deleting the first row and the last $n - \nu - 1$ rows, and the last $n - \nu$ columns. But this is the same as the leading principal minor determinant of order $\nu + 1$ of V. Thus *it is necessary and sufficient for the Weber-Voetter algorithm to go through* that every leading principal minor determinant of Y be nonvanishing. But this hypothesis will be satisfied "in general," since there is nothing special about a Krylov sequence of vectors. In fact, suppose

$$V = (v_1, v_2, \cdots, v_n)$$

is any nonsingular matrix, and v_{n+1} any vector. Then there exists a matrix A for which the vectors $v_1, \cdots, v_{n+1}$ are the first $n + 1$ vectors of a Krylov sequence. This is equivalent to saying that there exists a matrix A for which

$$A V = (v_2, \cdots, v_{n+1}),$$

and since V is nonsingular the matrix is

$$A = (v_2, \cdots, v_{n+1}) V^{-1}.$$

The vanishing of a particular minor determinant does not imply that the vectors from which these are formed are linearly dependent. Consequently, even when the algorithm fails at some stage it may be possible to permute the indices, as suggested above, and continue. But if $e_j{}^T P_i(\lambda) e_i = 0$ for every $j > i$, then the vectors are linearly dependent and it is not possible to continue. But in that event, $e_1{}^T P_i(\lambda) e_i$ is the minimal polynomial for the vector $v = e_1$ and hence is a divisor of the characteristic polynomial of A. In practice, however, these numbers may fall below the machine's level of precision without being zero in fact.

In the next section another method for obtaining the characteristic polynomial will be described which can also be related to the Krylov method. This one, due to Danilevskiĭ, applies a sequence of similarity transformations to A which ultimately reduce it to Frobenius form. Still other applications of the Krylov method will be developed thereafter which use similarity transformations but make only a partial reduction.

6.3. The Method of Danilevskiĭ.   The method of Danilevskiĭ applies the Gauss-Jordan method in solving (6.1.3) for f, multiplying V by a sequence of elementary matrices

(1) $$M_i = I - \sigma_i m_i e_i{}^T$$

such that the product

$$M = M_m \cdots M_2 M_1$$

satisfies

$$M V = \begin{pmatrix} I \\ 0 \end{pmatrix}.$$

Then

$$(MAM^{-1})MV = MVF,$$

hence

(2) $$MAM^{-1} = \begin{pmatrix} F & * \\ 0 & * \end{pmatrix},$$

where the asterisks designate matrices that are, for the present at least, of no interest. Evidently

$$M_1 v_1 = e_1,$$

$$M_2 M_1 v_2 = e_2, \qquad M_2 e_1 = e_1,$$

$$M_3 M_2 M_1 v_3 = e_3, \qquad M_3 e_2 = e_2, \qquad M_3 e_1 = e_1,$$

. . . .

The Danilevskiĭ method proper takes $v_1 = e_1$, whence $M_1 = I$, and the reduction starts, in that case, with M_2. In any event, A is replaced successively by matrices

$$A_1 = M_1 A M_1^{-1},$$

(3) $$A_2 = M_2 A_1 M_2^{-1},$$

$$A_3 = M_3 A_2 M_3^{-1},$$

. . . .

and the matrices V by

$$V_1 = M_1 V,$$

(4) $$V_2 = M_2 V_1,$$

$$V_3 = M_3 V_2,$$

. . . .

Now observe that

$$A_2 V_2 - V_2 F,$$

and that the first two columns of V_2 are e_1 and e_2, respectively; hence the first column on the right is e_2. Evidently, then,

$$A_2 e_1 = e_2,$$

which is to say that the first column of A_2 is e_2.

    Next,

$$A_3 V_3 = V_3 F,$$

the first three columns of V_3 are e_1, e_2, and e_3; the second column in the right is e_3, hence

$$A_3 e_2 = e_3,$$

which is to say that the second column of A_3 is e_3. Each matrix M_{i+1}, therefore, while reducing column $i + 1$ of V to e_{i+1}, reduces column i of A to the same.

    This demonstrates that by a sequence of reductions of the form (3) by elementary matrices (1), any matrix A can be reduced to the form (2) where F is a Frobenius matrix. The matrix V need not enter explicitly, and does not in the Danilevskiĭ method as ordinarily carried out. In fact, consider the determination of M_{i+1} and A_{i+1}, given A_i. It is required that

$$M_{i+1} A_i = A_{i+1} M_{i+1}$$

where the ith column of A_{i+1} is e_{i+1}, hence that

$$M_{i+1}A_i e_i = e_{i+1},$$

and therefore that

$$\sigma_{i+1}m_{i+1} = (A_i e_i - e_{i+1})/e_{i+1}^{\mathrm{T}}A_i e_i.$$

This is always possible if $e_{i+1}^{\mathrm{T}}A_i e_i \neq 0$.

If the condition fails, but $e_j^{\mathrm{T}}A_i e_i \neq 0$ for some $j > i + 1$, an interchange of rows and of columns can be made and the algorithm continued. Even if $e_{i+1}^{\mathrm{T}}A_i e_i$ is only small but nonvanishing an interchange may be desirable. But if every $e_j^{\mathrm{T}}A_i e_i = 0$, $j \geq i + 1$, then the matrix decomposes. In that event A_i can be relabeled A_{i+1} and the algorithm again continued.

6.4. The Hessenberg and the Lanczos Reductions. Since the diagonalization of the Krylov matrix V leads to the Frobenius form of the matrix A, it is natural to consider the effect of triangularizing V. This provides the Hessenberg form of A, and, in fact, the Hessenberg form results whether the triangularization is carried out by Gaussian elimination or by the application of elementary Hermitian matrices. Although again in practice the matrix V does not appear explicitly, it will be shown first that the triangularization of V does lead to the Hessenberg form. After that some consideration will be given to the steps required in practice.

The original method of Hessenberg amounts to the application of elimination to V, but the method will first be stated in somewhat more general terms. As in the Lanczos reduction, the simple Krylov sequence of iterates v_i is replaced by a sequence of vectors b_i, where $b_1 = v_1$ and each b_i is a linear combination of the vectors $v_1, \cdots, v_i$. Hence the matrix V is replaced by a matrix

$$(1) \qquad\qquad B = VQ,$$

where Q is unit upper triangular. In its most general form, the Hessenberg criterion for forming Q is to select a matrix C and require that

$$(2) \qquad\qquad C^{\mathrm{H}}B = M$$

be lower triangular and nonsingular and of maximal order. Evidently the columns of C must be linearly independent and the maximal order is equal to m, the degree of the minimal polynomial of v_1. (The matrix M is not, of course, the matrix M of the last section.) This is possible for "almost all" matrices C of m linearly independent columns. The Lanczos reduction is therefore a specialized Hessenberg reduction in which the columns of B and of C are formed by a parallel construction. Evidently

$$M^{-1}C^{\mathrm{H}}B = I,$$

and, in particular, if $m = n$, then

$$B^{-1} = M^{-1}C^{\mathrm{H}}.$$

Now from (1) and (6.1.9),

(3) $$AB = B(Q^{-1}FQ),$$

and it is readily verified that, Q and Q^{-1} being triangular,

(4) $$H = Q^{-1}FQ$$

is in Hessenberg form. Moreover, if $H = (\eta_{ij})$, then

(5) $$\eta_{i+1,i} = 1.$$

The requirement that M be lower triangular means that b_2 is orthogonal to c_1 in the plane of v_1 and v_2; that b_3 is orthogonal to both c_1 and c_2 in the space of v_1, v_2, and v_3; $\cdots$. Taking Q to be unit upper triangular means that in forming b_i as a linear combination of $v_1, \cdots, v_i$, the coefficient of v_i is unity. These conditions determine uniquely the elements of Q above the diagonal and these can be formed a column at a time. It is natural to take for C the columns of I, and to take $v_1 = b_1 = e_1$, and this is what Hessenberg does. Hence e_1 is the first column of B.

In practice, however, neither V nor Q is obtained explicitly. Given b_1, b_2 is the linear combination of b_1 and Ab_1 that is orthogonal to c_1; b_3 the linear combination of b_1, b_2, and Ab_2 that is orthogonal to c_1 and c_2; $\cdots$. Thus

(6)′
$$Ab_1 = b_1\eta_{11} + b_2,$$
$$c_1{}^H Ab_1 = c_1{}^H b_1\eta_{11};$$
$$Ab_2 = b_1\eta_{12} + b_2\eta_{22} + b_3,$$
$$c_1{}^H Ab_2 = c_1{}^H b_1\eta_{12},$$
$$c_2{}^H Ab_2 = c_2{}^H b_1\eta_{12} + c_2{}^H b_2\eta_{22};$$
$$\cdots\cdots\cdots\cdots\cdots\cdots\cdots\cdots\cdots$$

Let T be any upper triangular matrix. Then

$$T^H C^H B = T^H M$$

is lower triangular. But M does not occur explicitly in the calculation of H and B. Hence if C is replaced by CT, the same matrices B and H will result. This is to say that *if each c_i is replaced by any linear combination of $c_1, c_2, \cdots, c_i$, the result will be the same.* This result can be verified in detail by an examination of Eqs. (6).

Since the choice of the vector c_1 does not affect that of b_1, except for the requirement that $c_1{}^H b_1 \neq 0$, and, more generally, each b_i is independent of c_i, it should be possible to choose c_2 as that linear combination of c_1 and $A^H c_1$ that is orthogonal to b_1; c_3 as the linear combination of c_1, c_2, and $A^H c_2$ that is orthogonal to both b_1 and b_2, etc. Thus the vectors $c_1, c_2, c_3, \cdots$ are formed with A^H and the b_i in the same way as $b_1, b_2, b_3, \cdots$ are formed with A and the c_i. This is, in fact, the *Lanczos reduction* and leads to a Jacobi form of the matrix.

The requirement that Q be unit upper triangular implies a particular normalization of the columns b_i of B and leads to (5). Any other normalization would be equivalent to the use of a diagonal matrix D in which $\delta_1 = 1$, with B replaced by

$$BD = VQD,$$

$$A(BD) = (BD)(D^{-1}HD),$$

where $D^{-1}HD$ is also in Hessenberg form, but does not satisfy condition (5). Hence each vector b_i can be scaled arbitrarily and independently of the others, and this is often advantageous in practice, since the unscaled vectors b_i defined by (6) tend to become progressively smaller. In particular, D could be chosen so that

$$C^H BD = MD = L$$

is a unit triangular matrix.

In case the columns of C are the first m columns of I, B is trapezoidal (triangular when $m = n$). In this case, if the last $n - m$ columns of I are adjoined to BD to form a nonsingular matrix (in fact it is unit lower triangular), the matrix L is of order n and expressible as a product

$$L = L_1 L_2 \cdots L_m$$

of elementary triangles:

$$L_i = I - l_i e_i^T, \qquad e_j^T l_i = 0, \qquad j \le i \le m.$$

Then

$$AL = L \begin{pmatrix} D^{-1}HD & * \\ 0 & * \end{pmatrix},$$

where again the asterisks designate submatrices which do not vanish in general, and $L^{-1}AL$ can be regarded as the result A_m of a sequence of transformations

$$L_1^{-1}AL_1 = A_1,$$

$$L_2^{-1}A_1 L_2 = A_2,$$

$$\cdots \cdots \cdots \cdots \cdots$$

$$L_m^{-1}A_{m-1}L_m = A_m.$$

The first one can be disregarded since, as in the last section, it does not modify the form of A. The second yields A_2 whose first column is null below the second element:

$$A_1(I - l_2 e_2^T) = (I - l_2 e_2^T)A_2.$$

On equating first columns, the result is

$$A_1 e_1 = A_2 e_1 - l_2(e_2^T A_2 e_1),$$

the quantity in parentheses being a scalar. Since l_2 vanishes in its first two elements, and $A_2 e_1$ is to vanish everywhere but in these elements, this defines $A_2 e_1$ and l_2 uniquely provided

$$e_2{}^T A_1 e_1 = e_2{}^T A_2 e_1 \neq 0.$$

In case this condition fails, a permutational transformation can be applied to A_1 to interchange the second row and column with another row and column, and, in fact, by means of such interchanges it can always be arranged that

$$|\, l_2 \,| \leq e.$$

Subsequent transformations, L_3, L_4, $\cdots$ can be determined analogously. Direct verification shows that if BD^{-1} represents the first m columns of L, then

$$B = VQ$$

where Q is unit upper triangular and V is formed from a Krylov sequence.

The vectors c_i are arbitrary except for the requirement of linear independence, together with the requirement that $c_i{}^H b_i = 0$ implies $b_i = 0$. Hence a possible choice is

$$c_i = b_i,$$

fulfilling the latter requirement automatically. Hence

$$C = B, \qquad B^H B = M = \Delta^2,$$

where Δ^2 is diagonal. In particular, when $m = n$, $B\Delta^{-1}$ is unitary.

In case $m < n$, let $B\Delta^{-1} = W_1$, and let W_2 be a matrix such that (W_1, W_2) is unitary. Then $A W_1 = W_1 K_{11}$, where $K_{11} = \Delta H \Delta^{-1}$ is in Hessenberg form and

$$A(W_1, W_2) = (W_1, W_2)\begin{pmatrix} K_{11} & K_{12} \\ 0 & K_{22} \end{pmatrix}.$$

By applying the same type of transformation to K_{22}, and repeating if necessary, there will be formed eventually a unitary matrix W such that

(7) $$A W = W K,$$

where K is in Hessenberg form and is of order n. Equations (6) with $c_i = b_i$ apply directly in simplified form since $b_i{}^H b_j = 0$ for $i \neq j$.

This shows that *any matrix can be reduced to Hessenberg form by a unitary transformation*. Moreover, if A is real, W can be taken orthogonal. But if such a transformation is known to exist, it is in order to consider other possible ways of generating it. Two such methods may be mentioned in particular. One method is to generate W as a product of plane rotations, which amounts to the triangularization of the matrix V by the *method of Givens* described at the end of Section 5.4. For real A, if $v_1 = e_1$, the first step is to apply a plane rotation in the plane $(2, 3)$ to annihilate the element α_{31}; next a rotation in the plane $(2, 4)$ to annihilate α_{41}; $\cdots$; a rotation in the plane $(3, 4)$ to annihilate the element $\alpha_{42}; \cdots$; and finally a rotation in the plane $(n - 1, n)$ to annihilate the element $\alpha_{n.n-2}$.

The other method is the use of elementary Hermitians. Let

$$W_i = I - 2w_i w_i^H, \qquad w_i^H w_i = 1,$$
$$e_j^T w_i = 0, \qquad j < i.$$

The transformation W_1 does not affect the form of the matrix A, but only adjusts v_1:

$$W_1 v_1 = w e_1.$$

If e_1 is considered suitable as the vector v_1, this step may be omitted. In any event, if

$$W_1 A = A_1 W_1,$$
$$W_2 A_1 = A_2 W_2,$$

then w_2 is to be chosen so that the first column of A_2 is null below the second element. Hence, since $w_2^H e_1 = 0$, it follows that

$$A_1 e_1 - 2 w_2 (w_2^H A_1 e_1) = A_2 e_1.$$

This shows first that

$$e_1^T A_1 e_1 = e_1^T A_2 e_1,$$

that is, $A_1 e_1$ and $A_2 e_1$ agree in their first elements. It is sufficient, therefore, to consider only elements below the first in the vectors w_2, $A_1 e_1$, and $A_2 e_1$. But then the problem is identical with the first step in a triangularization as discussed in Section 5.4.

The next step in the reduction is made by deleting the first row and column of A_2 and proceeding as with A_1.

In case A is Hermitian, then each of

$$A_1 = W_1 A W_1,$$
$$A_2 = W_2 A_1 W_2,$$
$$\cdots\cdots\cdots\cdots$$

is also Hermitian. In A_2 all elements of the first column are null except for the first two; hence all elements of the first row are null except for the first two. Finally the matrix

$$K = W^H A W$$

is both Hermitian and of Hessenberg form, hence is of Jacobi form. Hence *the unitary reduction of a Hermitian matrix is equivalent to a Lanczos reduction.*

Returning to the general case of an arbitrary matrix A, the columns of the unitary matrix W that results are the orthonormalized vectors of a Krylov sequence, hence are the vectors p_i of (1.6.6) after normalization, and are the same as would be obtained by a direct orthonormalization considered as a special modification of the method of Hessenberg. In case the initial vector $v = v_1$ has degree $m < n$, the sequence breaks off, but a new one can be started as indicated above. In this case the final Hessenberg matrix is block-triangular, with each block along the diagonal itself of Hessenberg form.

The Hessenberg form can be reduced to Jacobi form quite directly. Suppose A is in lower Hessenberg form, with zeros above the superdiagonal. This can be brought about by applying a Hessenberg reduction to the conjugate transpose of the original matrix. Let the matrix C be any upper triangular matrix (or trapezoidal when $m < n$). It follows from (2) that B must be lower triangular (or trapezoidal). Then in

$$AB = BH,$$

since B is lower triangular and A lower Hessenberg in form, also H must be lower Hessenberg in form, which is to say that the zeros above the superdiagonal are not affected. But the reduction leads to a matrix H that is upper Hessenberg in form. Hence H must be in Jacobi form.

This result requires only that C be upper triangular or trapezoidal. Moreover, all matrices of that form lead to the same B and H, as shown above, with a given $b_1 = v_1$. Hence the identity I, or its first m columns, provide a possible choice for C. Also it may be noted that if the Lanczos process were applied with $c_1 = e_1$, the resulting matrix C would be upper triangular and the same B and H would be obtained. Thus if the simple Hessenberg process is applied to a matrix that is already in lower Hessenberg form, and with $C = (e_1, \cdots, e_m)$, the result is the same tridiagonal matrix H that would be obtained by applying the Lanczos process to that matrix with $c_1 = e_1$. The reduction in this case is especially simple, since in (6), η_{11} is chosen so that the first element of b_2 is null; η_{12} is formed by comparing initial elements of Ab_2 and b_1; η_{22} is chosen so that the second element of b_3 is null; $\cdots$.

6.5. Proper Values and Vectors. Iterative methods to be described in the next chapter can be applied to the transformed form of the matrix in order to obtain the characteristic roots and vectors, and the characteristic vectors for the original and the transformed matrices are related by means of the transforming matrix [see below]. If the characteristic polynomial, or a divisor of it, has been obtained explicitly and solved, then for any root λ, the homogeneous system

$$(\lambda I - A)x = 0$$

can be solved, preferably by factorization. It will be shown in the next chapter how an approximate vector so obtained can be improved by iteration.

When A is Hermitian and has been reduced to tridiagonal form H, a nested sequence of leading principal minor determinants of $\lambda I - H$ forms a Sturm sequence (Exercise 14) and the fact is used by Givens in solving for the roots. A similar sequence can be formed for any Hessenberg matrix, and this will now be developed, but the sequence is not in general a Sturm sequence. It should be emphasized that the relations now to be developed are of *theoretical interest only* and are not to be regarded as providing a useful computational scheme.

Suppose the matrix is in Hessenberg form and that there are no zeros along the subdiagonal. Then a diagonal matrix D can always be found such that every subdiagonal element in $D^{-1}HD$ has the value 1. Hence there is no restriction in supposing at the outset that

$$(1) \qquad\qquad \eta_{i+1,i} = 1.$$

Let

$$
\begin{aligned}
\psi_m(\lambda) &= 1, \\
\psi_{m-1}(\lambda) &= (\eta_{mm} - \lambda)\psi_m(\lambda), \\
(2) \qquad \psi_{m-2}(\lambda) &= (\eta_{m-1,\,m-1} - \lambda)\psi_{m-1}(\lambda) - \eta_{m-1,m}\psi_m(\lambda), \\
&\; \cdots\cdots\cdots\cdots\cdots\cdots\cdots\cdots\cdots\cdots\cdots \\
\psi_0(\lambda) &= (\eta_{11} - \lambda)\psi_1(\lambda) - \eta_{12}\psi_2(\lambda) + \cdots \pm \eta_{1m}\psi_m(\lambda).
\end{aligned}
$$

Then it is easy to verify that $\psi = \psi_0(\lambda)$ is the characteristic polynomial of H. In case H is in Jacobi or Frobenius form the equations simplify accordingly. By differentiating (2), analogous sequences for the derivatives can be written down.

Let λ be any zero of ψ, and form the vector

$$(3) \qquad q^{\mathrm{T}} = [\pm\psi_1(\lambda),\ \mp\psi_2(\lambda),\ \cdots,\ +\psi_m(\lambda)].$$

Then

$$Hq = \lambda q,$$

which is to say that q is a proper vector belonging to λ. If

$$AB = BH,$$

then

$$ABq = BHq = \lambda Bq,$$

which is to say that Bq is a proper vector of A belonging to λ.

If all proper values of H are distinct, there are m linearly independent proper vectors belonging to them, and hence m linearly independent proper vectors of A. Suppose, however, that λ is a double root of the characteristic equation. Then $\psi'(\lambda) = 0$. Let

$$q'^{\mathrm{T}} = [\pm\psi_1'(\lambda),\ \mp\psi_2'(\lambda),\ \cdots,\ -\psi_{m-1}'(\lambda),\ 0].$$

Then

$$Hq' = q + \lambda q'.$$

If λ is a triple root, $\psi''(\lambda) = 0$, and if

$$q''^{\mathrm{T}} = [\pm\psi_1''(\lambda),\ \cdots,\ +\psi_{m-2}''(\lambda),\ 0,\ 0],$$

then

$$Hq'' = q' + \lambda q''.$$

Now let the proper values be arranged in any order, $\lambda_1, \lambda_2, \cdots, \lambda_m$, except that equal roots are consecutive, and form a matrix Q as follows: If $\lambda_1 = \lambda_2 = \cdots = \lambda_\nu \neq \lambda_{\nu+1}$, then the first ν columns of Q are the vectors $q_1^{(\nu-1)}, \cdots, q_1$ formed from λ_1; if $\lambda_{\nu+1} = \cdots = \lambda_{\nu+\mu} \neq \lambda_{\nu+\mu+1}$, the next μ vectors are $q_2^{(\mu-1)}, \cdots, q_2$ formed from $\lambda_{\nu+1}; \cdots$. Then

$$HQ = Q\Lambda,$$

where

$$\Lambda = \mathrm{diag}\,(\Lambda_1, \Lambda_2, \cdots), \qquad \Lambda_i = \lambda_i I + J,$$

and

$$A(BQ) = (BQ)\Lambda.$$

The columns of BQ are therefore principal and proper vectors of A.

The row vectors are more simply expressed. If λ is a simple root, let

$$r^H = [1, \varphi_1(\lambda), \varphi_2(\lambda), \ldots],$$

where the $\varphi_i(\lambda)$ are the orthogonal polynomials defined in Section 1.6 for $D^{-1}HD$ with the vector e_1. Then it is verified directly that r^H is a proper row vector belonging to λ. If the reduction to Hessenberg form has been made by an orthogonal transformation, then the polynomials $\varphi_i(\lambda)$ are orthogonal polynomials for the original matrix A formed from a suitable starting vector v. In fact, the reduction can be made by (1.6.11) and

$$A P_m = P_m\,(D^{-1}\,HD),$$

$$D^{-1}\,HD = \begin{pmatrix} \sigma_{11} & \sigma_{12} & \sigma_{13} & \cdots \\ 1 & \sigma_{22} & \sigma_{23} & \cdots \\ 0 & 1 & \sigma_{33} & \cdots \\ & & \cdots & \end{pmatrix}.$$

The method just described for forming principal and proper vectors is, as remarked above, primarily of theoretical interest, and cannot be recommended in practice since the computations tend to be unstable.

6.6. The Method of Samuelson and Bryan. The method of Samuelson and Bryan utilizes the Krylov sequence in a different way, and expresses the coefficients of the characteristic polynomial of the matrix itself in terms of those of the polynomial of a principal submatrix. Hence, the characteristic polynomial can be built up recursively starting with that for a matrix of order one.

The simplest and most elegant derivation is due to Bryan. Suppose the characteristic polynomial $\phi_m(\lambda)$ is known for a matrix A_m of order m, and let A_{m+1} be the bordered matrix

$$A_{m+1} = \begin{pmatrix} A_m & a_m \\ a_m{}^* & \alpha_m \end{pmatrix}.$$

Let the adjoint of $\lambda I - A_{m+1}$ be represented

$$(1) \qquad (\lambda I - A_{m+1})^{\mathbf{A}} = \begin{pmatrix} F_m(\lambda) & f_m(\lambda) \\ f_m{}^*(\lambda) & \phi_m(\lambda) \end{pmatrix},$$

where the asterisks are merely distinguishing marks. If $\lambda I - A_{m+1}$ is multiplied by the last column of its adjoint, the result is

$$(2) \qquad \begin{aligned} (\lambda I - A_m)f_m(\lambda) - a_m\phi_m(\lambda) &= 0, \\ -a_m{}^*f_m(\lambda) + (\lambda - \alpha_m)\phi_m(\lambda) &= \phi_{m+1}(\lambda), \end{aligned}$$

since the entire product is equal to $\phi_{m+1}(\lambda)I$. The first of (2) can be written

$$\lambda f_m(\lambda) = a_m\phi_m(\lambda) + A_m f_m(\lambda).$$

But A_m and a_m are independent of λ, ϕ_m is of degree m and presumed known, and the elements of $f_m(\lambda)$ are of degree $m - 1$ at most. Hence by comparing coefficients of like powers of λ, the vector coefficients in $f_m(\lambda)$ can be obtained in sequence, starting with that of λ^{m-1} which is simply a_m. When $f_m(\lambda)$ is found, $\phi_{m+1}(\lambda)$ is obtainable from the last equation in (2).

For any proper value λ_i of A_{m+1}, $\phi_{m+1}(\lambda_i) = 0$, and therefore the last column of (1) (in fact, any nonnull column) is a proper vector belonging to λ_i. In case the column does vanish for a given λ_i, special steps can be taken for finding proper and principal vectors, but present concern is with the polynomial itself.

The usual derivation of Samuelson's method proceeds along different lines. With indeterminates $\zeta = \zeta_1$ and $z = z_1$, form the Krylov sequence

$$\begin{pmatrix} A_m & a_m \\ a_m{}' & \alpha_m \end{pmatrix}\begin{pmatrix} z_i \\ \zeta_i \end{pmatrix} = \begin{pmatrix} z_{i+1} \\ \zeta_{i+1} \end{pmatrix},$$

and from these relations eliminate $z_2, \cdots, z_{m+2}$. This leaves $m + 1$ equations involving $\zeta_1, \cdots, \zeta_{m+2}$, and the vectors $z_1, A_m z_1, \cdots, A_m{}^m z_1$. But application of the fact that $\phi_m(A_m) = 0$ permits elimination of z_1 and leaves a single relation in the ζ's. Since these are indeterminates, the coefficients of the ζ's are coefficients in $\phi_{m+1}(\lambda)$. In spite of the different derivations, it can be verified in detail that the method of forming ϕ_{m+1} is formally identical with that of Bryan.

6.7. The Method of Leverrier. The method of Leverrier antedates the oldest of these methods (that of Krylov) by nearly a century. Let

$$(1) \qquad \delta(\lambda I - A) = \phi(\lambda) = \lambda^n - \gamma_1\lambda^{n-1} + \gamma_2\lambda^{n-2} - \cdots \pm \gamma_n,$$

where the γ_i are the elementary symmetric polynomials in the roots λ_i. In particular

$$\gamma_1 = \tau(A), \qquad \gamma_n = \delta(A).$$

Let

$$(2) \qquad \sigma_\nu = \sum \lambda_i^\nu$$

be the sums of the powers of λ_i. Then

$$(3) \qquad \sigma_\nu = \tau(A^\nu).$$

The original method of Leverrier forms the σ_ν by (3), and utilizes the Newton identities [Exercise 16],

$$\sigma_1 - \gamma_1 = 0,$$
$$\sigma_2 - \sigma_1\gamma_1 + 2\gamma_2 = 0,$$
$$(4) \qquad \sigma_3 - \sigma_2\gamma_1 + \sigma_1\gamma_2 - 3\gamma_3 = 0,$$
$$\cdots\cdots\cdots\cdots\cdots\cdots\cdots$$
$$\sigma_n - \sigma_{n-1}\gamma_1 + \cdots + (-1)^n n\gamma_n = 0,$$

to obtain the γ_i. It may be observed that, since only traces of the powers of A are required, traces of higher powers can be obtained by computing only diagonal elements in

$$A^i A^j = A^{i+j}.$$

To improve the method somewhat, write the adjoint

$$(5) \qquad (\lambda I - A)^A = C(\lambda) = C_0\lambda^{n-1} - C_1\lambda^{n-2} + \cdots \pm C_{n-1}.$$

Since

$$(6) \qquad C(\lambda)(\lambda I - A) = \phi(\lambda)I,$$

it follows from a comparison of coefficients of like powers of λ that

$$C_0 = I,$$
$$C_1 + C_0 A = \gamma_1 I,$$
$$C_2 + C_1 A = \gamma_2 I,$$
$$(7) \qquad \cdots\cdots\cdots\cdots$$
$$C_{n-1} + C_{n-2} A = \gamma_{n-1} I,$$
$$C_{n-1} A = \gamma_n I.$$

By a simple induction, these identities imply

$$(8) \qquad C_\nu = \gamma_\nu I - \gamma_{\nu-1}A + \gamma_{\nu-2}A^2 - \cdots \pm A^\nu.$$

On multiplying through in (8) by A, taking traces, and comparing with (4), it follows that

$$(9) \qquad \tau(C_\nu A) = (\nu + 1)\gamma_{\nu+1}.$$

Hence with γ_1, C_1 can be obtained from the second of (7); from $C_1 A$ can be obtained γ_2 and thence C_2; $\cdots$. Thus the characteristic polynomial can be obtained completely.

Proper and principal vectors are obtainable from the matrix $C(\lambda)$ in rather obvious ways. In the simplest case, if λ_i is any proper value for which $C(\lambda_i) \neq 0$, then, as with Bryan's method, any nonnull column of $C(\lambda_i)$ is a proper vector belonging to λ_i. But $C(\lambda_i)$ *cannot vanish unless* λ_i *is at least a double root.* For by differentiation of (6) it follows that

$$C'(\lambda)(\lambda I - A) + C(\lambda) = \phi'(\lambda)I;$$

hence, if $C(\lambda_i) = 0$,

$$C'(\lambda_i)(\lambda_i I - A) = \phi'(\lambda_i)I,$$

but the matrix product on the left is singular, and that on the right nonsingular if λ_i is a simple root. Consequently when λ_i is a simple root, $C(\lambda_i)$ has at least one nonvanishing column.

When λ_i is a double root, $C(\lambda_i)$ may or may not vanish. If it does not, then a nonvanishing column is again a proper vector belonging to λ_i, and a nonvanishing column of $C'(\lambda_i)$ is a principal vector belonging to λ_i. If, however, $C(\lambda_i) = 0$, then $C'(\lambda_i)$ can be shown to have rank 2, and any nonvanishing column is a proper vector. For roots of higher multiplicity the situation is analogous.

6.8. Deflation. When one or more roots of a matrix have been found along with the invariant subspace belonging to it, it is possible, and often advantageous, to obtain a matrix of lower order whose roots are the yet unknown roots of the original matrix. This is analogous to the process of "dividing out" known roots of an algebraic equation, and it is known as "deflation." Deflations by similarity transformations have already been indicated, e.g., in (6.3.2). A deflation is also implicit in the stagewise Lanczos reduction, Section 1.5. The deflation to be described here does not depend upon similarity transformations, and it will be useful primarily in conjunction with the use of iterative methods.

Before describing these methods of deflation, certain preliminary lemmas will be needed. First observe that if X and Y are any two matrices of the same dimensions, then (Exercise 1.14)

(1) $$\delta(I - X^H Y) = \delta(I - Y X^H).$$

Suppose then that P is a matrix of ρ linearly independent columns forming a basis for an invariant subspace of A, hence that for some matrix M,

(2) $$AP = PM.$$

Let V be any matrix of ρ linearly independent columns satisfying

(3) $$V^H P = I,$$

and form

(4) $$B = A - PMV^H = A(I - PV^H).$$

Then it will be shown that

(5) $$\delta(\lambda I - A)/\delta(\lambda I - M) = \lambda^{-\rho}\,\delta(\lambda I - B).$$

This proves incidentally that for given A and M, a matrix P can exist satisfying (2) only if $\delta(\lambda I - M)$ divides $\delta(\lambda I - A)$. But the point of interest here is that *the nonnull roots of B are precisely the yet unknown roots of A.*

To establish (5), note first that

$$\lambda I - B = \lambda I - A + PMV^H$$
$$= (\lambda I - A)[I + (\lambda I - A)^{-1}PMV^H]$$
$$= (\lambda I - A)[I + P(\lambda I - M)^{-1}MV^H].$$

But (1) and (3) imply that

$$\delta[I + P(\lambda I - M)^{-1}MV^H] = \delta[I + (\lambda I - M)^{-1}M]$$
$$= \lambda^\rho\,\delta(\lambda I - M)^{-1}.$$

Obvious substitutions now complete the proof.

Since the matrix B is not obtained from A by a similarity transformation (otherwise all roots of B would agree with those of A), the invariant subspaces for the two matrices are related in a somewhat more complicated way. This will now be investigated.

From the form of the Jordan normal form, it is easy to see that *invariant row and column subspaces, for a given matrix, belonging to distinct proper values are orthogonal.* However, an independent proof can be given very easily. In fact, let

(6) $$S^H A = \Lambda S^H,$$

where Λ and M have no proper values in common. Then (6) implies that $S^H A P = \Lambda S^H P$, while (2) implies that $S^H A P = S^H PM$. Hence

$$\Lambda S^H P = S^H PM,$$

and since Λ and M have no roots in common, it follows (Exercise 1.16) that

$$S^H P = 0.$$

It will be convenient to suppose that A is nonsingular, which is a trivial restriction since A could be replaced by $\mu I + A$, with any convenient scalar μ. Then B is of rank $n - \rho$, since its roots, apart from the known ρ null roots, are $n - \rho$ of the roots of A (counting possible multiplicities). Suppose that none of these $n - \rho$ roots agree with those of M. This is to say that the space of P contains the complete invariant subspace belonging to every root of A that is a root of M. The roots of A therefore fall into two distinct sets, Σ_1, the set of roots of M, and Σ_2, the roots of A that are not roots of M. There should be no ambiguity in considering the set Σ_1 for B to consist of ρ zeros. Also, a principal (or proper) row or column vector for A or B will be said to belong to Σ_1 or Σ_2 if it belongs to a root in Σ_1 or in Σ_2.

The columns of the matrix P were assumed at the outset to provide a basis for the column space Σ_1 of A. From (3) and (4), it follows that

$$BP = 0,$$

whence the column space Σ_1 for B is the same as for A. Since the row spaces Σ_2 are orthogonal, these also coincide for the two matrices. Not only that, but *every principal vector in Σ_2 is of the same grade for B as for A*. The proof will be made by induction on the grade.

If y^H is a principal vector in Σ_2 for either A or B, then

$$y^H P = 0.$$

If y^H is a proper vector of A, then

$$y^H A = \lambda y^H, \qquad \lambda \in \Sigma_2.$$

Then

$$y^H B = y^H(A - PMV^H) = y^H A = \lambda y^H.$$

Conversely, if

$$y^H B = \lambda y^H, \qquad \lambda \in \Sigma_2,$$

then

$$y^H A = \lambda y^H.$$

Now suppose it established that any principal row vector of A of grade ν in Σ_2 is also a principal row vector of B of grade ν in Σ_2, and let y^H be a principal vector of A of grade $\nu + 1$. That is to say that

$$y^H(A - \lambda I)^{\nu+1} = 0, \qquad \lambda \in \Sigma_2.$$

Then

$$y^H(A - PMV^H - \lambda I)^{\nu+1} = y^H(A - \lambda I)(B - \lambda I)^{\nu}.$$

But $y^H(A - \lambda I)$ is a principal vector of A of grade ν, hence by the hypothesis of the induction a principal vector of B of grade ν. Hence the right member vanishes. A similar argument applies when y^H is assumed a principal vector of grade $\nu + 1$ for B. Hence the proof is complete.

The choice of V is by no means uniquely determined by (3), but if W is any matrix of ρ linearly independent columns such that $W^H P$ is nonsingular, then V can be chosen so that

(7) $$W^H B = 0.$$

In other words, *the row vectors for B in Σ_1 are almost arbitrary*, but when they are chosen, they then determine V uniquely. In fact, with (4), (7) implies that

(8) $$W^H A = (W^H P) M V^H,$$

and $W^H P$ and M are both assumed nonsingular. Hence (8) determines V uniquely. Ordinarily W^H is chosen to be ρ of the rows of I, and then (7) implies that the same ρ rows of B are null. Then the $n - \rho$ nonnull roots of B are those of a principal submatrix of B of order $n - \rho$.

There remain for consideration the column invariant subspaces Σ_2 of the two matrices. Suppose, for the moment, that M is in Jordan normal form. Then the Jordan normal form of A is a direct sum of M and a matrix Λ, also in Jordan normal form. The considerations already made establish that the Jordan normal form of B is the direct sum of a null matrix of order ρ, and the same matrix Λ. Hence there exists a matrix Q such that

(9) $$BQ = Q\Lambda.$$

The columns of Q are principal and proper vectors of B in Σ_2. It will be assumed that these are now known, and that it is required to find a matrix R of principal and proper vectors of A, which must satisfy

(10) $$AR = R\Lambda.$$

It is required that R be expressed in terms of Q.

It has already been established that if (6) is satisfied, then also

$$S^H B = \Lambda S^H,$$

for Λ as here defined, since S^H is a matrix of principal and proper row vectors for both A and B in Σ_2. With proper normalization, it can be supposed that

$$S^H P = 0, \qquad S^H R = S^H Q = I,$$

since the reciprocal of a nonsingular matrix of proper and principal column vectors is a matrix of proper and principal row vectors; and conversely. Hence R and Q must be related by

(11) $$R = Q + PZ,$$

for some matrix Z. But (10), (4), (9), (3), together with the fact that P is of rank ρ, imply that

(12) $$Z\Lambda - MZ = MV^H Q.$$

Conversely, if Z satisfies (12), and R is given by (11), then R satisfies (10). Moreover, (12) is known to have a unique solution (Exercise 1.16).

Actually it is not required that either M or Λ be in Jordan normal form, although the solution is simplest in that case. Moreover, it is not required that the matrices Q and Λ be known completely. Suppose Λ is itself expressible as a direct sum of Λ_1 and Λ_2, where Λ_1 and Λ_2 have no common root, and let

$$BQ_1 = \Lambda_1 Q_1, \qquad\qquad BQ_2 = \Lambda_2 Q_2.$$

Then
$$R_1 = Q_1 + PZ_1$$
and Z_1 satisfies
$$Z_1 \Lambda_1 - M Z_1 = M V^H Q_1.$$

Thus R_1 can be adjoined to P, Λ_1 to M, and a new matrix B of lower rank can be defined and employed for finding the remaining roots. Thus all Σ_2 vectors of A can be formed when those of B are known.

Other forms of deflation amount to the progressive reduction of A to triangular form. Let x be any proper vector:

$$Ax = \lambda x,$$

and let E be an elementary matrix, possibly an elementary Hermitian (Section 1.1) such that
$$x = Ee_1.$$
Then
$$AEe_1 = \lambda E e_1,$$
$$E^{-1}AEe_1 = \lambda e_1;$$

hence e_1 is a proper vector of $E^{-1}AE$, and therefore its first column is λe_1. Therefore computations may be continued with a submatrix of order $n - 1$. In case A is Hermitian, and $E = H$ is an elementary Hermitian, the resulting matrix is also Hermitian since $H^{-1} = H$.

REFERENCES

The earliest practical method of finding the characteristic polynomial is that of Leverrier (1840); next came the method of Krylov (1931), which, as shown here, represents an entire family of methods when properly interpreted. Leverrier's method was rediscovered several times, sometimes in slightly improved form, more than a century later by Horst (1935), Souriau (1948), Frame (1949), and Faddeev and Sominskii (1949). The method of Givens (1953, 1954) extends beyond the reduction to tridiagonal form and includes specifications for finding the roots of the reduced matrix. Upper and lower bounds for the roots are found; then by successive bisection and use of the Sturm properties of the characteristic polynomials of a nested sequence of principal submatrices, the roots are obtained to any desired degree of precision. The author's modification (Householder and Bauer 1959), applies only to the reduction, and was first proposed at the 1958 meeting of the Association for Computing Machinery.

The fact that the Weber-Voetter method can be reduced to a Krylov form was first shown by Paul (1957), but the method of reduction was quite different from the one used here.

For practical computational techniques reference is made again to Wilkinson. For general treatments see Faddeev and Faddeeva (1960), Durand (1961), and Korganoff (1961).

PROBLEMS AND EXERCISES

1. Let

$$P(\lambda) = \lambda^\nu I + \lambda^{\nu-1} P_1 + \cdots + P_\nu.$$

Show that

$$A = \begin{pmatrix} 0 & 0 & \cdots & 0 & -P_\nu \\ I & 0 & \cdots & 0 & -P_{\nu-1} \\ \cdots & \cdots & \cdots & \cdots & \cdots \\ 0 & 0 & \cdots & I & -P_1 \end{pmatrix}$$

has $\delta[P(\lambda)]$ as its characteristic polynomial. Also if $x^H = (x_1^H, x_2^H, \cdots, x_\nu^H)$ and $Ax = \lambda x$, then

$$P(\lambda)x_\nu = 0.$$

Hence, in general, if Q_0 is nonsingular and

$$Q(\lambda) = \lambda^\nu Q_0 + \lambda^{\nu-1} Q_1 + \cdots + Q_\nu,$$

then the problem of finding scalars λ and vectors y to satisfy $Q(\lambda)y = 0$ is reduced to the ordinary characteristic value and vector problem for a matrix of order νn. A similar reduction is possible if Q_0 is singular but Q_ν nonsingular.

2. Given that the vectors $v_1, \cdots, v_m$ of Section 6.1 are linearly independent, but that the vectors $v_1, \cdots, v_{m+1}$ are linearly dependent, state why it must be true that the polynomials $\psi(\lambda)$ of (6.1.7), $\chi_m(\lambda)$ of (1.5.10), and $\phi_m(\lambda)$ of (1.6.4) are identical.

3. Let the matrix B commute with A, let $w_1 = Bv_1$, and form the Krylov sequence (6.1.1) starting with w_1 and v_1. If V is defined by (6.1.2) and W is defined analogously show that along with (6.1.9), (6.1.10) is also satisfied. Conversely, if (6.1.9) and (6.1.10) are satisfied and $m \equiv n$, then there exists a matrix B commuting with A and such that $w_1 = Bv_1$.

4. Let $A = J + \epsilon^n e_1 e_n^T$. Show that the roots are $\omega^r \epsilon$, where ω is a primitive nth root of unity (Forsythe).

5. Show that application of the Weber-Voetter method to $F - \lambda I$ leads to

$$\begin{pmatrix} -\lambda & -\lambda^2 & \cdots & -\lambda^{n-1} & -\phi(\lambda) \\ 1 & 0 & \cdots & 0 & 0 \\ \cdots & \cdots & \cdots & \cdots & \cdots \\ 0 & 0 & \cdots & 1 & 0 \end{pmatrix}.$$

6. Let

$$A = \begin{pmatrix} \alpha_1 & \beta_1 & 0 & \cdots \\ \gamma_1 & \alpha_2 & \beta_2 & \cdots \\ 0 & \gamma_2 & \alpha_3 & \cdots \\ & \cdots\cdots & & \end{pmatrix}$$

and define

$$\phi_1(\lambda) = \lambda - \alpha_1,$$
$$\phi_2(\lambda) = (\lambda - \alpha_2)\phi_1(\lambda) - \beta_1\gamma_1,$$
$$\phi_\nu(\lambda) = (\lambda - \alpha_\nu)\phi_{\nu-1}(\lambda) - \beta_{\nu-1}\gamma_{\nu-1}\phi_{\nu-2}(\lambda).$$

Show that application of the Weber-Voetter method to $A - \lambda I$ leads to

$$\begin{pmatrix} -\phi_1(\lambda) & -\phi_2(\lambda) & -\phi_3(\lambda) & \cdots \\ \gamma_1 & 0 & 0 & \cdots \\ 0 & \gamma_1\gamma_2 & 0 & \cdots \\ 0 & 0 & \gamma_1\gamma_2\gamma_3 & \cdots \\ & \cdots\cdots\cdots\cdots\cdots & & \end{pmatrix}.$$

7. Define the vector l by (6.1.4), where λ is any root of

$$\psi(\lambda) = 0.$$

Then

$$l^{\mathrm{T}}F = \lambda l^{\mathrm{T}}.$$

Let

$$\nu!\, l^{(\nu)} = d^\nu l/d\lambda^\nu, \qquad l^{(0)} = l.$$

Then if λ is a root of multiplicity ν,

$$l^{(\nu-1)\mathrm{T}}F = \lambda l^{(\nu-1)\mathrm{T}} + l^{(\nu-2)\mathrm{T}}, \qquad \nu = 2, 3, \cdots.$$

Thus

$$LF = \Lambda L,$$

where Λ is a Jordan normal form, and L is a nonsingular matrix whose rows are vectors $l^{\mathrm{T}}, l'^{\mathrm{T}}, \cdots$, formed from the distinct roots of F in accordance with their multiplicities (i.e., if λ is a root of multiplicity ν, there will be rows

$$l^{\mathrm{T}}, l'^{\mathrm{T}}, \cdots, l^{(\nu-1)\mathrm{T}}$$

formed with that λ).

8. Develop the scalar equations for the rotational (Givens) reduction of a real matrix A to Hessenberg form (6.4.7) (i) for A symmetric, and (ii) for A nonsymmetric.

9. Do likewise for A complex and Hermitian.

10. Obtain the scalar equations for reducing a lower Hessenberg matrix to Jacobi form as at the end of Section 6.4.

11. Let $AB = BH$, where H is in Hessenberg form. Show that B can be related to a Krylov sequence in accordance with (6.4.1).

12. Suppose H is in Hessenberg form with $\eta_{i+1,i} = 1$. Form the Krylov sequence for H beginning with $v_1 = e_1$ and thence obtain the characteristic polynomial $\delta(\lambda I - H)$.

13. If the Hermitian matrix A has a pair of equal roots, in its Jacobi form there will be at least one pair of zeros in subdiagonal and superdiagonal positions, hence the Jacobi form is the direct sum of Jacobi matrices of lower orders.

14. Let

$$ H = \begin{pmatrix} \alpha_1 & \beta_1 & 0 & \cdots \\ \beta_1 & \alpha_2 & \beta_2 & \cdots \\ 0 & \beta_2 & \alpha_3 & \cdots \\ & & \cdots\cdots \end{pmatrix}, $$

let $\phi_0 = 1$, $\phi_1(\lambda) = \lambda - \alpha_1$, and $\phi_i(\lambda)$ the characteristic polynomial of the leading principal submatrix of order i. Assuming every $\beta_i \neq 0$, obtain the three-term recursion satisfied by the $\phi_i(\lambda)$. For any λ' show that the number of variations in sign in the sequence

$$ \phi_0, \phi_1(\lambda'), \phi_2(\lambda'), \cdots, \phi_n(\lambda') \neq 0, $$

is equal to the number of roots of H which exceed λ'. Relate the polynomials $\phi_i(\lambda)$ to those in Section 1.6, and to the polynomials $\chi_i(\lambda)$ in Section 1.5.

15. Any step in the Givens reduction of a symmetric matrix A affects elements in only two rows and the corresponding two columns. Thus the error matrix, after a permutational transformation, has the form

$$ E = \begin{pmatrix} \gamma_{11} & \gamma_{12} & a^T \\ \gamma_{21} & \gamma_{22} & b^T \\ a & b & 0 \end{pmatrix}. $$

If $|\gamma_{ij}| \leq \gamma$, show that the error introduced into any root of A cannot exceed the positive root μ of the quadratic

$$ \mu^2 - 2\gamma\mu - (a^T a + b^T b) = 0. $$

16. Expand both sides of the identity

$$ \phi'(\lambda) = \phi(\lambda) \sum (\lambda - \lambda_i)^{-1} $$

formally in powers of λ^{-1}, compare coefficients, and hence obtain Newton's identities required in Section 6.6.

17. Prove the assertions made in the last paragraph of Section 6.7.

18. Consider the equation (Wilkinson)

$$\phi(\lambda) \equiv (\lambda - 1)(\lambda - 2) \cdots (\lambda - 20) \equiv \lambda^{20} - \alpha_1\lambda^{19} + \alpha_2\lambda^{18} - \cdots + \alpha_{20} = 0$$

and compute approximately $\partial\phi/\partial\alpha_1/\phi'$ evaluated at $\lambda = 15$; hence conclude that a small change in α_1 will introduce a large error into this and neighboring roots.

19. Let $y' = Ay$ (prime denoting differentiation) be a homogeneous system of ordinary differential equations with constant coefficients. By elimination of $n - 1$ of the elements, there results a differential equation of the form

$$\eta^{(n)} + \alpha_1\eta^{(n-1)} + \cdots + \alpha_n\eta = 0.$$

Show that the characteristic equation of A is

$$\lambda^n + \alpha_1\lambda^{n-1} + \cdots + \alpha_n = 0,$$

hence deduce the Samuelson-Bryan method.

20. Verify in detail the assertion made at the end of Section 6.6.

21. Let $Ax = \alpha x$, $y^H A = \alpha y^H$, $y^H x = 1$. For arbitrary v with $e_1^T v \neq 0$, and

$$u = (e_1 - x)/v^H e_1,$$

let (Section 1.1)

$$E = E(u, v; 1).$$

Then

$$A_1 = E^{-1}(A - \alpha xy^H)E$$

has e_1 in its first column and when its first row and column are deleted, the roots of the remaining matrix are those other than α of A (Feller and Forsythe).

22. If A is Hermitian, consider

$$A_1 = H(A - \alpha xx^H)H,$$

where $Ax = \alpha x$, $x^H x = 1$, and $H = H(w)$ such that

$$He_1 = x.$$

23. Generalize (Exercise 22) as follows: Let

$$AX = XW, \qquad Y^H A = WY^H,$$

where X and Y are matrices of ρ columns and rank ρ. Let V and P be such that $V^H P$ and $I - V^H P$ are nonsingular of order ρ, and let

$$U = (P - X)(V^H P)^{-1}.$$

Let $E = I - UV^H$ and form

$$E^{-1}(A - XWY^H)E.$$

24. Rutishauser's deflation of a Hermitian band matrix, say tridiagonal, proceeds as follows: If

$$Ax = \alpha x, \qquad x^H x = 1,$$

the unitary matrix

$$H = h_1 e_1^T + h_2 e_2^T + \cdots + h_n e_n^T$$

is Hessenberg in form, with

$$h_n = x$$

$$\rho_{n-1} h_{n-1} = h_n - \eta_{n-1,n} e_n,$$

$$\rho_{n-2} h_{n-2} = h_{n-1} - \eta_{n-2,n-1} e_{n-1} - \eta_{n-2,n} e_n,$$

$$\rho_{n-3} h_{n-3} = h_{n-2} - \eta_{n-3,n-2} e_{n-2} - \eta_{n-3,n-1} e_{n-1},$$

$$\cdots \cdots \cdots \cdots \cdots \cdots \cdots$$

where the ρ_i are normalizing factors. The Hessenberg form requires that

$$e_n^T h_{n-1} = \eta_{n-2,n}, \qquad e_{n-1}^T h_{n-2} = \eta_{n-3,n-1}, \cdots$$

and the unitary properties fix the other coefficients. Show that if

$$AH = HB,$$

then the last column of B is αe_n, and that B retains the band form of A.

25. If $\alpha_{jk} = \beta_{jk} + i\gamma_{jk}$, let

$$A_{jk} = \begin{pmatrix} \beta_{jk} & -\gamma_{jk} \\ \gamma_{jk} & \beta_{jk} \end{pmatrix}$$

and form the matrix

$$M = \begin{pmatrix} A_{11} & \cdots & A_{1n} \\ \cdots\cdots\cdots\cdots \\ A_{n1} & \cdots & A_{nn} \end{pmatrix}.$$

Hence apply Exercise 1.22 to show that

$$\delta(M) = |\,\delta(A)\,|^2;$$

hence apply a permutational transformation to show that if

$$H = \begin{pmatrix} B & -C \\ C & B \end{pmatrix},$$

then

$$\delta(H) = |\,\delta(H)\,|^2,$$

hence that for every root λ of A, λ and $\bar{\lambda}$ are both roots of H (Brenner, 1961; see also Bodewig, 1959).

26. If x is a proper vector of A belonging to λ, then $\begin{pmatrix} x \\ -ix \end{pmatrix}$ is a proper vector of H belonging to λ and $\begin{pmatrix} x \\ ix \end{pmatrix}$ is a proper vector belonging to $\bar{\lambda}$.

CHAPTER 7

Proper Values and Vectors:
Successive Approximation

7.0. Methods of Successive Approximation. Methods of successive approximation may be characterized, first, as being self-correcting or not; second, with respect to the order of convergence. In a method that is not self-correcting, the matrix is subjected to a sequence of transformations, and, in practice, the transforms may deviate progressively from the original. Nevertheless, if convergence is sufficiently rapid, reasonable convergence may be obtained before the deviations become too serious, and in any event the approximations so obtained may be subsequently improved by the use of a method that is self-correcting.

The order of convergence of a method is an index of the number of steps required to reduce the error to within tolerance. But the computations required per step vary considerably from method to method so that this is not in itself an adequate index of efficiency. Moreover, the computations per step for a given method depend strongly upon the particular form of the matrix, so that the method may be quite unsuitable for a general matrix, but relatively efficient for one in some special form, Hessenberg, Jacobi, or, conceivably, other. Hence it would not be unreasonable to consider, for example, first a transformation to Jacobi form, then application of method A, possibly not self-correcting, followed by application of method B, probably self-correcting, to the original matrix, to purify the approximations to required accuracy.

Most of the known methods apply to arbitrary matrices, with certain simplifications in the Hermitian case, and are based upon asymptotic properties of powers of a matrix. In the simplest form it provides the dominant root and the vector belonging to it, but the method can be extended to provide simultaneously all roots of the matrix, or any number of the roots of larger modulus. Two methods, however, are based upon different principles, and these will be discussed first. One, the method of Jacobi, will be described for Hermitian matrices only, although its applicability to arbitrary normal matrices is fairly obvious. The other applies to more general matrices, but provides only improvements of approximations to roots and vectors already at hand, and already sufficiently close.

178

7.1. The Method of Jacobi. The method of Jacobi was developed to apply only to Hermitian matrices, and in practice is used almost exclusively for real symmetric matrices. When suitably modified it can be applied also to any normal matrix, but so far attempts to extend the method to apply to a general matrix have not proved very successful.

Let $\eta(A)$ represent the Euclidean norm of A:

$$\eta^2(A) = \tau(A^H A) = \tau(AA^H).$$

This norm is unitarily invariant (Exercise 2.48), which is to say that if W is any unitary matrix, then

$$\eta(WA) = \eta(AW) = \eta(A).$$

If A is Hermitian, let W be unitary with

$$A = W\Lambda W^H,$$

where Λ is diagonal. Then

$$\eta^2(A) = \eta^2(\Lambda) = \sum \lambda_i^2$$

which is to say that the Euclidean norm of a Hermitian matrix is equal to the square root of the sum of the squares of its roots. Evidently, therefore, if α_{ii} is the ith diagonal element of A, then

$$\sum \alpha_{ii}^2 \le \sum \lambda_i^2$$

and equality holds if and only if A is diagonal. Therefore, if a sequence of unitary transformations can be performed, each having the effect of increasing the sum of the squares of the diagonal elements, hence of decreasing the sum of the squares of the moduli of the off-diagonal elements, and if in the limit the transforms become diagonal, then this limit is Λ, possibly with the diagonal elements permuted. But such a sequence can certainly be found.

The existence of the required sequence of transformations follows from the fact that any principal submatrix of a Hermitian matrix is itself Hermitian, and hence diagonalizable by a unitary transformation. This is true in particular of a principal submatrix of order 2. It is sufficient to consider the second-order submatrix at the upper left, since by a permutational transformation any element α_{ij}, $i \ne j$, can be brought into the position $(1, 2)$, and this, at the same time, brings α_{ji} into the position $(2, 1)$. Let

$$(1) \qquad A = \begin{pmatrix} A_{11} & A_{12} \\ A_{21} & A_{22} \end{pmatrix}, \qquad A_{11} = \begin{pmatrix} \alpha_{11} & \alpha_{12} \\ \alpha_{21} & \alpha_{22} \end{pmatrix},$$

let W_1 be a unitary matrix such that

$$(2) \qquad W_1^H A_{11} W_1 = M_1 = \begin{pmatrix} \mu_1 & 0 \\ 0 & \mu_2 \end{pmatrix},$$

and let

(3) $$W = \text{diag } (W_1, I).$$

Then W is also unitary and

(4) $$W^H A W = \begin{pmatrix} M_1 & W_1{}^H A_{12} \\ A_{21} W_1 & A_{22} \end{pmatrix}.$$

In view of the invariance properties of the Euclidean norm,

(5) $$\eta(A_{12}) = \eta(W_1{}^H A_{12}), \qquad \eta(A_{21}) = \eta(A_{21} W_1),$$
$$\eta^2(A_{11}) = \eta^2(M_1) = \mu_1{}^2 + \mu_2{}^2.$$

Moreover,

(6) $$\eta^2(A) = \eta^2(A_{11}) + \eta^2(A_{12}) + \eta^2(A_{21}) + \eta^2(A_{22}).$$

Hence the sum of the squares of the diagonal elements of A is increased by $2 \mid \alpha_{12} \mid^2$.

In the Jacobi method, strictly so-called, at each step the off-diagonal element α_{ij} of largest modulus is annihilated (in practice, the permutational transformation is not necessary).

For machine computation the search for the largest element is costly in time, and in practice it is customary to annihilate the elements in some preassigned cyclic order. Since the method is not self-correcting, this has the obvious disadvantage of permitting greater accumulation of rounding errors because of requiring a larger number of transformations. A less obvious disadvantage is that special precautions are required for making certain of convergence, and, in fact, the proof of convergence is considerably more complicated. Some amelioration can be obtained by annihilating only elements of modulus exceeding some assigned lower bound that is progressively reduced. This is the so-called *threshold method*.

The new diagonal elements μ_1 and μ_2 satisfy

(7) $$(\mu - \alpha_{11})(\mu - \alpha_{22}) = \alpha_{12}\alpha_{21},$$

with the right member nonnegative. But if $\alpha_{12} = \alpha_{21} = 0$, there is nothing to be done, hence it may be supposed that

$$\alpha_{12}\alpha_{21} > 0.$$

Hence each of μ_1 and μ_2 either exceeds both α_{11} and α_{22} or is exceeded by both. Since

$$(8) \qquad \mu_1 + \mu_2 = \alpha_{11} + \alpha_{22},$$

one must exceed, the other be exceeded by, α_{11} and α_{22}. For definiteness, it is customary to require that if α_{ij} and α_{ji} are being annihilated, then μ_i and μ_j are selected with $\mu_i > \mu_j$ for $i < j$.

The Jacobi method is thus seen to be a method of forming a unitary matrix V as the limit of an infinite product of plane rotational matrices, such that

$$V^H A V = \Lambda$$

is diagonal. Evidently the columns of V are the proper vectors of A, and hence can be obtained by forming the products as the computations proceed. The method is essentially not self-correcting.

7.2. The Method of Collar and Jahn. This method, like that of Jacobi, transforms the entire matrix, but unlike that of Jacobi it requires that a nonsingular matrix X of approximate proper vectors be available or at least that there be an approximate separation into invariant subspaces. Suppose first that the matrix A has no multiple roots. Then it is normalizable. If X is a nonsingular matrix of approximate proper vectors, then

$$(1) \qquad X^{-1} A X = M + R,$$

where M is a diagonal matrix, and R is a matrix of small elements with vanishing diagonal. A matrix E of small elements is required such that $X(I + E)$ is a matrix of better approximations. This is equivalent to saying that the off-diagonal elements of

$$(I + E)^{-1}(M + R)(I + E)$$

are smaller than those in R.

On the assumption that the elements in E and in R are small enough so that powers and products of the matrices can be neglected, this leads to the requirement that

$$(2) \qquad EM - ME = R,$$

which determines off-diagonal elements of E uniquely if the diagonal elements of M are distinct.

If A is Hermitian, X may be unitary, and a unitary factor would be desired. Hence, in place of $I + E$, take $(I + S)(I - S)^{-1}$ where S is skew-Hermitian. This leads to

$$2(SM - MS) = R,$$

which is equivalent up to terms of the second order.

To return to the general case, the method can be extended to improve the separation of invariant subspaces of any dimensionality, but belonging to distinct sets of proper values. It is sufficient to consider sets Σ_1 and Σ_2 of proper values of A, with no common member. Let M and R have the forms

$$M = \begin{pmatrix} M_1 & 0 \\ 0 & M_2 \end{pmatrix}, \quad R = \begin{pmatrix} 0 & R_{12} \\ R_{21} & 0 \end{pmatrix},$$

where the elements of R_{12} and R_{21} are small. It is required to find

$$E = \begin{pmatrix} 0 & E_{12} \\ E_{21} & 0 \end{pmatrix}$$

so that (2) is satisfied. Then

(3)
$$E_{12}M_2 - M_1E_{12} = R_{12},$$
$$E_{21}M_1 - M_2E_{21} = R_{21}.$$

But the roots of M_1 are approximately those in Σ_1, and the roots of M_2 are approximately those of Σ_2. Hence M_1 and M_2 can be assumed to have no common root, and Eqs. (3) can be solved uniquely.

A similar modification is possible when the matrix is Hermitian. If

$$S = \begin{pmatrix} 0 & Q \\ -Q^H & 0 \end{pmatrix}, \quad R = \begin{pmatrix} 0 & P \\ P^H & 0 \end{pmatrix}$$

and

$$P = QM_2 - M_1Q,$$

then

$$X(I + S)(I - S)^{-1}$$

will provide an improved separation of the invariant subspaces.

7.3. Powers of a Matrix. Other methods are based upon the properties of high powers of the matrix A itself and of its compounds, hence matrix powers will now be considered. If J (Section 1.0) is of order n, $J^n = 0$, and for $\nu \geq n$, $\lambda \neq 0$,

$$(\lambda I - J)^\nu = \lambda^\nu[I - \nu\lambda^{-1} + \cdots \pm C_{n-1,\nu}\lambda^{-n+1}J^{n-1}].$$

For ν sufficiently large, λ being fixed, the last coefficient is the one of greatest magnitude, hence

$$C_{n-1,\nu}\,|\,\lambda\,|^{\nu-n+1} < \mathrm{lub}_e\,[(\lambda I - J)^\nu] < n\,C_{n-1,\nu}\,|\,\lambda\,|^{\nu-n+1}.$$

But

$$C_{n-1,\nu} = \nu^{n-1}(1 - \nu^{-1})(1 - 2\nu^{-1})\cdots[1 - (n-2)\nu^{-1}]/(n-1)!;$$

hence the coefficient of ν^{n-1} has the limit $1/(n-1)!$, and, for sufficiently large ν,

$$1/(n-1)! < \mathrm{lub}_e\,[(\lambda I - J)^\nu]\,|\,\lambda\,|^{-\nu+n-1}\nu^{-n+1} < n/(n-1)!.$$

Also, for any norm whatsoever, there exists a constant κ such that

$$\kappa^{-1} \operatorname{lub}_e (A) \le \operatorname{lub} (A) \le \kappa \operatorname{lub}_e (A),$$

independently of A (Section 2.1). On combining the two results, it follows that *for any norm there exist constants κ' and κ'' such that for all positive integers ν,*

(1) $$\kappa' \le \operatorname{lub} [(\lambda I - J)^\nu] \,|\, \lambda \,|^{-\nu} \nu^{-n+1} \le \kappa''.$$

Now let A be any matrix with $\rho(A) > 0$, and let

$$A = P \Lambda P^{-1},$$

where Λ is the Jordan normal form. The matrix Λ is block diagonal, with diagonal blocks of the form $\lambda_i I - J$ of order n_i. Let $|\,\lambda_1\,| = \rho$, and if $|\,\lambda_i\,| = \rho$ for $i > 1$, let $n_i \le n_1$. Evidently

$$\kappa^{-1}(P) \operatorname{lub} (\Lambda^\nu) \le \operatorname{lub} (A^\nu) \le \kappa(P) \operatorname{lub} (\Lambda^\nu), \qquad \kappa(P) = \operatorname{lub} (P) \operatorname{lub} (P^{-1}).$$

Hence there exist scalars κ_1 and κ_2 such that for every positive integer ν,

(2) $$\kappa_1 \le \rho^{-\nu} \nu^{-n_1+1} \operatorname{lub} (A^\nu) \le \kappa_2.$$

From this it follows immediately that

(3) $$\lim_{\nu \to \infty} \operatorname{lub}^{1/\nu}(A^\nu) = \rho(A),$$

whatever the norm (Exercise 8).

The trace $\tau(A)$ is known to satisfy

$$\tau(A) = \sum \lambda_i,$$

the sum being taken over all characteristic roots. Hence for any ν,

(4) $$\tau_\nu = \tau(A^\nu) = \sum \lambda_i^\nu.$$

If

$$|\,\lambda_1\,| > |\,\lambda_2\,| \ge \cdots \ge |\,\lambda_n\,|,$$

then for increasing ν,

$$\tau_{\nu+1}/\tau_\nu = \lambda_1[1 + O \,|\, \lambda_2/\lambda_1 \,|^\nu].$$

where the symbol O represents a sum that vanishes with $|\,\lambda_2/\lambda_1\,|^\nu$. In particular, if A is positive definite Hermitian, then

$$\lambda_1^\nu \le \tau_\nu \le n\lambda_1^\nu,$$
$$\lambda_1 \le \tau_\nu^{1/\nu} \le \lambda_1 n^{1/\nu}.$$

Let

$$\tau_{\nu, 2} = \delta \begin{pmatrix} \tau_\nu & \tau_{\nu+1} \\ \tau_{\nu+1} & \tau_{\nu+2} \end{pmatrix}$$
$$= \lambda_1^\nu \lambda_2^\nu (\lambda_2 - \lambda_1)^2 + \lambda_1^\nu \lambda_3^\nu (\lambda_3 - \lambda_1)^2 + \cdots.$$

Then if

$$| \lambda_1 | \geq | \lambda_2 | > | \lambda_3 | \geq \cdots, \qquad \lambda_1 \neq \lambda_2,$$

it follows that for increasing ν

$$\tau_{\nu+1,2}/\tau_{\nu,2} = \lambda_1 \lambda_2 [1 + O \,| \, \lambda_3/\lambda_2 \,|^\nu].$$

More generally, if

$$(5) \qquad \tau_{\nu,\sigma} = \delta \begin{pmatrix} \tau_\nu & \cdots & \tau_{\nu+\sigma-1} \\ \cdots\cdots\cdots\cdots\cdots\cdots \\ \tau_{\nu+\sigma-1} & \cdots & \tau_{\nu+2\sigma-2} \end{pmatrix}$$

$$= \lambda_1^\nu \cdots \lambda_\sigma^\nu \delta^2 \begin{pmatrix} 1 & \cdots & 1 \\ \cdots\cdots\cdots\cdots \\ \lambda_1^{\sigma-1} & \cdots & \lambda_\sigma^{\sigma-1} \end{pmatrix} + \cdots,$$

if

$$(6) \qquad | \lambda_1 | \geq | \lambda_2 | \geq \cdots \geq | \lambda_\sigma | > | \lambda_{\sigma+1} | \geq \cdots \geq | \lambda_n |,$$

and if

$$(7) \qquad \delta \begin{pmatrix} 1 & \cdots & 1 \\ \cdots\cdots\cdots\cdots \\ \lambda_1^{\sigma-1} & \cdots & \lambda_\sigma^{\sigma-1} \end{pmatrix} \neq 0,$$

then for increasing ν

$$(8) \qquad \tau_{\nu+1,\sigma}/\tau_{\nu,\sigma} = \lambda_1 \cdots \lambda_\sigma [1 + O(\,| \, \lambda_{\sigma+1}/\lambda_\sigma \,| \,)^\nu].$$

The roots of the σth compound, $A^{(\sigma)}$, of a matrix A are the products of the λ_i taken σ at a time. Moreover, the compound of a product of matrices is the product of the compounds. Hence the roots of

$$(A^\nu)^{(\sigma)} = (A^{(\sigma)})^\nu$$

are the products of the λ_i^ν taken σ at a time. From this it follows in partic-ular that *if* (6) *holds, then independently of* (7),

$$(9) \qquad \tau[(A^{\nu+1})^{(\sigma)}]/\tau[(A^\nu)^{(\sigma)}] = \lambda_1 \cdots \lambda_\sigma [1 + O \,| \, \lambda_{\sigma+1}/\lambda_\sigma \,|^\nu].$$

Considered as a function of the variable λ, the matrix $(\lambda I - A)^{-1}$ has elements that are rational fractions, having the characteristic polynomial as a common denominator. If x and y are arbitrary nonnull vectors, then

$$(10) \qquad \eta(\lambda) = y^H (\lambda I - A)^{-1} x$$

is also expressible as a rational fraction, whose denominator is the character-istic polynomial. Hence $\eta(\lambda)$ has no singularities except poles. Moreover, the expansion

$$(11) \qquad \begin{aligned} \eta(\lambda) &= \eta_0 \lambda^{-1} + \eta_1 \lambda^{-2} + \cdots, \\ \eta_i &= y^H A^i x \end{aligned}$$

converges and is analytic for $| \, \lambda \,| > \rho(A)$. A classical theorem of analysis can be applied to show that

(12) $$\lim \eta_{i+1}/\eta_i = \lambda_1,$$

but the fact will be proved otherwise below.

Indeed, let

(13) $$|\lambda_1| \geq |\lambda_2| \geq \cdots \geq |\lambda_\sigma| > |\lambda_{\sigma+1}| \geq \cdots,$$

and suppose that $y_1, \cdots, y_\sigma$ and $x_1, \cdots, x_\sigma$ are two sets of linearly independent vectors. Then the σth compounds of the matrices $(y_1, \cdots, y_\sigma)$ and $(x_1, \cdots, x_\sigma)$ are vectors of dimension $C_{n,\sigma}$ and

(14) $$[(y_1, \cdots, y_\sigma)^H A^\nu (x_1, \cdots, x_\sigma)]^{(\sigma)} = \delta \begin{pmatrix} y_1^H A^\nu x_1 & \cdots & y_1^H A^\nu x_\sigma \\ \cdots & \cdots & \cdots \\ y_\sigma^H A^\nu x_1 & \cdots & y_\sigma^H A^\nu x_\sigma \end{pmatrix}.$$

Then *as ν becomes infinite, the ratio of one of these determinants* (14) *to the preceding one is equal to $\lambda_1 \lambda_2 \cdots \lambda_\sigma$.*

The last two assertions will follow from the following considerations. Supposing that the first σ roots of A exceed all others in modulus, let V be any matrix such that

$$V^{-1}AV = \begin{pmatrix} A_1 & 0 \\ 0 & A_2 \end{pmatrix},$$

where the submatrix A_1 is of order σ with the roots $\lambda_1, \cdots, \lambda_\sigma$, and A_2 is of order $n - \sigma$ with roots $\lambda_{\sigma+1}, \cdots, \lambda_n$. In particular, the submatrices may be in Jordan normal form. Let $x_1, \cdots, x_\sigma$ be arbitrary linearly independent vectors, and

$$Vu_1 = x_1, \cdots, Vu_\sigma = x_\sigma.$$

Let $|\lambda_\sigma| \geq \rho > \rho(A_2)$. Then

$$(V^{-1}AV)^\nu u_i = V^{-1}A^\nu Vu_i = \begin{pmatrix} A_1^\nu u_i^{(1)} \\ A_2^\nu u_i^{(2)} \end{pmatrix}, \qquad i = 1, \cdots, \sigma,$$

where

$$u_i = \begin{pmatrix} u_i^{(1)} \\ u_i^{(2)} \end{pmatrix},$$

and

$$\rho^{-\nu}(V^{-1}AV)^\nu u_i = \rho^{-\nu} \begin{pmatrix} A_1^\nu u_i^{(1)} \\ A_2^\nu u_i^{(2)} \end{pmatrix} \to \rho^{-\nu} \begin{pmatrix} A_1^\nu u_i^{(1)} \\ 0 \end{pmatrix}$$

as ν becomes infinite, since

$$\rho^{-\nu} A_2^\nu \to 0.$$

Hence if

$$V = (V_1, V_2),$$

where V_1 is the matrix of the first σ columns of V, in the limit

$$A^\nu x_i \to V_1 A_1^\nu u_i^{(1)},$$

which is a vector in the invariant subspace of A belonging to the roots $\lambda_1, \cdots, \lambda_\sigma$. In particular, if the vectors x_i form a Krylov sequence for A, then the vectors $u_i^{(1)}$ form a Krylov sequence for the matrix A_1. But then if it happens that σ consecutive vectors of the sequence of vectors $u_i^{(1)}$ are linearly independent, a vanishing linear combination of $\sigma + 1$ of them can be formed with coefficients proportional to the coefficients of the characteristic polynomial of A_1. In the limit, then, $\sigma + 1$ consecutive vectors

$$x_\nu, \cdots, x_{\nu+\sigma}$$

are linearly dependent with coefficients of dependence proportional to the coefficients of a polynomial whose zeros are $\lambda_1, \cdots, \lambda_\sigma$. This implies that *if* (13) *is satisfied, and if the vectors* $y_1, y_2, \cdots, y_\sigma$ *are linearly independent, and also* $x_1, x_2, \cdots, x_\sigma$ *are linearly independent, then the roots of the equation*

$$(15) \qquad \delta \begin{pmatrix} 1 & \cdots & \lambda^\sigma \\ y_1^H A^\nu x_1 & \cdots & y_1^H A^{\nu+\sigma} x_{\sigma+1} \\ \cdots\cdots\cdots\cdots\cdots\cdots\cdots\cdots \\ y_\sigma^H A^\nu x_1 & \cdots & y_\sigma^H A^{\nu+\sigma} x_{\sigma+1} \end{pmatrix} = 0$$

are, in the general case, equal in the limit to $\lambda_1, \lambda_2, \cdots, \lambda_\sigma$. *In particular, this is true of the roots of the equation*

$$(16) \qquad \delta \begin{pmatrix} 1 & \cdots & \lambda^\sigma \\ \eta_\nu & \cdots & \eta_{\nu+\sigma} \\ \cdots\cdots\cdots\cdots\cdots\cdots \\ \eta_{\nu+\sigma-1} & \cdots & \eta_{\nu+2\sigma-1} \end{pmatrix} = 0,$$

where the η_i are defined by (11). This could fail to take place only in the very special circumstance that a vector y_i or a vector x_i would be in the invariant subspace (row or column, respectively) belonging to the roots $\lambda_{\sigma+1}, \lambda_{\sigma+2}, \cdots, \lambda_n$. The polynomials $\chi_\nu(\lambda)$ and $\phi_\nu(\lambda)$ of Sections 1.5 and 1.6 are, apart from constant factors, special cases of these. Evidently the product of the roots of (16) is the ratio of consecutive determinants (14), which proves the assertion made about their ratios.

An important special case occurs when $\bar\lambda_1 = \lambda_2$, λ_1 and λ_2 being complex conjugates. Then the iteration can only converge to vectors in the subspace spanned by the two proper vectors. If the roots are simple, and if

$$|\lambda_2| > |\lambda_3| \geq \cdots,$$

then the subspace is two-dimensional and a quadratic equation in λ_1 and λ_2 is obtained.

"In general," σ successive iterates $u_i^{(1)}$ will be linearly independent, as assumed, but should the matrix A_1 be derogatory (i.e., its minimal equation is of degree less than that of the characteristic equation), A would be also, and any σ iterates would be linearly dependent. If $\sigma' < \sigma$ are linearly independent, but $\sigma' + 1$ linearly dependent, then in (15) and (16) σ must be replaced by σ' (cf. Section 7.5).

7.4. Simple Iteration (the Power Method).   The simplest application of the foregoing results can be made when

$$|\lambda_1| > |\lambda_2| \geq \cdots,$$

and this can be made by continuing the Krylov sequence

$$x_{i+1} = Ax_i,$$

starting with any convenient vector x_0. For computational purposes scale factors can be introduced according to convenience, either at every step or intermittently. Then the modified iteration has the form

$$\rho_{i+1} x_{i+1} = Ax_i,$$

and common and convenient methods of scaling are to require that $\| x_i \| = 1$ for some norm. In any event, for any convenient y,

$$y^H Ax_i / y^H x_i \to \lambda_1,$$

and the method converges in the rate that

$$(\lambda_2/\lambda_1)^i \to 0.$$

In particular, if $y = x_i$, the estimate is the Rayleigh quotient, which is the center of the smallest inclusion circle.

If the iteration is carried out with some polynomial $\psi(A)$, in place of A itself, where

$$|\psi(\lambda_1)| > |\psi(\lambda_2)| \geq \cdots,$$

and if

$$|\psi(\lambda_2)/\psi(\lambda_1)| < |\lambda_2/\lambda_1|,$$

then the convergence will be more rapid. A very simple choice is the linear polynomial $A - \mu I$, with μ a convenient scalar, and sometimes it is possible to find a μ that gives some improvement.

In case the roots are all real, suppose that scalars μ_1 and μ_2 have been found in some way such that

$$\lambda_1 > \mu_1 \geq \lambda_2 \geq \cdots \geq \lambda_n \geq \mu_2.$$

There is no restriction in supposing that

$$\mu_1 = -\mu_2 = 1,$$

since this involves only replacing A by a suitable linear polynomial in A. Then

(1) $$\lambda_1 > 1 \geq \lambda_2 \geq \cdots \geq \lambda_n \geq -1.$$

Let

$$T_\nu(\lambda) = \cos{(\nu \cos^{-1} \lambda)}, \qquad -1 \leq \lambda \leq 1,$$
$$\cosh{(\nu \cosh^{-1} \lambda)}, \qquad |\lambda| \geq 1$$

be the Chebyshev polynomial of degree ν. Then

$$T_\nu(\lambda)/T_\nu(\lambda_1)$$

can be shown to be that polynomial of degree ν whose value is unity at λ_1, and whose maximum absolute value is least on the same interval. Consequently, in the absence of any information as to the roots $\lambda_2, \cdots, \lambda_n$, except that they lie on that interval, an iteration of the form

$$u_{i+1} = T_\nu(A)u_i.$$

has obvious advantages.

Returning to the iteration with A itself, suppose the iterates $x_0, \cdots, x_\nu$ have been formed. Then, it may be asked, what linear combination

$$y_\nu = x_\nu + \alpha_1 x_{\nu-1} + \cdots + \alpha_\nu x_0$$

approximates most closely the proper vector belonging to the proper value λ_1 of A? But evidently

$$y = \alpha_\nu(A)x_0,$$
$$\alpha_\nu(\lambda) = \lambda^\nu + \alpha_{\nu,1}\lambda^{\nu-1} + \cdots + \alpha_{\nu,\nu},$$

and therefore, in the sense already discussed, if the relations (1) are assumed the choice

$$\alpha_\nu(\lambda) = T_\nu(\lambda)$$

is again indicated. But since the Chebyshev polynomials can be generated by a 3-term recursion (Section 4.22) of the form

$$T_{\nu+1}(\lambda) = 2\lambda T_\nu(\lambda) - T_{\nu-1}(\lambda), \qquad \nu \geq 1,$$

it follows that the vectors y_ν can be formed in sequence by

$$y_{\nu+1} = 2Ay_\nu - y_{\nu-1}, \qquad \nu \geq 1.$$

When the process has been carried sufficiently far, calculation of λ_1 can be made by

$$\lambda_1 \doteq y^H A y_\nu / y^H y_\nu,$$

where y is any convenient vector, possibly y_ν.

When something is known about the distribution of the nondominant roots on the interval from -1 to $+1$, it may be possible to choose polynomials that are more effective than the Chebyshev polynomials. This possibility will not be discussed, since the cases would be quite special, and the choice would depend upon special properties of orthogonal polynomials.

The possibilities for accelerating convergence by use of polynomials in A are limited even when the roots of A are real. When the roots are not all real the limitations are even greater. But much more flexibility is provided by the use of rational fractions in A, even by the simplest such, in the form $(A - \mu I)^{-1}$.

In fact, given any scalar μ, if there is a particular λ_i such that $| \mu - \lambda_i |$ is less than any $| \mu - \lambda_j |$ for $j \neq i$, then $(\lambda_i - \mu)^{-1}$ is the root of greatest modulus of $(A - \mu I)^{-1}$. Hence the vectors of a Krylov sequence defined by $(A - \mu I)^{-1}$ approach the proper vector of A belonging to λ_i, and the rate of convergence is the rate at which

$$[(\lambda_i - \mu)/(\lambda_k - \mu)]^\nu \to 0,$$

where

$$| \lambda_k - \mu | = \min_{j \neq i} | \lambda_j - \mu |.$$

This is Wielandt's fractional iteration (inverse power method).

Suppose μ_1 is any approximation to λ_i in the above sense, and that x_1 is any approximation to the proper vector belonging to λ_i. Then

$$x_2 = (A - \mu_1 I)^{-1} x_1$$

is in general a closer approximation to that proper vector, and, in general, if

$$y^H x_1 / y^H x_2 = \mu_2 - \mu_1,$$

shen μ_2 should be a closer approximation to λ_i than is μ_1. Hence the sequence

$$x_{\nu+1} = (A - \mu_\nu I)^{-1} x_\nu,$$
$$y^H x_\nu / y^H x_{\nu+1} = \mu_{\nu+1} - \mu_\nu,$$

where y may or may not remain fixed, should yield optimal convergence. On the other hand, each step requires the solution of a system of equations of the form

$$(A - \mu_\nu I) x_{\nu+1} = x_\nu,$$

and it is probably not advantageous to change μ_ν at every step, since each time μ_ν is changed the matrix must be refactored.

If the approximation μ_1 is already close, having been obtained, possibly, by a method that is not self-correcting, and that does not provide an approximate vector, x_1 may be selected nearly at random, and x_2, or at worst x_3, may be an adequate approximation to the proper vector. Suppose that a factorization has been made,

$$A - \mu_1 I = L_1 R_1,$$

where L_1 is a unit lower triangular matrix, and R_1 an upper triangular matrix. If μ_1 is in fact a close approximation to a root, then

$$\delta(A - \mu_1 I) = \delta(R_1)$$

must be small, and there must be small elements along the diagonal of R_1. If interchanges have been carried out (search for pivot), there will be small elements at the lower right. Then

$$x_2 = R_1^{-1} L_1^{-1} x_1.$$

But a "random" selection for x_1 can be made by choosing $L_1^{-1} x_1$ instead of choosing x_1 explicitly.

7.5. Multiple Roots and Principal Vectors.
To return to the sequence

$$x_{\nu+1} = A x_\nu = A^\nu x_1,$$

if it should happen that

(1) $$|\lambda_1| = |\lambda_2| = \cdots = |\lambda_\sigma| > |\lambda_{\sigma+1}| \geq \cdots,$$

the vectors x_ν may not have a limit, but they will approach the invariant subspace belonging to $\lambda_1, \cdots, \lambda_\sigma$. This is to say that in the limit any $\sigma + 1$ consecutive iterates will be linearly dependent, and if it happens that σ of them are linearly independent, then the coefficients of the vanishing linear combination are coefficients of the polynomial whose zeros are $\lambda_1, \cdots, \lambda_\sigma$. This polynomial can be given in the form (7.3.18), or (7.3.19). However, it could happen that the maximal sequence of linearly independent vectors $x_\nu, x_{\nu+1}, \cdots$, has (in the limit) fewer than σ vectors. For example, if the matrix A is Hermitian, and

$$\lambda_1 = \lambda_2 = \cdots = \lambda_\sigma,$$

then A has an invariant subspace of dimension σ, each vector of which is a proper vector, hence in the limit x_ν approaches some vector in this invariant subspace, and x_ν and $x_{\nu+1}$ are linearly dependent in the limit. The same could happen for non-Hermitian matrices.

The number of possible cases that can arise when (1) holds is quite considerable, and no exhaustive treatment will be attempted. But by way of illustration, suppose that

(2) $$\lambda_1 = \lambda_2 = \lambda_3, \quad |\lambda_3| > |\lambda_4| \geq \cdots,$$

and that associated with λ, there is only a single proper vector. Then in the Jordan normal form there will occur the submatrix $\lambda_1 I + J$ of order 3. Let u_1, u_2, u_3 be principal vectors such that if

(3)
$$U = (u_1, u_2, u_3)$$

then

(4)
$$AU = U(\lambda_1 I + J).$$

This is to say that

(5)
$$(A - \lambda_1 I)U = UJ,$$

and therefore

(6)
$$(A - \lambda_1 I)^3 U = 0.$$

For sufficiently large ν, all vectors $x_\nu, x_{\nu+1}, \cdots$ can be considered, with sufficient approximation, to lie in the space of U, hence for some vector u, the relation

$$x_\nu = Uu$$

must hold at least approximately. Then

$$(A - \lambda_1 I)^3 x_\nu = 0,$$

or

$$x_{\nu+3} - 3\lambda_1 x_{\nu+2} + 3\lambda_1^2 x_{\nu+1} - \lambda_1^3 x_\nu = 0.$$

Therefore (7.3.18) or (7.3.19) becomes for $\sigma = 3$, and after dividing out the leading coefficient,

$$(\lambda - \lambda_1)^3 = 0.$$

Given λ_1, the choice

(7)
$$u_1 = x_\nu,$$
$$u_2 = (A - \lambda_1 I)x_\nu = x_{\nu+1} - \lambda_1 x_\nu,$$
$$u_3 = (A - \lambda_1 I)^2 x_\nu = x_{\nu+2} - 2\lambda_1 x_{\nu+1} + x_\nu,$$

is effective for (3) and (4).

Further consideration of roots of equal or nearly equal moduli will be given in the next sections, where it will be shown how to obtain a complete invariant subspace.

7.6. Staircase Iteration (Treppeniteration). If the characteristic roots of A satisfy

(1)
$$|\lambda_1| \geq |\lambda_2| \geq \cdots \geq |\lambda_\nu| > |\lambda_{\nu+1}| \geq \cdots,$$

then in the limit at most ν consecutive iterates

$$x_i = A^i x$$

of the single vector x will be linearly independent. But if X is a matrix of ν

linearly independent columns, then the columns of each matrix

$$(2) \qquad\qquad X_i = A^i X$$

will in general remain linearly independent. It could happen, in principle, that proper vectors belonging to a λ_i, $i \leq \nu$, could have no component along any column of X, but in general rounding errors would introduce such components. Hence such a theoretical possibility will be left out of account, and then it can be said that in the limit the columns of the X_i form a basis for the invariant subspace belonging to $\lambda_1, \cdots, \lambda_\nu$. The same would be true of the columns of

$$Y_i = X_i M_i,$$

where M_i is any nonsingular matrix. *Staircase iteration provides a sequence of matrices, each assured of having linearly independent columns, and such that in the limit the first ν columns are a basis for the invariant subspace belonging to the ν roots of A of largest modulus.*

Let C and B be matrices which, for the moment, will be supposed square and nonsingular. Let

$$(3) \qquad\qquad B = \Lambda_0 R_0,$$

and

$$(4) \qquad\qquad A\Lambda_i = \Lambda_{i+1} R_{i+1},$$

where each R_i is upper triangular, and $C\Lambda_i$ is unit lower triangular. Then

$$(5) \qquad\qquad A^i B = \Lambda_i P_i, \qquad P_i = R_i R_{i-1} \cdots R_0,$$

and

$$(6) \qquad\qquad (CAC^{-1})^i (CB) = C\Lambda_i P_i.$$

If the matrices Λ_i and R_i have limits Λ and R, then

$$A\Lambda = \Lambda R,$$

and if all matrices are of order n, the matrices A and R are similar and have the same proper values. But R is triangular and hence its proper values appear on its own diagonal.

Suppose (1) is satisfied. Since each R_i is upper triangular, the first ν columns of Λ_i are formed independently of the others. Form the νth compounds of all matrices, with lexicographic ordering of the rows and columns. Then

$$(C\Lambda_i)^{(\nu)} = C^{(\nu)} \Lambda_i^{(\nu)}$$

is unit lower triangular, and $R_i^{(\nu)}$ is upper triangular. Equation (4) implies

$$(7) \qquad\qquad A^{(\nu)} \Lambda_i^{(\nu)} = \Lambda_{i+1}^{(\nu)} R_{i+1}^{(\nu)}.$$

Since the roots of $A^{(\nu)}$ are the products ν at a time of those of A, (1) implies that $A^{(v)}$ has $\lambda_1 \cdots \lambda_v$ as the root of maximal modulus, therefore the first column of $\Lambda_i^{(v)}$ has a limit, and the element of $R_i^{(v)}$ in the upper left has the limit $\lambda_1 \cdots \lambda_v$. But this is the determinant of the upper left submatrix of order v from R_i, which is the product of the first v diagonal elements. Hence if (1) is satisfied, the product of the first v diagonal elements of R_i has a limit, even if the elements themselves do not.

It is easy to see that *if the first column of $\Lambda_i^{(v)}$ has a limit, then the vth column of Λ_i has a limit.* For consider the first $n - v + 1$ elements of the first column of $(C\Lambda_i)^{(v)}$. These are the values of the determinants of the lower triangular submatrices which are formed from the first v columns of $C\Lambda_i$ by using first $v - 1$ rows from the first $v - 1$ rows and one other from $C\Lambda_i$. Hence the first $n - v + 1$ elements of the first column of $(C\Lambda_i)^{(v)}$ are the nonnull elements in the vth column of $C\Lambda_i$.

The linear independence of the first v columns of each Λ_i is assured by the fact that $C\Lambda_i$ is unit lower triangular. Hence, *if (1) is satisfied, then for sufficiently large i, the first v columns of Λ_i are a basis for the invariant subspace belonging to the v roots of largest modulus.* This means that if $\Lambda_i(1, v)$ is the matrix of the first v columns of Λ_i, and if C_v is the matrix of the first v rows of C, then the matrix $\Gamma_i(1, v)$ defined by

(8)
$$C_v \Lambda_i(1, v) \Gamma_{i+1}(1, v) - C_v \Lambda_{i+1}(1, v)$$

is such that for large i,

(9)
$$\Lambda_i(1, v) \Gamma_{i+1}(1, v) \doteq \Lambda_{i+1}(1, v),$$

approximately. But $\Gamma_i(1, v)$ is unit lower triangular, since the matrices $C_v \Lambda_i(1, v)$ are of this form. The argument can be repeated for every v' with $|\lambda_{v'}| > |\lambda_{v'+1}|$, hence *there exists a unit lower triangular matrix Γ_i for every i, which is a direct sum of lower triangular matrices, such that for large i,*

(10)
$$\Lambda_i \Gamma_{i+1} \doteq \Lambda_{i+1}.$$

Each triangular block, $\Gamma_i(v + 1, v')$, of order $v' - v$, if not further reducible, is such that

(11)
$$|\lambda_{v+1}| = \cdots = |\lambda_{v'}| > |\lambda_{v'+1}|$$

and that the product of the diagonal elements of R_i, from $v + 1$ to v', is equal, in the limit, to $\lambda_{v+1} \cdots \lambda_{v'}$. In particular, if all roots are distinct in modulus, limits Λ and R of Λ_i and R_i exist, and Γ_i is then equal to I.

Otherwise stated, let

(12)
$$\Lambda_i L_{i+1} = \Lambda_{i+1},$$

where the equality is strict, and not merely approximate as in (10). Then L_{i+1} is unit lower triangular, and if (11) is satisfied for any $v' > v \geq 0$, then in the L_i all elements which lie in columns from $v + 1$ to v' and rows $v' + 1$ and below must vanish in the limit.

The most natural choice for C is a permutation matrix so chosen that in the triangular factorization a largest pivot is selected from each column. If not all roots of A are required, but only the ν roots of greatest modulus, then it is sufficient to take for B and C matrices of ν columns and ν rows, respectively, since the iteration on these is unaffected by the presence or absence of additional columns and rows.

In the most usual, and most important, special case, the roots will differ in modulus except for pairs of complex conjugates. In the limit, then, the matrix L_i will differ from the identity only by individual elements below the diagonal to mark the presence of such pairs.

In any event, it follows from (4) and (12) that

$$(13) \qquad A\Lambda_i = \Lambda_i L_{i+1} R_{i+1}.$$

For sufficiently large i, $L_{i+1}R_{i+1}$ is block triangular, and the characteristic polynomials of the diagonal blocks will have roots of equal modulus.

7.7. The LR-Transformation. From (7.6.4)

$$A\Lambda_{i-1} = \Lambda_i R_i,$$

and from (7.6.12)

$$A\Lambda_i = \Lambda_i R_i L_i.$$

But by comparing this with (7.6.13) it can be seen that

$$(1) \qquad R_i L_i = L_{i+1} R_{i+1},$$

while

$$(2) \qquad A\Lambda_0 = \Lambda_0 L_1 R_1$$

with Λ_0 arbitrary nonsingular. Thus if

$$(3) \qquad A_1 = \Lambda_0^{-1} A \Lambda_0 = L_1 R_1,$$
$$(4) \qquad A_i = R_{i-1} L_{i-1} = L_i R_i,$$

these matrices form a sequence of matrices similar to A which can be obtained by factorization followed by multiplication in reverse order. The behavior of the matrices L_i for increasing i was discussed in the last section.

From (3) and (7.6.3) it follows that

$$A_1 = R_0(B^{-1}AB)R_0^{-1}$$

and from (4) that

$$(5) \qquad A_{i+1} = R_i A_i R_i^{-1}.$$

Hence, by (7.6.5),

$$(6) \qquad A_{i+1} = P_i(B^{-1}AB)P_i^{-1}.$$

The formation of the sequence of matrices A_i by (4) is Rutishauser's *LR*-transformation. The method is, unfortunately, not self-correcting, while the staircase iteration is self-correcting.

As indicated in the last section, the matrices L_i, although they may not approach a limit, nevertheless approach a limiting form, which is a direct sum of lower triangular matrices, and in the usual case these triangular submatrices are at worst of second order. Hence the matrices A_i approach a limiting form in which elements vanish below triangular blocks that correspond to those in the limiting form of the L_i. When these elements have become sufficiently small, subsequent convergence can be accelerated by interrupting the normal sequence to introduce transformations by elementary triangular matrices.

Suppose that in the jth column of A_i, the elements below the kth, for some $k > j$, have become small, and consider the matrix

$$E = I - l_j e_j^{\mathrm{T}},$$

where l_j has nonnull elements only in positions below the kth. Form the matrix

$$A_{i,j} = (I - l_j e_j^{\mathrm{T}}) A_i (I + l_j e_j^{\mathrm{T}}),$$

similar to A_i, by choosing l_j so as to reduce as far as possible the elements below the kth in the jth column. The jth column of $A_{i,j}$ is

$$A_{i,j} c_j = A_i c_j \mid A_i l_j \quad l_j c_j^{\mathrm{T}} A_i c_j \quad l_j c_j^{\mathrm{T}} A l_j.$$

If the nonvanishing elements of l_j are small, the last term can be neglected, and then the requirement that the last $n - k$ elements of the left member vanish provides $n - k$ linear equations in the $n - k$ elements of l_j. Since only elements from the jth column and beyond in A_i appear in these equations, a cycle can be carried out to reduce elements in all columns taken in sequence before continuing with the factorization.

Another method of acceleration is also possible (Rutishauser and Bauer). If $B = C = I$, then (7.6.6)

(7) $$A^i = \Lambda_i P_i$$

and hence

$$A^{2i} = \Lambda_{2i} P_{2i} = \Lambda_i P_i \Lambda_i P_i.$$

Therefore if

$$P_i \Lambda_i = \tilde{\Lambda}_i \tilde{P}_i$$

where the factorization on the right is into unit lower triangular and upper triangular matrices, respectively, then

(8) $$\Lambda_{2i} = \Lambda_i \tilde{\Lambda}_i \qquad P_{2i} = \tilde{P}_i P_i.$$

Thus the Λ_i and P_i can be formed in steps for i given as powers of 2. This process can be followed by a single application of (7.6.4) to produce R_{i+1}.

This algorithm, however, has the obvious disadvantage that the diagonal elements of P_i, being the ith powers of the roots of A, become either very large or very small; hence the algorithm is unstable. Let

$$(9) \qquad\qquad P_i = \Delta_i S_i,$$

where Δ_i is diagonal, and S_i unit upper triangular, and factor

$$(10) \qquad\qquad S_i \Lambda_i = H_i D_i U_i,$$

where the factors are unit lower triangular, diagonal, and unit upper triangular, respectively. Then

$$A^{2i} = \Lambda_{2i} \Delta_{2i} S_{2i} = \Lambda_i \Delta_i S_i \Lambda_i \Delta_i S_i$$
$$= \Lambda_i (\Delta_i H_i \Delta_i^{-1}) \Delta_i D_i \Delta_i (\Delta_i^{-1} U_i \Delta_i) S_i.$$

But $\Delta_i H_i \Delta_i^{-1}$ and $\Delta_i^{-1} U_i \Delta_i$ are unit lower and upper triangular, respectively. Hence

$$\Lambda_{2i} = \Lambda_i \Delta_i H_i \Delta_i^{-1},$$
$$(11) \qquad\qquad \Delta_{2i} = \Delta_i^2 D_i,$$
$$S_{2i} = \Delta_i^{-1} U_i \Delta_i S_i.$$

7.8. Bi-iteration. Bi-iteration provides a self-correcting alternative [cf. (9) below]. Let $B = B_0$ and $C = C_0$ be, again, any two nonsingular matrices, and let

$$(1) \qquad\qquad C_i B_i = M_i V_i,$$

where M_i is a unit lower triangular matrix and V_i is upper triangular. From (7.6.3), unique factorization implies that

$$M_0 = C \Lambda_0, \qquad\qquad V_0 = R_0.$$

Let

$$(2) \qquad\qquad B_i V_i^{-1} = S_i, \qquad\qquad M_i^{-1} C_i = Q_i,$$
$$AS_i = B_{i+1}, \qquad\qquad Q_i A = C_{i+1}.$$

Then

$$(3) \qquad\qquad Q_i S_i = I,$$

and

$$(4) \qquad\qquad A^i B = S_i V_i V_{i-1} \cdots V_0,$$
$$CA^i = M_0 M_1 \cdots M_i Q_i.$$

Hence, because of (3),

$$(5) \qquad\qquad CA^{2i} B = M_0 \cdots M_i V_i \cdots V_0.$$

Comparison of this with (7.6.6) shows that

$$(6) \qquad M_0 M_1 \cdots M_i = C \Lambda_{2i}, \qquad V_i V_{i-1} \cdots V_0 = P_{2i},$$

in view of the fact that the factorization of a matrix into a unit lower triangular matrix and an upper triangular matrix is unique. Hence

(7) $V_0 = R_0,$ $V_i = R_{2i}R_{2i-1},$ $i \geq 1,$

and

(8) $M_0 = L_0,$ $M_i = \Lambda_{2i-2}^{-1}\Lambda_{2i} = L_{2i-1}L_{2i},$ $i \geq 1.$

Since

$$CA^{2i+1}B = (CA^i)A(A^iB),$$

(7.6.6), (7.7.5), (7.7.12), (4), (6), and (8) imply that

(9) $S_i^{-1}AS_i = Q_iAS_i = L_{2i+1}R_{2i+1} = A_{2i+1}.$

These are among the matrices that occur in the LR-transformation.

7.9. The QR-Transformation. As in Section 7.7, let

(1) $| \lambda_1 | \geq | \lambda_2 | \geq \cdots \geq | \lambda_\nu | > | \lambda_{\nu+1} | \geq \cdots,$

and consider the iteration

(2) $AW_i = W_{i+1}R_{i+1},$ $i = 0, 1, 2, \cdots,$

where

(3) $W_i^H W_i = I$

is of order ν, and where each R_i is upper triangular. By taking νth compounds, the argument of Section 7.6 can be paraphrased to show that

$$\delta(R_i) \to \lambda_1\lambda_2 \cdots \lambda_\nu.$$

In fact, the discussion in Section 7.3 shows that the columns of the W_i
approach the invariant subspace of A belonging to $\lambda_1, \lambda_2, \cdots, \lambda_\nu$, and since
they are linearly independent by virtue of their orthogonality, in the limit
the columns of any W_i form an orthonormal basis for this subspace. Thus
for sufficiently large i, there exists a unitary matrix Ω_i such that approximately

(4) $W_i\Omega_{i+1} \doteq W_{i+1}.$

Then

(5) $AW_i \doteq W_i\Omega_{i+1}R_{i+1}.$

Now suppose the matrices W_i are square matrices, hence unitary, and
define the unitary matrix Ω_{i+1} to satisfy (4) exactly. Hence (5) will be
satisfied exactly. In the limit, the matrices Ω_i will become block diagonal,
each diagonal block being unitary. The product Ω_iR_i will be block triangular
in the limit, and the roots of each diagonal block will be of equal modulus.

Along with (5) and (4), consider

$$AW_{i-1} = W_iR_i,$$
$$AW_i = W_iR_i\Omega_i.$$

Hence

(6)
$$R_i\Omega_i = \Omega_{i+1}R_{i+1}.$$

Hence if

$$A = A_0 = \Omega_1 R_1,$$

$$A_1 = R_1\Omega_1 = \Omega_2 R_2,$$

(7)
$$\cdots\cdots\cdots\cdots\cdots\cdots$$

$$A_i = R_i\Omega_i = \Omega_{i+1}R_{i+1},$$

there results a sequence similar to the LR-transformation (Section 7.7) of repeated factorization followed by multiplication in reverse order, but here the factorization is into a unitary matrix and an upper triangular matrix. In the limit the matrices A_i become block triangular. The method is not self-correcting.

REFERENCES

For the careful analytic treatment of asymptotic properties of powers of matrices see papers by Ostrowski and by Gautschi. Müntz (1913) gives an early, possibly the earliest, treatment of simple iteration as a practical computational device. The idea of iterating on a matrix to obtain more than one root at a time is perhaps fairly natural. It was suggested by Horst (1937), but the paper was read only by a limited group. Perron (1908) provides the basis for a rigorous justification. More modern treatments are listed under the names of Rutishauser, of Bauer, and of Francis.

The great name of von Neumann, and the simplicity of the method, combined to make popular the Jacobi method [Goldstine, Murray, and von Neumann (1959)], and a description and an error analysis were circulated privately for some time before actual publication. Meanwhile, Forsythe, Henrici, Tompkins, and others were considering the possible pitfalls of proceeding cyclically, rather than by search as the method properly required. Goldstine and Horwitz (1959) adapted the method to normal matrices; Greenstadt, and others, attempted, with no great success, to find a proper and effective generalization for triangularizing or diagonalizing matrices of more general form. However, even for symmetric matrices the methods of Lanczos and of Givens (Chapter 6) appear to be superior, in general, and it seems unlikely that generalizing the method of Jacobi can have other than theoretical interest.

The method of fractional iteration was first proposed by Wielandt (1944). In a class of related methods one can iterate with $(A - \mu I)^{-1}$ for some fixed μ, changing only the vector each time; one can hold the vector fixed and modify μ each time, or one can modify both: Unger (1950), Crandall (1951). Methods of one type or another have been rediscovered a number of times, but the most complete and rigorous development is given in a series of papers by Ostrowski.

For carrying out simple iteration in practice, see Wilkinson (1954). For general treatments see Faddeev and Faddeeva (1960), Korganoff (1961), Durand (1961), and Householder (1953).

The LR-algorithm is due to Rutishauser; the QR-algorithm was developed independently by Francis and Kublanovskaja.

PROBLEMS AND EXERCISES

1. Verify in detail the expansion (7.3.5).

2. If (7.3.7) fails because a single pair of the roots coincide say $\lambda_1 = \lambda_2$, but the others remain distinct, what can be said in place of (7.3.8)?

3. For applying the Jacobi method, obtain explicit expressions for the elements of the plane rotation matrix W, that annihilates α_{12} and α_{21}.

4. Show that the Jacobi method can be applied to normal matrices.

5. If y and z are fixed vectors, let

$$\mu_\nu = z^H A x_\nu / z^H x_\nu, \qquad (A - \mu_\nu I) x_{\nu+1} = y.$$

If, for some simple root λ, $\epsilon = \lambda - \mu_0$ is sufficiently small in magnitude, and if z^H and y are sufficiently close to proper vectors belonging to λ, then $\lim_{\nu \to \infty} \mu_\nu = \lambda$ [for the proof there is no restriction in supposing A to have the form

$$A = \begin{pmatrix} \lambda & 0 \\ 0 & B \end{pmatrix}$$

with $B - \lambda I$ nonsingular].

6. If y is a fixed vector, let

$$\mu_\nu = x_\nu^H A x_\nu / x_\nu^H x_\nu, \qquad (A - \mu_\nu I) x_{\nu+1} = y.$$

If $\lambda - \mu_\nu = \epsilon$ is sufficiently small in magnitude, and λ is a root corresponding to simple elementary divisors, then $\lambda - \mu_{\nu+1}$ is approximately proportional to ϵ^2.

7. Let

$$M = \begin{pmatrix} A & B \\ B^H & C \end{pmatrix}$$

be Hermitian and let A, C, and M have the roots

$$\alpha_1 \geq \alpha_2 \geq \cdots \geq \alpha_m, \qquad \gamma_1 \geq \gamma_2 \geq \cdots \geq \gamma_{n-m}, \qquad \mu_1 \geq \mu_2 \geq \cdots \geq \mu_r.$$

If $\mu_1 > \lambda_0 > \max(\alpha_1, \gamma_1)$, then the sequence in which $\lambda_{\nu+1}$ is the largest root of

$$A - B(\lambda_\nu I - C)^{-1} B^H$$

converges to μ_1 with

$$\lambda_0 < \lambda_2 < \cdots < \mu_1 < \cdots < \lambda_3 < \lambda_1.$$

Likewise if $\mu_n < \lambda_0 < \min(\alpha_m, \gamma_{n-m})$, and the smallest root is taken for $\lambda_{\nu+1}$, then the sequence converges to μ_n and

$$\lambda_0 > \lambda_2 > \cdots > \mu_n > \cdots > \lambda_3 > \lambda_1$$

(Shishov).

8. Show by an example that the sequence $\mathrm{lub}^{1/\nu}(H^\nu)$ (7.3.3) need not be monotonic.

9. If $A \geq 0$ is irreducible, form the Krylov sequence starting with $g_0 = g > 0$, and let

$$\underline{\alpha}_\nu g_\nu \leq g_{\nu+1} \leq \bar{\alpha}_\nu g_\nu.$$

Then the sequences $\underline{\alpha}_\nu$ and $\bar{\alpha}_\nu$ are monotonically increasing and decreasing, respectively, and have the common limit $\rho(A)$. Apply this to obtain an alternative proof to the Frobenius theorem (Section 2.4).

10. Prove (7.8.9) in detail.

11. Complete the proof of the assertion that the matrices (7.9.7) become block-triangular in the limit.

12. Prove that (Section 7.8)

$$Q_i S_{i+1} = L_{2i+1} R_{2i+2}^{-1},$$
$$Q_{i+1} S_i = L_{2i+2}^{-1} R_{2i+1}.$$

Show that (i) an LR-transformation applied to a Hessenberg matrix produces a Hessenberg matrix, even when interchanges are made to obtain the largest pivot; (ii) a QR-transformation applied to a Hessenberg matrix produces a Hessenberg matrix; (iii) an LR-transformation applied to a tridiagonal produces a tridiagonal matrix if and only if interchanges are not made.

13. If A is positive definite it is unnecessary to search for the largest pivot in triangular factorization. Show that the LR-transformation can be replaced by a sequence of Choleski factorizations

$$A_i = P_i^H P_i, \qquad P_i P_i^H = A_{i+1},$$

retaining symmetry, with no effect on convergence. Hence if A is merely Hermitian, the same transformation can be applied to $A + \mu I$ for suitable μ.

14. Show that for A Hermitian, convergence can be accelerated by use of the modified sequence in which

$$A_{i+1} = P_i P_i^H - \mu_i I,$$

where μ_i is just large enough to assure positive definiteness.

15. If A is Hermitian, then every matrix A_ν produced by the QR-transformation (Section 7.9) is Hermitian; hence if A is also tridiagonal every A_ν is tridiagonal.

16. Let A be positive definite and tridiagonal, and, in the application of either the LR-algorithm or the QR-algorithm, let A_ν have diagonal elements

$$\alpha_{\nu 1}, \cdots, \alpha_{\nu n},$$

and off-diagonal elements

$$\beta_{\nu 1}, \cdots, \beta_{\nu, n-1}.$$

Show that

$$\alpha_{\nu+1, 1}, \cdots, \alpha_{\nu+1, n}, \beta^2_{\nu+1, 1}, \cdots, \beta^2_{\nu+1, n-1},$$

can be obtained from

$$\alpha_{\nu, 1}, \cdots, \alpha_{\nu, n}, \beta^2_{\nu, 1}, \cdots, \beta_{\nu, n-1},$$

by rational processes (Ortega and Kaiser).

17. Let the Hermitian matrix A be partitioned in the form

$$\begin{pmatrix} \beta & b^H \\ b & B \end{pmatrix},$$

let u be a unit vector of $n - 1$ dimensions such that $u^H b \neq 0$. Let β' be the larger root of the matrix

$$\begin{pmatrix} \beta & b^H u \\ u^H b & u^H B u \end{pmatrix},$$

and let r be the normalized vector belonging to it:

$$r^H - (\bar{\rho}, \bar{\sigma}), \; |\rho|^2 \; | \; |\sigma|^2 = 1.$$

Then $\beta' > \beta$, and if

$$v^H = (\bar{\rho}, \bar{\sigma} \, u^H),$$

and V is any unitary matrix for which v is the first column, the matrix $V^H A V$ will have β' in the upper left. In particular, V can be an elementary Hermitian, and the process can be repeated until, in the limit, the off-diagonal elements in the first row and first column vanish. This Jacobi-like method isolates the roots one by one. Alternatively the smaller root could be taken at each step [Kaiser].

Bibliography

ABRAMOV, A. A. (1950): "Ob odnom sposobe uskorenija iteracionnyh processov," Dokl. Akad. Nauk SSSR **74,** 1051–1052.

——, AND NEUHAUS, M. (1958): "Bemerkungen über Eigenwertprobleme von Matrizen höherer Ordnung," Les mathématiques de l'ingénieur, Mém. Publ. Soc. Sci. Arts Lett. Hainaut Vol. hors série, pp. 176–179.

AFRIAT, S. N. (1951): "An iterative process for the numerical determination of characteristic values of certain matrices," Quart. J. Math. Oxford Ser. (2) **2,** 121–122.

——, (1954): "Composite matrices," Quart. J. Math. Oxford Ser. (2) **5,** 81–98.

——, (1957): "Orthogonal and oblique projectors and the characteristics of pairs of vector spaces," Proc. Cambridge Philos. Soc. **53,** 800–816.

——, (1959): "Analytic functions of finite dimensional linear transformations," Proc. Cambridge Philos. Soc. **55,** 60–61.

AITKEN, A. C. (1931): "Further numerical studies in algebraic equations and matrices," Proc. Roy. Soc. Edinburgh Sect. A **51,** 80–90.

——, (1932): "On the evaluation of determinants, the formation of their adjugates, and the practical solution of simultaneous linear equations," Proc. Edinburgh Math. Soc. (2) **3,** 207–219.

——, (1936–7a): "Studies in practical mathematics. I. The evaluation with application of a certain triple product matrix," Proc. Roy. Soc. Edinburgh Sect. A **57,** 172 181.

——, (1936–7b): "Studies in practical mathematics. II. The evaluation of the latent roots and latent vectors of a matrix," Proc. Roy. Soc. Edinburgh Sect. A **57,** 269–304.

——, (1950): "Studies in practical mathematics. V. On the iterative solution of a system of linear equations," Proc. Roy. Soc. Edinburgh Sect. A **63,** 52–60.

AKAIKE, HIROTOGU (1958): "On a computation method for eigenvalue problems and its application to statistical analysis," Ann. Inst. Statist. Math. **10,** 1–20.

ALBRECHT, J. (1961): "Bemerkungen zum Iterationsverfahren von Schulz zur Matrixinversion," Z. Angew. Math. Mech. **41,** 262–263.

ALLEN, D. N. DE G. (1954): *Relaxation Methods.* McGraw-Hill Publishing Company, Inc., New York, ix + 257 pp.

AMIR-MOEZ, ALI R. (1956): "Extreme properties of eigenvalues of a Hermitian transformation and singular values of a sum and product of linear transformations," Duke Math. J. **23,** 463–476.

——, and HORN, ALFRED (1958): "Singular values of a matrix," Amer. Math. Monthly **65,** 742–748.

205

ANDERSEN, EINAR (1947): "Solution of great systems of normal equations together with an investigation of Andrae's dot-figure. An arithmetical-technical investigation," Geodaet. Inst. Skr. 3, Raekke 11, 65 pp.

ANDREE, R. V. (1951): "Computation of the inverse of a matrix," Amer. Math. Monthly 58, 87–92.

ANSORGE, R. (1960): "Über die Konvergenz der Iterationsverfahren zur Auflösung linearer Gleichungssysteme im Falle einer singulären Koeffizientenmatrix," Z. Angew. Math. Mech. 40, 427.

APARO, ENZO (1957): "Sulle equazioni algebriche matriciali," Atti Accad. Naz. Lincei. Rend. Cl. Sci. Fis. Mat. Nat. (8) 22, 20–23.

ARMS, R. J., GATES, L. D., and ZONDEK, B. (1956): "A method of block iteration," J. Soc. Indust. Appl. Math. 4, 220–229.

ARNOLDI, W. E. (1951): "The principle of minimized iterations in the solution of the matrix eigenvalue problem," Quart. Appl. Math. 9, 17–29.

ARŽANYH, I. S. (1951): "Rasprostranenie metoda A. N. Krylova na polinomial'nye matricy," Dokl. Akad. Nauk. SSSR 81, 749–752.

ASCHER, MARCIA, and FORSYTHE, G. E. (1958): "SWAC experiments on the use of orthogonal polynomials for data fitting," J. Assoc. Comput. Mach. 5, 9–21.

ASPLUND, EDGAR (1959): "Inverses of matrices (a_{ij}) which satisfy $a_{ij} = 0$ for $j > i + p$," Math. Scand. 7, 57–60.

AZBELOV, N., and VINOGRAD, R. (1952): "Process posledovatel'nyh približeniĭ dlja otyskanija sobstvennyh čisel i sobstvennyh vektorov," Dokl. Akad. Nauk SSSR 83, 173–174.

BANACHIEWICZ, T. (1937a): "Zur Berechnung der Determinanten, wie auch der Inversen, und zur darauf basierten Auflösung der Systeme linearer Gleichungen," Acta Astr. 3, 41–72.

——, (1937b): "Calcul des déterminants par la méthode des cracoviens," Bull. Iternat. Acad. Polon. Sci. Lett. A, 109–120.

——, (1937c): "Sur la résolution numérique d'un système d'équations linéaires," Bull. Internat. Acad. Polon. Sci. Lett. A, 350–354.

——, (1942): "An outline of the cracovian algorithm of the method of least squares," Astronom J., 38–41.

BANDEMER, HANS (1956/57): "Berechnung der reellen Eigenwerte einer reellen Matrix mit dem Verfahren von Rutishauser," Wiss. Z. Martin-Luther-Univ. Halle-Wittenberg, Math.-Nat. Reihe 6, 807–814.

BARANKIN, E. W. (1945): "Bounds for characteristic roots of a matrix," Bull. Amer. Math. Soc. 51, 767–770.

BARGMANN, V., MONTGOMERY, D., and VON NEUMANN, J. (1946): *Solution of Linear Systems of High Order*, Bureau of Ordnance, Navy Department, Princeton Institute for Advanced Study, Contract NORD-9596.

BARTLETT, M. S. (1951): "An inverse matrix adjustment arising in discriminant analysis," Ann. Math. Statist. 22, 107–111.

BARTSCH, HELMUT (1953): "Ein Einschliessungssatz für die charakteristischen

Zahlen allgemeiner Matrizen-Eigenwertaufgaben," Arch. Math. **4**, 133–136.

——, (1954): "Abschätzung für die kleinste charakeristische Zahl einer positivdefinit hermitschen Matrix," Z. Angew. Math. Mech. **34**, 72–74.

BAUER, FRIEDRICH L. (1954): "Beiträge zur Entwicklung numerischer Verfahren für programmgesteuerte Rechenanlagen. I. Quadratisch konvergente Durchführung der Bernoulli-Jacobischen Methoden zur Nullstellenbestimmung von Polynomen," Akad. Wiss. München, Math.-Nat. Kl. S.-B, 275–303.

——, (1955a): *Beiträge zum Danilewski-Verfahren.* Internat. Koll. über Probl. d. Rechentechnik, Dresden, 133–139.

——, (1955b): "Der Newton-Prozess als quadratisch konvergente Abkürzung des allgemeinen linearen stationären Iterationsverfahrens 1. Ordnung (Wittmeyer-Prozess)," Z. Angew. Math. Mech. **35**, 469–470.

——, (1956a): "Beiträge zur Entwicklung numerischer Verfahren für programmgesteuerte Rechenanlagen. II. Direkte Faktorisierung eines Polynoms," Bayer. Akad. Wiss. Math.-Nat. Kl. S.-B., 163–203.

——, (1956b): "Iterationsverfahren der linearen Algebra vom Bernoullischen Konvergenztyp," Nachr. Techn. Fachber. **4**, 171–176.

——, (1956c): "Zur numerischen Behandlung von algebraischen Eigenwertproblemen höherer Ordnung," Z. Angew. Math. Mech. **36**, 244–245.

——, (1956d): "Das Verfahren der abgekürzten Iteration," Z. Angew. Math. Phys. **7**, 17.

——, (1957a): *Zusammenhänge zwischen einigen Iterationsverfahren der linearen Algebra,* Bericht über das Internat. Mathematiker-Kolloquium, Dresden, 22 bis 27 November 1955, 99–111.

——, (1957b): "Das Verfahren der Treppeniteration und verwandte Verfahren zur Lösung algebraischer Eigenwertprobleme," Z. Angew. Math. Phys. **8**, 214–235.

——, (1958): "On modern matrix iteration processes of Bernoulli and Graeffe type," J. Assoc. Comput. Mach. **5**, 246–257.

——, (1959): "Sequential reduction to tridiagonal form," J. Soc. Indust. Appl. Math. **7**, 107–113.

——, (1960a): *On the Definition of Condition Numbers and On Their Relation to Closed Methods for Solving Linear Systems,* Information Processing Proc. ICIP, Unesco, Paris, 109–110.

——, (1960b): "Wertevorrat eines Hauptabschnitts einer normalen Matrix," Jber. Deutsch. Math.-Verein. **63**, 180–182.

——, (1961): "A further generalization of the Kantorovic inequality," Numer. Math. **3**, 117–119.

——, (1962): "On the field of values subordinate to a norm," Numer. Math. **4**, 103–114.

——, and FIKE, C. T. (1960): "Norms and exclusion theorems," Numer. Math. **2**, 137–141.

208   BIBLIOGRAPHY

——, and HOUSEHOLDER, A. S. (1960a): "Moments and characteristic roots," Numer. Math. **2**, 42–53.

——, and ——, (1960b): "Some inequalities involving the Euclidean condition of a matrix," Numer. Math. **2**, 308–311.

——, and ——, (1961): "Absolute norms and characteristic roots," Numer. Math. **3**, 241–246.

——, and SAMELSON, KLAUS (1957): "Polynomkerne und Iterationsverfahren," Math. Z. **67**, 93–98.

——, STOER, J., and WITZGALL, C. (1961): "Absolute and monotonic norms," Numer. Math. **3**, 257–264.

BECKMAN, F. S. (1960): *The Solution of Linear Equations by the Conjugate Gradient Method. Mathematical Methods for Digital Computers.* John Wiley and Sons, Inc., New York, pp. 62–72.

BELLMAN, RICHARD (1953): *Stability Theory of Differential Equations,* McGraw-Hill Book Company, Inc., New York, xiii + 166 pp.

——, (1955): "Notes on matrix theory, IV (an equality due to Bergström)," Amer. Math. Monthly **62**, 172–173.

——, (1959a): "Kronecker products and the second method of Lyapunov," Math. Nachr. **20**, 17–19.

——, (1959b): "Representation theorems and inequalities for Hermitian matrices," Duke Math. J. **26**, 485–490.

——, (1960): *Introduction to Matrix Analysis.* McGraw-Hill Book Company, Inc., New York, Toronto, London, xx + 328 pp.

BENDIXSON, IVAR (1902): "Sur les racines d'une équation fondamentale," Acta Math. **25**, 359–365.

BENOIT, COMMANDANT (1924): "Sur la méthode de résolution des équations normales, etc. (Procédés du commandant Cholesky)," Bull. Géodésique **2**, 67–77.

DI BERARDINO, VINCENZO (1956): "Risoluzione dei sistemi di equazioni algebriche lineari per incrementi successivi delle incognite, Nuovo metodo di calcolo," Riv. catasto e ser. tecn. erariali **11**, 304–308.

——, (1957): "Il metodo di Hardy-Cross e la sua giustificazione mediante un nuovo procedimento di risoluzione dei sistemi di equazioni algebriche lineari," Ingegneria ferroviaria **12**, 821–831.

——, and GIRARDELLI, LAMBERTO (1957): "Nuovo metodo di risoluzione di particulari sistemi di equazioni algebriche lineari. Sistemo a catena," Riv. catasto e ser. tecn. erariali **12**, 46–50.

BEREZIN, F. A., and GEL'FAND, I. M. (1956): "Neskol'ko zamečanii k teorii sferičeskih funkciĭ simmetričeskih rimanovyh mnogoobrazijah," Trudy Moskov. Mat. Obšč. **5**, 311–351.

BEREZIN, I. S., and ŽIDKOV, N. P. (1959): *Metody vyčisleniĭ.* Gosudarstv. Izdat. Fiz.-Mat. Lit. Moscow, Vol. I, 464 pp.; Vol. 2, 620 pp.

BESSMERTNYH, G. A. (1959): "Ob odnovremennom otyskanii dvuh sobstvennyh čisel samosoprjažennago operatora," Dokl. Akad. Nauk SSSR

128, 1106–1109.

BIALY, H. (1960): "Iterative Lösung einer Matrizengleichung," Z. Angew. Math. Mech. **40,** 130–132.

BICKLEY, W. G., and McNAMEE, JOHN (1960): "Matrix and other direct methods for the solution of systems of linear difference equations," Philos. Trans. Roy. Soc. London Ser. A. **252,** 69–131.

BIRKHOFF, GARRETT, and MacLANE, SAUNDERS (1953): *A Survey of Modern Algebra* (revised edition). The Macmillan Company, New York, 472 pp.

——, and VARGA, RICHARD S. (1958): "Reactor criticality and nonnegative matrices," J. Soc. Indust. Appl. Math. **6,** 354–377.

——, and ——, (1959a): "Errata: Reactor criticality and nonnegative matrices," J. Soc. Indust. Appl. Math. **7,** 343.

——, and ——, (1959b): "Implicit alternating direction methods," Trans. Amer. Math. Soc. **92,** 13–24.

BIRMAN, M. Š. (1950): "Nekotorye ocenki dlja metoda naiskoreĭsego spuska," Uspehi Mat. Nauk **5,** 3(37) 152-155.

——, (1952): "O vyčislenii sobstvennyh čisel metodom naiskoreĭsego spuska," Zap. Leningrad. Gorn. Inst. **27,** (1), 209–216.

BJERHAMMAR, ARNE (1958): "A generalized matrix algebra," Kungl. Tekn. Högsk. Handl. Stockholm **124,** 32 pp.

BLACK, A. N., and SOUTHWELL, R. V. (1938): "Relaxation methods applied to engineering problems, II. Basic theory, with application to surveying and to electrical networks, and an extension to gyrostatic systems," Proc. Roy. Soc. London Ser. A. **164,** 447–467.

BLAIR, A., METROPOLIS, N., VON NEUMANN, J., TAUB, A. H., and TSINGOU, M. (1959): "A study of a numerical solution of a two-dimensional hydrodynamical problem," Math. Tables Aids Comput. **13,** 145–184.

BLANC, C., and LINIGER, W. (1956): "Erreurs de chute dans la résolution de systèmes algébriques linéaires," Comment. Math. Helv. **30,** 257–264.

BODEWIG, E. (1959): *Matrix Calculus* (second revised and enlarged edition). Interscience Publishers, Inc., New York; North Holland Publishing Company, Amsterdam, xi + 452 pp.

BONFERRONI, C. F. (1942): "Una disuguaglianza sui determinante e il teorea di Hadamard," Boll. Un. Mat. Ital. (II) **4,** 158–165.

BONNESEN, T., and FENCHEL, W. (1934): *Theorie der konvexen Körper.* Chelsea Publishing Company, 1948; Julius Springer, Berlin, vii + 164 pp.

BORSCH-SUPAN, W., and BOTTENBRUCH, H. (1958): "Eine Methode zur Eingrenzung sämtlicher Eigenwerte einer hermiteschen Matrix mit überwiegender Hauptdiagonale," Z. Angew Math. Mech. **38,** 13–71.

BOTTEMA, O. (1957): "A geometrical interpretation of the relaxation method," Quart. Appl. Math. **7,** 422–423.

——, (1958): "Zur Stabilitätsfrage bei Eigenwertproblemen," Neder. Akad. Wetensch. Proc. Ser. A. **61,** 501–504.

BOURGIN, D. G. (1939): "Positive determinants, "Amer. Math. Monthly

46, 225–226.

Bowie, O. L. (1947): "Least-square application to relaxation methods," J. Appl. Phys. **18,** 830–833.

Bowker, Albert H. (1947): "On the norm of a matrix," Ann. Math. Statist. **18,** 285–288.

Brauer, Alfred (1946): "Limits for the characteristic roots of a matrix," Duke Math. J. **13,** 387–395.

——, (1947): "Limits for the characteristic roots of a matrix, II," Duke Math. J. **14,** 21–26.

——, (1948): "Limits for the characteristic roots of a matrix, III," Duke Math. J. **15,** 871–877.

——, (1953): "Bounds for characteristic roots of matrices," Nat. Bur. Standards Appl. Math. Ser. **29,** 101–106.

——, (1957): "A method for the computation of the greatest root of a positive matrix," J. Soc. Indust. Appl. Math. **5,** 250–253.

Brenner, J. L. (1954a): "A bound for a determinant with dominant main diagonal," Proc. Amer. Math. Soc. **5,** 631–634.

——, (1954b): "Bounds for determinants," Proc. Nat. Acad. Sci. U.S.A. **40,** 452–454.

——, (1957): "Bounds for determinants," Proc. Amer. Math. Soc. **8,** 532–534.

——, (1959): "Relations among the minors of a matrix with dominant principal diagonal," Duke Math. J. **26,** 563–567.

——, (1961): "Expanded matrices from matrices with complex elements," SIAM Rev. **3,** 165–166.

Brickman, Louis (1961): "On the field of values of a matrix," Proc. Amer. Math. Soc. **12,** 61–66.

Bromwich, J. T. I'A (1906): "On the roots of the characteristic equation of a linear substitution," Acta Math. **30,** 297–304.

Browne, Edward Tankard (1918): "The characteristic equation of a matrix," Bull. Amer. Math. Soc. **34,** 363–368.

——, (1930a): "The characteristic roots of a matrix," Bull. Amer. Math. Soc. **36,** 705–710.

——, (1930b): "On the separation property of the roots of the secular equation," Amer. J. Math. **52,** 843–850.

——, (1958): *Introduction to the Theory of Determinants and Matrices.* The University of North Carolina Press, Chapel Hill, xi + 270 pp.

de Bruijn, N. G. (1956): "Inequalities concerning minors and eigenvalues," Nieuw Arch. Wisk. (3) **4,** 18–35.

——, and van Dantzig, D. (1952): "Inequalities concerning determinants and systems of linear equations," Neder. Akad. Wetensch. Proc. Ser. A. **50,** 315–321.

——, and Szekeres, G. (1955): "On some exponential and polar representations of matrices," Nieuw Arch. Wisk. (3) **3,** 20–32.

BRYAN, JOSEPH G. (1950): *A Method for the Exact Determination of the Characteristic Equation and Latent Vectors of a Matrix with Applications to the Discriminant Function for More Than Two Groups*, Thesis, Harvard University.

BÜCKNER, HANS (1952): *Die praktische Behandlung von Integralgleichungen*. Springer Verlag, Berlin-Göttingen-Heidelberg, vi + 126 pp.

——, (1949/50): "Über ein unbeschränkt anwendbares Iterationsverfahren für Systeme linearer Gleichungen," Arch. Math. **2**, 5–9.

BURDINA, V. I. (1958): "K odnomu metodu rešenija sistem lineĭnyh algebraičeskyh uravneniĭ," Dokl. Akad. Nauk SSSR **120**, 235–238.

CAPRIOLI, LUIGI (1953): "Sulla risoluzione dei sistemi di equazioni lineari con il metodo di Cimmino," Boll. Un. Mat. Ital. (3) **8**, 260–265.

CARRÉ, B. A. (1961): "The determination of the optimum accelerating factor for successive over-relaxation," Comput. J. **4**, 73–78.

CASSINA, U. (1948): "Sul numero delle operazioni elementari necessarie per la risoluzione dei sistemi di equazioni lineari," Boll. Un. Mat. Ital. (3) **3**, 142–147.

CASSINIS, GINO (1944): "I metodi di H. Boltz per la risoluzione dei sistemi di equazioni lineari e il loro impiego nella compenzazione delle triangolazione," Riv. catasto e dei Servizi tecn. erariali No. 1.

——, (1946): "Risoluzione dei sistemi de equazioni algebrichi lineari," Rend. Sem. Mat. Fis. Milano **17**, 62–78.

CASTALDO, D. (1957): "Matrici cicliche e relativi determinanti," Rev. Math. Pures Appl. (2) **7**, 29–40.

CAUCHY, AUGUSTIN L. (1812): "Mémoire sur les fonctions qui ne peuvent obtenir que deux valeurs égales et de signes contraires par suite des transpositions opérées entre les variables qu'elles renferment," J. de l'Ecole Polytech **10**, Cah. 17, 29–112; Oeuvres Complètes (2) **1**, 91–169.

——, (1821): "Cours d'analyse de l'École Royale Polytechnique," Oeuvres Complètes (2) **3**.

——, (1829): "Sur l'équation à l'aide de laquelle on détermine les inégalités séculaires des mouvements des planètes," Ex. de Math. **4**, 140–160; Oeuvres Complètes (2) **9**, 174–195.

——, (1847): "Méthode générale pour la résolution des systèmes d'équations simultanées," C. R. Acad. Sci. Paris **25**, 536–8; Oeuvres Complètes (1) **10**, 399–402.

CAUSEY, ROBERT L. (1958a): "On some error bounds of Givens," J. Assoc. Comput. Mach. **5**, 127–131.

——, (1958b): "Computing eigenvalues of non-Hermitian matrices by methodes of Jacobi type," J. Soc. Indust. Appl. Math. **6**, 172–181.

——, and GREGORY, R. T. (1961): "On Lanczos' algorithm for tridiagonalizing matrices," SIAM Rev. **3**, 322–328.

——, and HENRICI, PETER (1960): "Convergence of approximate eigenvectors in Jacobi methods," Numer. Math. **2**, 67–78.

CAYLEY, ARTHUR (1857): "A memoir on the theory of matrices," Philos. Trans. Roy. Soc. London **148**, 17–37; Coll. Works **2**, 475–496.

ČEREPKOV, F. S. (1936): "O rešenii sistem lineĭnyh uravneniĭ metodom iteraciĭ," Akad. Nauk SSSR i Moskov. Mat. Obšč. Mat. Sb. **1**, 953–960.

CESARI, LAMBERTO (1937a): "Sulla risoluzione dei sistemi di equazioni lineari per approssimazioni successive," Atti Accad. Naz. Lincei Rend. Cl. Sci. Fis. Mat. Nat. (6), **25**, 422–428.

———, (1937b): "Sulla risoluzione dei sistemi di equazioni lineari per approssimazioni successive," Ricerca Sci. II, **8**, I, 512–522.

CHARTRES, B. A. (1960): *Computing Extreme Eigenvalues of Real Symmetric Matrices.* Basser Computing Department, School of Physics, The University of Sydney, Technical Report No. 9.

CHEN, T. C., and WILLOUGHBY, R. A. (1958): "A note on the computation of eigenvalues and vectors of Hermitian matrices," IBM J. Res. Develop. **2**, 169–170.

CHIÓ, F. (1853): *Mémoire sur les fonctions connues sous le nom de résultants ou de déterminants.* Turin, 32 pp.

CIMMINO, GIANFRANCO (1938): "Calcolo approssimato per le soluzioni dei sistemi di equazioni lineari," Ricerca Sci. II, **9**, I, 326–333.

CLASEN, B. J. (1888): "Sur une nouvelle méthode de résolution des équations linéaires et sur l'application de cette méthode au calcul des déterminants," Ann. Soc. Sci. Bruxelles (2) **12**, 251–81; Mathesis **9** (Suppl. 2), 1–31.

CLEMENT, PAUL A. (1959): "A class of triple-diagonal matrices for test purposes," SIAM Rev. **1**, 50–52.

COLLAR, A. R. (1948): "Some notes on Jahn's method for the improvement of approximate latent roots and vectors of a square matrix," Quart. J. Mech. Appl. Math. **1**, 145–148.

———, (1951): "On the reciprocal of a segment of a generalized Hilbert matrix," Proc. Cambridge Philos. Soc. **47**, 11–17.

COLLATZ, LOTHAR (1939): "Genäherte Berechnung von Eigenwerte," Z. Angew. Math. Mech. **19**, 224–249.

———, (1942a): "Fehlerabschätzung für das Iterationsverfahren zur Auflosüng linearer Gleichungssysteme," Z. Angew. Math. Mech. **22**, 357–361.

———, (1942b): "Einschliessungssatz für die characteristischen Zahlen von Matrizen," Math. Z. **48**, 221–226.

———, (1950a): "Über die Konvergenzkriterien bei Iterationsverfahren fur lineare Gleichungssysteme," Math. Z. **53**, 149–161.

———, (1950b): "Zur Herleitung von Konvergenzkriterien für Iterationsverfahren bei linearen Gleichungssystemen," Z. Angew. Math. Mech. **30**, 278–280.

———, (1952a): "Aufgaben monotoner Art," Arch. Math. **3**, 366–376.

———, (1952b): "Einschliessungssätze bei Iteration und Relaxation," Z. Angew. Math. Mech. **32**, 76–84.

COOPER, J. L. B. (1948): "The solution of natural frequency equations by relaxation methods," Quart. Appl. Math. **6**, 179–182.

COUFFIGNAL, LOUIS (1944): "La résolution numérique des systèmes d'équations linéaires. Premier Mémoire: L'opération fondamentale de réduction d'un tableau," Rev. Sci. **82**, 67–78.

———, (1951): "Sur la résolution numérique des systèmes d'équations linéaires. Deuxieme partie: Précision de la solution," Rev. Sci. **89**, 3-10.

COURANT, RICHARD, and HILBERT, DAVID (1953): *Methods of Mathematical Physics*. Interscience Publishers, New York, Vol. I, xv + 561 pp.

CRAIG, EDWARD J. (1955): "The N-step iteration procedures," J. Math. Phys. **34**, 65–73.

CRANDALL, S. H. (1951): "Iterative procedures related to relaxation methods for eigenvalue problems," Proc. Roy. Soc. London Ser. A **207**, 416–423.

———, (1951/52): "On a relaxation method for eigenvalue problems," J. Math. Phys. **30**, 140–145.

CROUT, PRESCOTT D. (1941): "A short method for evaluating determinants and solving systems of linear equations with real or complex coefficients," Trans. Amer. Inst. Elect. Engrs. **60**, 1235–1240.

CURTISS, J. H. (1954): "A generalization of the method of conjugate gradients for solving systems of linear algebraic equations," Math. Tables Aids Comput. **8**, 189–193.

CUTHILL, ELIZABETH H., and VARGA, RICHARD S. (1959): "A method of normalized block iteration," J. Assoc. Comput. Mach. **6**, 236–244.

DANILEVSKII, A. (1937): "O čislennom rešenii vekovogo uravnenija," Mat. Sb. **2** (44), 169–171.

DAVIS, P. J., HAYNSWORTH, E. V., and MARCUS, M. (1961): "Bound for the P-condition number of matrices with positive roots," J. Res. Nat. Bur. Standards **65B**, 13–14.

DEBREU, GERARD, and HERSTEIN, I. N. (1953): "Nonnegative square matrices," Econometrica **21**, 597–607.

DEDEKIND, R. (1901): "Gauss in seiner Vorlesung über die Methode der kleinsten Quadrate," Gesammelte Math. Werke **2** (1931), 293–306.

DENT, BENJAMIN A., and NEWHOUSE, ALBERT (1959): "Polynomials orthogonal over discrete domains," SIAM Rev. **1**, 55–59.

DERWIDUÉ, L. (1954): "La méthode de L. Couffignal pour la résolution numérique des systèmes des équations linéaires," Mathesis **63**, 9–12.

———, (1955): "Une méthode mécanique de calcul des vecteurs propres d'une matrice quelconque," Bull. Soc. Roy. Sci. Liège 149–171.

———, (1957): *Introduction à l'algèbre supérieure et au calcul numérique algébrique*. Masson et Cie, Éditeurs, Paris, 432 pp.

DESPLANQUES, J. (1887): "Théorème d'algébre," J. Math. Speciales (3) **1**, 12–13.

DIMSDALE, BERNARD (1958): "The nonconvergence of a characteristic root method" J. Soc. Indust. Appl. Math. **6,** 23–25.

DMITRIEV, N., and DYNKIN, E. (1945): "On the characteristic numbers of a stochastic matrix," Dokl. Akad. Nauk SSSR **49,** 159–162.

——, and ——, (1946): "Harakterističeskie korni stohastičeskih matric," Izv. Akad. Nauk SSSR Ser. Mat. **10,** 167–184.

DODGSON, C. L. (1866): "Condensation of determinants," Proc. Roy. Soc. London **15,** 150–155.

DRAZIN, M. P. (1951): "On diagonable matrices and normal matrices," Quart. J. Math. (2) **2,** 189–198.

——, DUNGEY, J. W., and GRUENBERG, K. W. (1951): "Some theorems on commutative matrices," J. London Math. Soc. **26,** 221–228.

DUNCAN, W. J. (1944): "Some devices for the solution of large sets of simultaneous linear equations (with an appendix on the reciprocation of partitioned matrices)," Philos. Mag. (7) **35,** 660–670.

——, and COLLAR, A. R. (1934): "A method for the solution of oscillation problems by matrices," Philos. Mag. (7) **17,** 865–909.

——, and ——, (1935): "Matrices applied to the motions of damped systems," Philos. Mag. (7) **19,** 197–219.

VAN DEN DUNGEN, F. H. (1928): "Über die Biegungsschwingungen einer Welle," Z. Angew. Math. Mech. **8,** 225–231.

DURAND, ÉMILE (1961a): "Recherche des valeurs propres d'une matrice à éléments complexes, par une méthode de triangularisation," C. R. Acad. Sci. Paris **252,** 1267–1269.

——, (1961b): *Solutions numériques des équations algébriques. Tome II. Systèmes de plusieurs équations. Valeurs propres des matrices.* Masson et Cie, Éditeurs, Paris, viii + 445 pp.

DWYER, PAUL S. (1941a): "The solution of simultaneous equations," Psychometrika **6,** 101–129.

——, (1941b): "The evaluation of linear forms," Psychometrika **6,** 355–365.

——, (1941c): "The Doolittle technique," Ann. Math. Statist. **12,** 449–458.

——, (1944): "A matrix presentation of least squares and correlation theory with matrix justification of improved methods of solution," Ann. Math. Statist. **15,** 82–89.

——, (1945): "The square root method and its use in correlation and regression," J. Amer. Statist. Assoc. **40,** 493–503.

——, (1951): *Linear Computations.* John Wiley and Sons, Inc., New York, xi + 344 pp.

——, (1953): "Errors of matrix computations," Nat. Bur. Standards Appl. Math. Ser. **29,** 49–58.

——, and WAUGH, FREDERICK V. (1953): "On errors in matrix inversion," J. Amer. Statist. Assoc. **48,** 289–319.

EASTERFIELD, T. E. (1957): "Matrix norms and vector measures," Duke Math. J. **24,** 663–669.

ECKART, CARL, and YOUNG, GALE (1939): "A principal axis transformation for non-Hermitian matrices," Bull. Amer. Math. Soc. **45**, 118–121.

EGERVÁRY, EUGEN (1953): "On a property of the projector matrices and its application to the canonical representation of matrix functions," Acta Sci. Math. Szeged. **15**, 1–6.

———, (1954): "On a lemma of Stieltjes on matrices," Acta Sci. Math. Szeged. **15**, 99–103.

———, (1955): "Über die Faktorisation von Matrizen und ihre Anwendung auf die Lösung von linearen Gleichungssystemen," Z. Angew Math. Mech. **35**, 111–118.

———, (1959): "Über eine konstruktive Methode zur Reduktion einer Matrix auf die Jordansche Normalform," Acta Math. Acad. Sci. Hungar. **10**, 31–54.

———, (1960): "On rank-diminishing operations and their applications to the solution of linear equations," Z. Angew. Math. Phys. **11**, 376–386.

EGGLESTON, H. G. (1958): *Convexity*, Cambridge tracts in Mathematics and Mathematical Physics. Cambridge University Press, viii + 136 pp.

EISEMANN, KURT (1957): "Removal of ill-conditioning for matrices," Quart. Appl. Math. **15**, 225–230.

ENGELI, M. (1959): "Overrelaxation and related methods," Mitt. Inst. Angew. Math. Zurich **8**, 79–101.

———, GINSBURG, T., RUTISHAUSER, H., and STIEFEL, E. (1959): *Refined Iterative Methods for Computations of the Solution and the Eigenvalues of Self-Adjoint Boundary Value Problems*. Birkhäuser Verlag, Basel/ Stuttgart, 107 pp.

EPSTEIN, MARVIN, and FLANDERS, HARLEY (1955): "On the reduction of a matrix to diagonal form," Amer. Math. Monthly **62**, 168–170.

ERDELYI, A. (1938): "Eigenfrequenz inhomogener Saiten," Z. Angew. Math. Mech. **18**, 177–185.

ERDÖS, PAUL (1959): "Elementary divisors of normal matrices," IBM J. Res. Develop. **3**, 197.

ERŠOV, A. P. (1955): "Ob odnom metode obraščenija matric," Dokl. Akad. Nauk SSSR **100**, 209–211.

FADDEEV, D. K. (1958a): "K voprosu o verhneĭ relaksacii pri rešenii sistem lineĭnyh uravneniĭ," Izv. Vysš. Uč. Zaved. Matematika (**5**), 122–125.

———, (1958b): "O nekotoryh posledovatel'nostjah polinomov poleznyh dlja postroenija iteracionnyh metodov rešenija sistem lineĭnyh algebraičeskih uravneniĭ," Vestnik Leningrad. Univ. **13**, 155–159.

———, and FADDEEVA, V. N. (1960): *Vyčislitel'nye metody lineĭnoĭ algebry*, Gosudarstv. Izdat. Fiz.-Mat. Lit., Moscow, 656 pp.

———, and SOMINSKIĬ, I. S. (1949): *Sbornik zadač po vysšeĭ algebra*, 2nd ed. Gosudarstv. Izdat. Tehn.-Teoret. Lit., Moscow-Leningrad, 308 pp.

FADDEEVA, V. N. (1950): *Vyčislitel'nye metody lineĭnoĭ algebry*. Gosudarstv. Izdat. Tehn.-Teoret. Lit., Moscow-Leningrad, 240 pp.

———, (1959): *Computation Methods of Linear Algebra* (authorized translation from the Russian by Curtis D. Benster). Dover Publishing Company, New York, 252 pp.

FAGE, M. K. (1946): "Obobščenie neravenstva adamara ob opredeliteljah," Dokl. Akad. Nauk SSSR **54**, 765–768.

——, (1951): "O simmetrizuemyh matricah," Uspehi Mat. Nauk **6**, 3(43) 153–156.

FALK, SIGURD (1954a): "Neue Verfahren zur direkten Lösung des allgemeinen Matrixeneigenwertproblemes," Z. Angew. Math. Mech. **34**, 289–291.

——, (1954b): "Neue Verfahren zur direkten Lösung algebraischer Eigenwertprobleme," Abh. Braunschweigischen Wiss. Ges. **6**, 166–194.

——, (1956): "Das Ersatzwertverfahren als Hilfsmittel bei der iterativen Bestimmung von Matrizeneigenwerten," Abh. Braunschweigischen Wiss. Ges. **8**, 99–110.

——, (1961): "Einschliessungssätze für Eigenwerte und vektoren normaler Matrizenpaare," Wiss. Z. Techn. Hochsch. Dresden **10**, 1033–1039.

——, and LANGEMEYER. P. (1960): "Das Jacobische Rotationsverfahren für reellsymmetrische Matrizenpaare, Teil 2," Elektr. Daten **8**, (Dec.) 35–43.

FAN, KY (1949): "On a theorem of Weyl concerning eigenvalues of linear transformations, I," Proc. Nat. Acad. Sci., U.S.A. **35**, 653–655.

——, (1950): "On a theorem of Weyl concerning eigenvalues of linear transformations, II," Proc. Nat. Acad. Sci. U.S.A. **36**, 31–35.

——, (1951): "Maximum properties and inequalities for the eigenvalues of completely continuous operators," Proc. Nat. Acad. Sci. U.S.A. **37**, 760–766.

——, (1954): "Inequalities for eigenvalues of Hermitian matrices," Nat. Bur. Standards Appl. Math. Ser. **39**, 131–139.

——, (1955a): "A comparison theorem for eigenvalues of normal matrices," Pacific J. Math. **5**, 911–913.

——, (1955b): "Some inequalities concerning positive-definite matrices," Proc. Cambridge Philos. Soc. **51**, 414–421.

——, (1958a): "Topological proofs for certain theorems on matrices with nonnegative elements," Monatsh. Math. **62**, 219–237.

——, (1958b): "Note on circular disks containing the eigenvalues of a matrix," Duke Math. J. **25**, 441–445.

——, (1960): "Note on M-matrices," Quart. J. Math. Oxford Ser. (2) **11**, 43–49.

——, and HOFFMAN, A. J. (1954): "Lower bounds for the rank and location of the eigenvalues of a matrix," Nat. Bur. Standards Appl. Math. Ser. **39**, 117–130.

——, and ——, (1955): "Some metric inequalities in the space of matrices," Proc. Amer. Math. Soc. **6**, 111–116.

——, and HOUSEHOLDER, A. S. (1959): "A note concerning positive matrices and M-matrices," Monatsh. Math. **63**, 265–270.

——, and PALL, GORDON (1957): "Imbedding conditions for Hermitian and normal matrices," Canad. J. Math. **9**, 298–304.

FARNELL, A. B. (1944): "Limits for the characteristic roots of a matrix," Bull. Amer. Math. Soc. **50**, 788–794.

——, (1960): "Characteristic root bounds of Gerschgorin type," SIAM Rev. **2**, 36–38.

FELLER, WILLIAM, and FORSYTHE, GEORGE E. (1951): "New matrix transformations for obtaining characteristic vectors," Quart. Appl. Math. **8**, 325–331.

FETTIS, HENRY E. (1950): "A method for obtaining the characteristic equation of a matrix and computing the associated modal columns," Quart. Appl. Math. **8**, 206–212.

FIEDLER, MIROSLAV (1960a): "Poznámka o positivné definitních maticích," Časopis Pěst. Mat. **85**, 75–77.

——, (1960b): *Some Estimates of Spectra of Matrices. Symposium on the Numerical Treatment of Ordinary Differential Equations, Integral and Integro-Differential Equations (Rome).* Birkhäuser, Basel, pp. 33–36.

——, (1961): "Über eine Ungleichung für positiv definite Matrizen," Math. Nachr. **23**, 197–199.

——, and PTÁK, VLASTIMIL (1956): "Über die Konvergenz des verallgemeinerten Seidelschen Verfahrens zur Lösung von Systemen linearer Gleichungen," Math. Nachr. **15**, 31–38.

——, and ——, (1960a): "Some inequalities for the spectrum of a matrix," Mat.-Fyz. Časopis Sloven. Akad. Vied **10**, 148–166.

——, and ——, (1960b): "O jedné iterační metodě diagonalisace symetrických matic," Časopis Pěst. Mat. **85**, 18–36.

FIKE, C. T. (1959): "Note on the practical computation of proper values," J. Assoc. Comput. Mach. **6**, 360–362.

FISCHER, E. (1908): "Ueber den Hadamardschen Determinantensatz," Arch. Math. Phys. (3) **13**, 32–40.

FISHER, MICHAEL E., and FULLER, A. T. (1958): "On the stabilization of matrices and the convergence of linear iterative processes," Proc. Cambridge Philos. Soc. **54**, 417–425.

FLANDERS, DONALD A., and SHORTLEY, GEORGE (1950): "Numerical determination of fundamental modes," J. Appl. Phys. **21**, 1326–1332.

FLOMENHOFT, H. I. (1950): "A method for determining mode shapes and frequencies above the fundamental by matrix iteration," J. Appl. Mech. **17**, 249–256.

FORSYTHE, GEORGE E. (1952): "Alternative derivations of Fox's escalator formulae for latent roots," Quart. J. Mech. Appl. Math. **5**, 191–195.

——, (1953a): "Tentative classification of methods and bibliography on solving systems of linear equations," Nat. Bur. Standards Appl. Math. Ser. **29**, 1–28.

——, (1953b): "Solving linear algebraic equations can be interesting," Bull. Amer. Math. Soc. **59**, 299–329.

———, (1957): "Generation and use of orthogonal polynomials for data-fitting with a digital computer," J. Soc. Indust. Appl. Math. **5,** 74–88.

———, and HENRICI, PETER (1960): "The cyclic Jacobi method for computing the principal values of a complex matrix," Trans. Amer. Math. Soc. **94,** 1–23.

———, and MOTZKIN, THEODORE S. (1952): "An extension of Gauss' transformation for improving the condition of systems of linear equations," Math. Tables Aids Comput. **6,** 9–17.

———, and ORTEGA, J. (1960): *Attempts to Determine the Optimum Factor for Successive Over-Relaxation.* Information Processing Proc. ICIP, Unesco, Paris, p. 110.

———, and STRAUS, E. G. (1955): "On best conditioned matrices," Proc. Amer. Math. Soc. **6,** 340–345.

———, and STRAUS, LOUISE W. (1955): "The Souriau-Frame characteristic equation algorithm on a digital computer," J. Math. Phys. **34,** 152–156.

———, and WASOW, WOLFGANG, (1960): *Finite-Difference Methods for Partial Differential Equations.* John Wiley and Sons, New York, x + 444 pp.

FOX, L. (1948): "A short account of relaxation methods," Quart. J. Mech. Appl. Math. **1,** 253–280.

———, (1950): "Practical methods for the solution of linear equations and the inversion of matrices," J. Roy. Statist. Soc. Ser. B **12,** 120–136.

———, (1952): "Escalator methods for latent roots," Quart. J. Mech. Appl. Math. **5,** 178–190.

———, and HAYES, J. C. (1951): "More practical methods for the inversion of matrices," J. Roy. Statist. Soc. Ser. B **13,** 83–91.

———, HUSKEY, H. D., and WILKINSON, J. H. (1948): "Notes on the solution of algebraic linear simultaneous equations," Quart. J. Mech. Appl. Math. **1,** 149–173.

FRAME, J. S. (1949): "A simple recursion formula for inverting a matrix" (abstract), Bull. Amer. Math. Soc. **55,** 1045.

FRANCIS, J., and STRACHEY, C. (1960): *Elimination Methods in Matrix Eigenvalue Problems.* Information Processing Proc. ICIP, Unesco, Paris, p. 112.

FRANCIS, J. F. G. (1961, 1962): "The QR transformation. A unitary analogue to the LR transformation," Comput. J. **4,** 265–271.

FRANK, WERNER L. (1958): "Computing eigenvalues of complex matrices by determinant evaluation and by methods of Danilewski and Wielandt," J. Soc. Indust. Appl. Math. **6,** 378–392.

———, (1960): "Solution of linear systems by Richardson's method," J. Assoc. Comput. Mach. **7,** 274–286.

FRANKEL, STANLEY P. (1950): "Convergence rates of iterative treatments of partial differential equations," Math. Tables Aids Comput. **4,** 65–75.

FRANKLIN, J. N. (1957): *Computation of Eigenvalues by the Method of Iteration.* CIT Computation Center, Technical Report No. 111.

FRAZER, R. A. (1947): "Note on the Morris escalator process for the solution of linear simultaneous equations," Philos. Mag. **38**, 287–289.

——, DUNCAN, W. J., and COLLAR, A. R. (1946): *Elementary Matrices and Some Applications to Dynamics and Differential Equations.* The Macmillan Company, New York, xvi + 416 pp.

FREEMAN, G. F. (1943): "On the iterative solution of linear simultaneous equations," Philos. Mag. (7) **34**, 409–416.

FROBENIUS, G. (1896): "Über die vertauschbaren Matrizen," S.-B. Deutsch. Akad. Wiss. Berlin Math.-Nat. Kl. (1896) 601–614.

——, (1908): "Über Matrizen aus positiven Elementen," S.-B. Deutsch. Akad. Wiss. Berlin (1908) 471–476.

——, (1909): "Über Matrizen aus positiven Elementen, II," S.-B. Deutsch. Akad. Wiss. Berlin (1909) 514–518.

——, (1912): "Über Matrizen aus nicht negative Elementen," S.-B. Deutsch. Akad. Wiss. Berlin (1912) 456–477.

FRÖBERG, CARL-ERIK (1958): "Diagonalization of Hermitian matrices," Math. Tables Aids Comput. **12**, 219–220.

FURTWÄNGLER, P. (1936): "Über einen Determinantensatz," S.-B. Akad. Wiss. Wien, Math.-Nat. Kl. IIa **145**, 527–528.

GANTMACHER, F. R. (1959a): *The Theory of Matrices.* Translated from the Russian, with further revisions by the author. Chelsea Publishing Company, New York, Vol. 1, x + 374 pp.; Vol. 2, x + 277 pp.

——, (1959b): *Applications of the Theory of Matrices.* Translated from the Russian and revised by J. L. Brenner. Interscience Publishers, Inc., New York, London, 319 pp.

GANTMAHER, F. R. (1954): *Teorija matric.* Gosudarstv. Izdat. Tehn.-Teoret. Lit., Moscow-Leningrad, 491 pp.

——, and KREIN, M. G. (1950): *Oscilljacionnye matricy i jadra i malye kolebanija mehaničeskih sistem.* Gosudarstv. Izdat. Tehn.-Teoret. Lit., Moscow-Leningrad, 359 pp.

GANTMAKHER, F. R., and KREIN, M. G. (1935): "Sur les matrices oscillatoires et complètement non négatives," Compositio Math. **4**, 445–476.

DE LA GARZA, A. (1951): *An Iterative Method for Solving Systems of Linear Equations.* Oak Ridge Gaseous Diffusion Plant, Report K-731, Oak Ridge, Tennessee.

——, (1953): "Error bounds on approximate solutions to systems of linear algebraic equations," Math. Tables Aids Comput. **7**, 81–84.

GASTINEL, NOËL (1958): "Procédé iteratif pour la résolution numérique d'un système d'équations linéaires," C. R. Acad. Sci. Paris **246**, 2571–2574.

——, (1959): "Conditionnement d'un système d'équations linéaires," C. R. Acad. Sci. Paris **248**, 2707–2709.

——, (1960a): "Utilisation de matrices vérifiant une équation de degré 2 pour la transmutation de matrices," C. R. Acad. Sci. Paris **250**, 1960–1961.

——, (1960b): *Matrices du second degré et normes générales en analyse numéri-.*

que linéaire. Thesis, Université de Grenoble, 137 pp.

GAUSS, C. F. (1826): "Supplementum theoriae combinationis observationum erroribus minimis obnoxiae," Werke **4**, 55–93.

GAUTSCHI, WERNER (1953): "The asymptotic behaviour of powers of matrices," Duke Math. J. **20**, 127–140; 375–379.

——, (1954a): "Bounds of matrices with regard to an Hermitian metric," Compositio Math. **12**, 1–16.

——, (1954b): On Norms of Matrices and Some Relations between Norms and Eigenvalues. Birkhäuser Verlag, Basel, 39 pp.

GAVURIN, M. K. (1950): "Primenenie polinomov nailučšego približenija dlja ulučšenija shodimosti iterativnyh processov," Uspehi Mat. Nauk **5**, 3(37) 156–160.

GERMANSKY, BORIS (1936): "Zur angenäherten Auflösung linearer Gleichungssysteme mittels Iteration," Z. Angew. Math. Mech. **16**, 57–58.

GERSCHGORIN, S. (1931): Über die Abgrenzung der Eigenwerte einer Matrix, Izv. Akad. Nauk SSSR Otd. Mat. Estest. 749–754.

GIVENS, WALLACE (1952): "Fields of values of a matrix," Proc. Amer. Math. Soc. **3**, 206–209.

——, (1953): "A method of computing eigenvalues and eigenvectors suggested by classical results on symmetric matrices," Nat. Bur. Standards Appl. Math. **29**, 117–122.

——, (1954): Numerical Computation of the Characteristic Values of a Real Symmetric Matrix. Oak Ridge National Laboratory, Report ORNL-1574.

——, (1957): "The characteristic value-vector problem," J. Assoc. Comput. Mach. **4**, 298–307.

——, (1958): "Computation of plane unitary rotations transforming a general matrix to triangular form," J. Soc. Indust. Appl. Math. **6**, 26–50.

——, (1959): The Linear Equations Problem. Applied Mathematics and Statistics Laboratories, Stanford University, Technical Report No. 3.

——, (1961): Elementary Divisors and Some Properties of the Lyapunov Mapping $X \to AX + XA^*$. Argonne National Laboratory, ANL 6456.

GLIDMAN, S. (1958): "Some theorems concerning determinants," Bull. Acad. Polon. Sci. Sér. Sci. Math. Astr. Phys. **6**, 275–280.

——, (1959): "Some theorems on the Gauss algorithm and the matrix orthogonalization process," Bull. Acad. Polon. Sci. Sér. Sci. Math. Astr. Phys. **7**, 373–379.

GLODEN, A. (1960): "Clasen's method for the solution of a system of linear equations," Scripta Math. **25**, 243–246.

GOLDSTINE, HERMAN H., and HORWITZ, L. P. (1959): "A procedure for the diagonalization of normal matrices," J. Assoc. Comput. Mach. **6**, 176–195.

——, MURRAY, F. J., and VON NEUMANN, JOHN (1959): "The Jacobi method for real symmetric matrices," J. Assoc. Comput. Mach. **6**, 59–96.

——, and VON NEUMANN, JOHN (1951): "Numerical inverting of matrices of high order, II," Proc. Amer. Math. Soc. **2**, 188–202.

GOLUB, GENE H., and VARGA, RICHARD S. (1961): "Chebyshev semi-iterative methods, successive over-relaxation iterative methods, and second order Richardson iterative methods," Numer. Math. **3**, 147–168.

GREENSTADT, JOHN (1955): "A method for finding roots of arbitrary matrices," Math. Tables Aids Comput. **9**, 47–52.

——, (1960): *The Determination of the Characteristic Roots of a Matrix by the Jacobi Method. Mathematical Methods for Digital Computers.* John Wiley and Sons, Inc., New York, pp. 84–91.

GREGORY, ROBERT T. (1958): "Results using Lanczos' method for finding eigenvalues of arbitrary matrices," J. Soc. Indust. Appl. Math. **6**, 182–188.

GREUB, WERNER, and RHEINBOLDT, WERNER (1959): "On a generalization of an inequality of L. V. Kantorovich," Proc. Amer. Math. Soc. **10**, 407–415.

GREVILLE, T. N. E. (1959): "The pseudoinverse of a rectangular or singular matrix and its application to the solution of systems of linear equations," SIAM Rev. **1**, 38–43.

——, (1960): "Some applications of the pseudoinverse of a matrix," SIAM Rev. **2**, 15–22.

GROSSMAN, D. P. (1950): "K zadače čislennogo rešenija sistem sovmestnyh lineĭnyh algebraičeskih uravneniĭ," Uspehi Mat. Nauk **5**, 3(37) 87–103.

——, (1956): "O rešenii pervoĭ kraevoĭ zadači dlja elliptičeskih uravneniĭ metodom setok," Dokl. Akad. Nauk **106**, 770–772.

GUDERLEY, KARL G. (1958): "On nonlinear eigenvalue problems for matrices," J. Soc. Indust. Appl. Math. **6**, 335–353.

GYIRES, BÉLA (1960): "Über die Schranken der Eigenwerte von Hypermatrizen," Publ. Math. Debrecen **7**, 374–381.

GUEST, J. (1955): "The solution of linear simultaneous equations by matrix iteration," Austral. J. Phys. **8**, 425–439.

HAMBURGER, H. L., and GRIMSHAW, M. E. (1951): *Linear Transformations in n-dimensional Vector Space. An Introduction to the Theory of Hilbert Space.* University Press, Cambridge, x + 195 pp.

HANSEN, ELDON R. (1962): "On quasicyclic Jacobi methods," J. Assoc. Comput. Mach. **9**, 118–135.

HAUSDORF, FELIX (1919): "Der Wertvorrat einer Matrix," Math. Z. 314–316.

HAYNSWORTH, EMILIE V. (1953): "Bounds for determinants with dominant main diagonal," Duke Math. J. **20**, 199–209.

——, (1955): "Quasi-stochastic matrices," Duke Math. J. **22**, 15–24.

——, (1957): "Note on bounds for certain determinants," Duke Math. J. **24**, 313–319.

——, (1959): "Application of a theorem on partitioned matrices," J. Res. Nat. Bur. Standards **62B**, 73–78.

——, (1960): "Bounds for determinants with positive diagonals," Trans. Amer. Math. Soc. **96**, 395–399.

——, (1961): "Special types of partitioned matrices," J. Res. Nat. Bur. Standards **65B**, 7–12.

HEINRICH, H. (1956): "Bemerkungen zu den Verfahren von Hessenberg und Voetter," Z. Angew. Math. Mech. **36**, 250–252.

HENRICI, PETER (1958a): "The quotient-difference algorithm," Nat. Bur. Standards Appl. Math. Ser. **49**, 23–46.

——, (1958b): "On the speed of convergence of cyclic and quasicyclic Jacobi methods for computing eigenvalues of Hermitian matrices," J. Soc. Indust. Appl. Math. **6**, 144–162.

HERMITE, CHARLES (1855): "Remarque sur un théorème de Cauchy," C.R. Acad. Sci. Paris **41**, 181–183.

HERSTEIN, I. N. (1954): "A note on primitive matrices," Amer. Math. Monthly **61**, 18–20.

HESSENBERG, K. (1941): *Auflösung linearer Eigenwertaufgaben mit Hilfe der Hamilton-Cayleyschen Gleichung.* Dissertation, Technische Hochschule, Darmstadt.

HESTENES, MAGNUS R. (1953): "Determination of eigenvalues and eigenvectors of matrices," Nat. Bur. Standards Appl. Math. Ser. **29**, 89–94.

——, (1955): "Iterative computational methods," Comm. Pure Appl. Math. **8**, 85–95.

——, (1956): *The Conjugate-Gradient Method for Solving Linear Systems.* Proc. Sixth Symposium Appl. Math., 83–102.

——, (1958): "Inversion of matrices by biorthogonalization and related results," J. Soc. Indust. Appl. Math. **6**, 51–90.

——, (1961): "Relative Hermitian matrices," Pacific J. Math. **11**, 225–245.

——, and KARUSH, WILLIAM (1951a): "A method of gradients for the calculation of the characteristic roots and vectors of a real symmetric matrix," J. Res. Nat. Bur. Standards **47**, 45–61.

——, and ——, (1951b): "Solutions of $Ax = \lambda Bx$," J. Res. Nat. Bur. Standards **47**, 471–478.

——, and STIEFEL, EDUARD (1952): "Methods of conjugate gradients for solving linear systems," J. Res. Nat. Bur. Standards **49**, 409–436.

HIRSCH, A. (1901): "Sur les racines d'une équation fondamentale," Acta Math. **25**, 367–370.

HLODOVSKII, I. N. (1933): "K teorii obščego slučaja preobrazovanija vekovogo uravnenija metodom Akademika A. N. Krylova," Izv. Akad. Nauk. Otd. Mat. Estest. 1077–1102.

HOFFMAN, ALAN J., and TAUSSKY, OLGA (1954): "A characterization of normal matrices," J. Res. Nat. Bur. Standards **52**, 17–19.

——, and WIELANDT, H. W. (1953): "The variation of the spectrum of a normal matrix," Duke Math. J. **20**, 37–40.

HÖLDER, O. (1913): "Über einige Determinanten," Ber. Verh. Königlich Sächs. Gesellschaft Wiss. Leipzig Math.-phys. Kl. **65**, 110–120.

HOLLADAY, JOHN C., and VARGA, RICHARD S. (1958): "On powers of nonnegative matrices," Proc. Amer. Math. Soc. **9,** 631–634.

VON HOLDT, RICHARD ELTON (1956): "An iterative procedure for the calculation of the eigenvalues and eigenvectors of a real symmetric matrix," J. Assoc. Comput. Mach. **3,** 223–238.

HOPF, EBERHARD (1959): "Zur Kennzeichnung der Euklidischen Norm," Math. Z. **72,** 76–81.

HORN, ALFRED (1950): "On the singular values of a product of completely continuous operators," Proc. Nat. Acad. Sci. **36,** 374–375.

——, (1954): "On the eigenvalues of a matrix with prescribed singular values," Proc. Amer. Math. Soc. **5,** 4–7.

——, and STEINBERG, ROBERT (1959): "Eigenvalues of the unitary part of a matrix," Pacific J. Math. **9,** 541–550.

HORST, PAUL (1935): "A method of determining the coefficients of a characteristic equation," Ann. Math. Statist. **6,** 83–84.

——, (1937): "A method of factor analysis by means of which all coordinates of the factor matrix are given simultaneously," Psychometrika **2,** 225–236.

HOTELLING, HAROLD (1943a): "Some new methods in matrix calculation," Ann. Math. Statist. **14,** 1–34.

——, (1943b): "Further points on matrix calculation and simultaneous equations," Ann. Math. Statist. **14,** 440–441.

HOUSEHOLDER, ALSTON S. (1953): *Principles of Numerical Analysis.* McGraw-Hill Book Company, Inc., New York, x + 274 pp.

——, (1954): *On Norms of Vectors and Matrices.* Oak Ridge National Laboratory, ORNL-1756, Physics, Oak Ridge, Tennessee.

——, (1955): "Terminating and nonterminating iterations for solving linear systems," J. Soc. Indust. Appl. Math. **3,** 67–72.

——, (1956): "On the convergence of matrix iterations," J. Assoc. Comput. Mach. **3,** 314–324.

——, (1957): "A survey of some closed methods for inverting matrices," J. Soc. Indust. Appl. Math. **5,** 155–169.

——, (1958a): "A class of methods for inverting matrices," J. Soc. Indust. Appl. Math. **6,** 189–195.

——, (1958b): "The approximate solution of matrix problems," J. Assoc. Comput. Mach. **5,** 204–243.

——, (1958c): "On matrices with nonnegative elements," Montash. Math. **62,** 238–242.

——, (1958d): "Unitary triangularization of a nonsymmetric matrix," J. Assoc. Comput. Mach. **5,** 339–342.

——, (1958e): "Generated error in rotational tridiagonalization," J. Assoc. Comput. Mach. **5,** 335–338.

——, (1959): "Minimal matrix norms," Monatsh. Math. **63,** 344–350.

——, (1961): "On deflating matrices," J. Soc. Indust. Appl. Math. **9,** 89–93.

——, and BAUER, F. L. (1958): "On certain methods for expanding the characteristic polynomial," Numer. Math. **1**, 29–37.

——, and ——, (1960): "On certain iterative methods for solving linear systems," Numer. Math. **2**, 55–59.

——, and YOUNG, GALE (1938): "Matrix approximation and latent roots," Amer. Math. Monthly **45**, 165–171.

INGRAHAM, MARK H. (1937): "A note on determinants," Bull. Amer. Math. Soc. **43**, 579–580.

IVANOV, V. K. (1939): "O shodimosti processov iteracii pri rešenii sistem lineĭnyh algebraičeskih uravneniĭ," Izv. Akad. Nauk SSSR, 477–483.

JACOBI, C. G. J. (1845): "Ueber eine neue Auflösungsart der bei der Methode der kleinsten Quadrate vorkommenden linearen Gleichungen," Astronom. Nachr. **22**, 297–306.

——, (1846): "Über ein leichtes Verfahren die in der Theorie der Säculär-störungen vorkommenden Gleichungen numerisch aufzulösen," Crelle's J. **30**, 51–94.

JAHN, H. A. (1948): "Improvement of an approximate set of latent roots and modal columns of a matrix by methods akin to those of classical perturbation theory," Quart. J. Mech. Appl. Math. **1**, 131–144.

JAMES, HUBERT M., and COOLIDGE, ALBERT SPRAGUE (1933): "The ground state of the hydrogen molecule," J. Chem. Phys. **1**, 825–835.

JENSEN, H. (1944): "Attempt at a systematic classification of some methods for the solution of normal equations," Geodaet. Inst. København Medd. No. 18.

JOHANSEN, DONALD E. (1961): "A modified Givens method for the eigenvalue evaluation of large matrices," J. Assoc. Comput. Mach. **8**, 331–335.

KACZMARZ, S. (1937): "Angenäherte Auflösung von Systemen linearer Gleichungen," Bull. Internat. Acad. Polon. Sci. Cl. A. 355–357.

KAHAN, W. (1958): *Gauss-Seidel Methods of Solving Large Systems of Linear Equations.* Thesis, University of Toronto.

KAMELA, C. (1943): "Die Lösung der Normalgleichungen nach der Methode von Professor Dr. T. Banachiewicz," Schweiz. Z. Vermessg. Kulturtech. Photogr. **41**, 225–232, 265–275.

KAMKE, E. (1939): "Weinsteins Einschliessungssatz," Math. Z. **45**, 788–790.

KANTOROVIČ, L. V. (1948): "Funkcional'nyĭ analiz i prikladnaja matematika," Uspehi Mat. Nauk **3**(28), 89–185.

KANTOROVITCH, L. V. (1939): "The method of successive approximations for functional equations," Acta Math. **71**, 62–97.

KARLQVIST, OLLE (1952): "Numerical solution of elliptic differential equations by matrix methods," Tellus **4**, 374–384.

KARPELEVIČ. F. I. (1951): "O harakteristiceskih kornjah matric s neotricatel'nymi elementami," Izv. Akad. Nauk SSSR **15**, 361–383.

KARUSH, W. (1951): "An iterative method for finding characteristic vectors of a symmetric matrix," Pacific J. Math. **1**, 233–248.

KASANIN, RADIVOYE (1952): "Geometriska interpretaciya Banahyeviceve sheme," Srpska Akad. Nauka, Zb. Rad. 18 Mat. Inst. **2**, 93–96.

KATO TOSIO, (1949): "On the upper and lower bounds of eigenvalues," J. Phys. Soc. Japan **4**, 334–339.

———, (1960): "Estimation of iterated matrices, with application of the von Neumann condition," Numer. Math. **2**, 22–29.

KELBASINSKIĬ, A. S. (1960): "Nekotoroe obobščenie teorem A. M. Ostrovskogo ob iteracionnyh processah," Vestnik Moskov. Univ. Ser. I Mat. Meh. **1**(5), 40–52.

KELLER, HERBERT B. (1958): "On some iterative methods for solving elliptic difference equations," Quart. Appl. Math. **16**, 209–226.

———, (1960): "Special block iterations with applications to Laplace and biharmonic difference equations," SIAM Rev. **2**, 277–287.

KELLOGG, R. B., and NODERER, L. C. (1960): "Scaled iterations and linear equations," J. Soc. Indust. Appl. Math. **8**, 654–661.

KINCAID, W. M. (1947): "Numerical methods for finding characteristic roots and vectors of matrices," Quart. Appl. Math. **5**, 320–345.

KIPPENHAHN, RUDOLPH (1951): "Über den Wertevorrat einer Matrix," Math. Nachr. **6**, 193–228.

KJELLBERG, GÖRAN (1958): "On the convergence of successive over-relaxation applied to a class of linear systems of equations with complex eigenvalues," Ericsson Technics **14**, 245–258.

———, (1961): "On the successive over-relaxation method for cyclic operators," Numer, Math. **3**, 87–91.

VON KOCH, H. (1913): "Über das Nichtverschwinden einer Determinante, nebst Bemerkungen über Systeme unendlich vieler linearer Gleichungen," Jber. Deutsch. Math. Verein. **22**, 285–291.

KOCH, J. J. (1926): "Bestimmung höherer kritischer Drehzahlen schnell laufender Wellen," Verh. internat. Kongr. techn. Mech., Zürich 213–218.

KOGBETLIANTZ, ERVAND GEORGE (1955): "Solution of linear equations by diagonalization of coefficients matrix," Quart. Appl. Math. **13**, 123–132.

KOHN, WALTER (1947): "A note on Weinstein's variational method," Phys. Rev. **71**, 902–904.

———, (1949): "A variational iteration method for solving secular equations," J. Chem. Phys. **17**, 670.

KOLMOGOROFF, A. (1934): "Zur Normierbarkeit eines allgemeinen topologischen linearen Raumes," Studia Math. **5**, 29–33.

KÖNIG, JULIUS (1884): "Ueber eine Eigenschaft der Potenzreihen," Math. Ann. **23**, 447–449.

KORGANOFF, A., et al. (1961): *Méthodes de calcul numérique.* Tome I. *Algèbre non linéaire.* Dunod, Paris, xxvii + 375 pp.

KOSTARČUK, V. N. (1954): "Ob odnom metode rešenija sistem lineĭnyh uravneniĭ i otyskanija sobstvennyh vektorov matricy," Dokl. Akad. Nauk SSSR **98**, 531–534.

KOTELJANSKIĬ, D. M. (1950): "K teorii neotricatel'nyh i oscilljacionnyh matric," Ukrain. Mat. Ž. 2(2), 94–101.

——, (1952): "O nekotoryh svoĭstvah matric s položitel'nymi élementami," Mat. Sb. 31, 497–506.

——, (1955a): "O nekotoryh dostatočnyh priznakah veščestvennosti i prostoty matričnogo spektre," Mat. Sb. 36(78), 163–168.

——, (1955b): "O raspoloženii toček matričnogo spektra," Ukrain. Mat. Ž. 7, 131–133.

——, (1955c): "O vliyanii preobrazovanija Gaussa na spektru matric," Uspehi Mat. Nauk 10, 1(163), 117–121.

KOWALEWSKI, GERHARD (1948): Einführung in die Determinantentheorie (third edition). Chelsea Publishing Company, New York, 320 pp.

KRASNOSEL'SKIĬ, M. A. (1960): "O rešenii metodom posledovatel'nyh približeniĭ uravneniĭ s samosopryažennymi operatorami," Uspehi Mat. Nauk 15, No. 3(93), 161–165.

——, and KREĬN, S. G. (1952): "Iteracionnyĭ process s minimal'nymi nevjazkami," Mat. Sb. 31, 315–334.

——, and ——, (1953): "Zamečanie o raspredelenii ošibok pri rešenii sistemy lineĭnyh uravneniĭ pri pomošči iteracionnogo processa," Uspehi Mat. Nauk 7, 4(50), 157–161.

KREIN, M. G. (1933): "Über eine neue Klasse von Hermiteschen Formen and über eine Verallgemeinerung des trigonometrischen Momentenproblems," Izv. Akad. Nauk SSSR. Otd. Mat. Estest. 1259–1275.

KREIN, S. G., and PROZOROVSKAJA, O. I. (1957): "Analog metoda zeidelja dlja operatornyh uravneniĭ," Ministerstvo Vyssego Obrazovaniya SSSR. Voronež. Gos. Univ. Trudy Sem. Funkcional. Anal. 5, 35–38.

KREYSZIG, ERWIN (1954): "Die Einschliessung von Eigenwerten hermitescher Matrizen beim Iterationsverfahren," Z. Angew. Math. Mech. 34, 459–469.

——, (1955): "Die Ausnutzung zusätzlicher Vorkenntnisse für die Einschliessung von Eigenwerten beim Iterationsverfahren," Z. Angew. Math. Mech. 35, 89–95.

——, (1958): "Einschliessung von Eigenwerten und Mohrsches Spannungsdiagramm," Z. Angew. Math. Phys. 9, 202–206.

KRON, GABRIEL (1953): "A set of principles to interconnect the solutions of physical systems," J. Appl. Phys. 24, 965–980.

KRULL, WOLFGANG (1958): "Über eine Verallgemeinerung der Hadamardschen Ungleichung," Arch. Math. 9, 42–45.

KRYLOV, A. N. (1931): "O čislennom rešenii uravnenija, kotorym v techničeskih voprasah opredeljajutsja častoty malyh kolebaniĭ material'nyh sistem," Izv. Akad. Nauk SSSR Otd. Mat. Estest. 491–539.

KUBLANOVSKAJA, V. N. (1961): "O nekotoryh algorifmah dlja rešenija polnoĭ problemy sobstvennyh značeniĭ," Dokl. Akad. Nauk SSSR 136, 26–28.

LAASONEN, PENTTI (1956): "Simultane Bestimmung mehrerer Eigenwerte mittels gebrochener Iteration," Ann. Acad. Sci. Fenn. Ser. A.I., 218, 8 pp.

——, (1958): "On the iterative solution of the matrix equation, $AX^2 - I = 0$," Math. Tables Aids Comput. **12**, 109–116.

——, (1959): "A Ritz method for simultaneous determination of several eigenvalues and eigenvectors of a big matrix," Ann. Acad. Sci. Fenn. Ser. A.I., 265, 16 pp.

LADERMAN, JACK (1948): "The square root method for solving simultaneous linear equations," Math. Tables Aids Comput. **3**, 13–16.

LANCASTER, P. (1960a): "Free vibrations of lightly damped systems by perturbation methods," Quart. J. Mech. Appl. Math. **13**, 138–155.

——, (1960b): "Inversion of lambda-matrices and applications to the theory of linear vibrations," Arch. Rational Mech. Anal. **6**, 105–114.

LANCZOS, CORNELIUS (1950): "An iteration method for the solution of the eigenvalue problem of linear differential and integral operators," J. Res. Nat. Bur. Standards **45**, 255–282.

——, (1952a): "Chebyshev polynomials in the solution of large-scale linear systems," Assoc. Comput. Mach. Proc. Toronto, 124–133.

——, (1952b): "Solution of systems of linear equations by minimized iterations," J. Res. Nat. Bur. Standards **49**, 33–53.

——, (1956): *Applied Analysis*. Prentice-Hall, Inc., Englewood Cliffs, New Jersey, 539 pp.

——, (1958a): "Iterative solution of large-scale linear systems," J. Soc. Indust. Appl. Math. **6**, 91–109.

——, (1958b): "Linear systems in self-adjoint form," Amer. Math. Monthly **65**, 665–679.

LAPPO-DANILEVSKY, I. A. (1933): *Mémoires sur la théorie des systèmes des équations differentielles linéaires*. Chelsea Publishing Company, New York, (1953), Vol. 1, 253 pp.; Vol. 2, 204 pp.

LAPPO-DANILEVSKIĬ, I. A. (1957): *Primenenie funkciĭ ot matric teorii lineĭnyh sistem obyknovennyh differencial'nyh uravneniĭ*. Gosudarstv. Izdat. Tehn.-Teoret. Lit., Moscow, 456 pp.

LÄUCHLI, PETER (1959): "Iterative Lösung und Fehlerabschätzung in der Ausgleichsrechnung," Z. Angew. Math. Phys. **10**, 245–280.

——, (1961): "Jordon-Elimination und Ausgleichung nach kleinsten Quadraten," Numer. Math. **3**, 226–240.

LAVRENT'EV, M. M. (1954): "O točnosti rešenija sistem lineĭnyh uravneniĭ," Mat. Sb. **34**, (76), 259–268.

LAVUT, A. P. (1952): "Raspoloženie sobstvennyh čisel preobrazovanii Zeĭdelja dlja sistem normal'nyh uravneniĭ," Uspehi Mat. Nauk **7**, 6(52): 197–202.

LAX, PETER D. (1958): "Differential equations, difference equations, and matrix theory," Comm. Pure Appl. Math. **11**, 175–194.

LEDERMANN, WALTER (1950): "Bounds for the greatest latent root of a positive matrix," J. London Math. Soc. **25,** 265–268.

LEHMANN, N. JOACHIM (1949–50a): "Berechnung von Eigenwertschranken bei linearen Problemen," Arch. Math. **2,** 139–147.

——, (1949–50b): "Beiträge zur numerischen Lösung linearer Eigenwertprobleme," Z. Angew. Math. Mech. **29,** 341–356; **30,** 1–16.

——, (1950): "Bemerkungen zu einem Einschliessungssatz für Eigenwerte," Z. Angew. Math. Mech. **30,** 223–225.

LE VERRIER, U. J. J. (1840): "Sur les variations séculaires des élements elliptiques des sept planètes principales," J. Math. Pures Appl. **5,** 220–254.

LEVY, LUCIEN (1881): "Sur la possibilité de l'équilibre électrique," C.R. Acad. Sci. Paris **93,** 706–708.

LEWIS, JR., DANIEL C., and TAUSSKY, OLGA (1960): "Some remarks concerning the real and imaginary parts of the characteristic roots of a finite matrix," J. Math. Phys. **1,** 234–236.

LIDSKIĬ, V. B. (1950): "O sobstvennyh značenijah summy i proizvedenija simmetričeskih matric," Dokl. Akad. Nauk SSSR **75,** 769–772.

LIEBMANN, HEINRICH (1918): "Die angenäherte Ermittelung harmonischer Funktionen und konformer Abbildungen (nach Ideen von Boltzmann und Jacobi)," S.-B. Math.-Nat. Kl. Bayerischen Akad. Wiss. München 385–416.

LIVŠIC, B. L. (1960): "Ob uskorennom "Utočnenii korneĭ vekovyh uravnenii" po metodu Majanca," Dokl. Akad. Nauk SSSR **132,** 1295–1298.

LJUSTERNIK, L. A. (1947): "Zamečanija k čislennomu rešeniju kraevyh zadač yravnenija Laplasa i vyčisleniju sobstvennyh značenii metodom setok," Trudy Mat. Inst. Steklov. **20,** 49–64.

——, (1956), *Vypuklye figury i mnogogranniki.* Gosudarstv. Izdat. Tehn.-Teoret. Lit., Moscow, 212 pp.

LOHMAN, JOHN B. (1949): "An iterative method for finding the smallest eigenvalue of a matrix," Quart. Appl. Math. **7,** 234.

LOMONT, J. S., and WILLOUGHBY, R. A. (1959): "Dominant eigenvectors of a class of test matrices," SIAM Rev. **1,** 64–65.

LONSETH, A. T. (1947): "The propagation of error in linear problems," Trans. Amer. Math. Soc. **62,** 193–212.

——, (1949): "An extension of an algorithm of Hotelling," Proc. Berkeley Symposium Math. Statist. Problems **1945, 1946,** 353–357.

LOPŠIC, A. M. (1949): "Čislennyĭ metod nahoždenija sobstvennyh značeniĭ i sobstvennyh ploskostei lineĭnogo operatora," Trudy Sem. Vektor. Tenzor. Anal. **7,** 233–259.

LOTKIN, MARK (1955): "A set of test matrices," Math. Tables Aids Comput. **9,** 153–161.

——, (1956): "Characteristic values of arbitrary matrices," Quart. Appl. Math. **14,** 267–275.

——, (1957): "The diagonalization of skew-Hermitian matrices," Duke Math.

J. **24**, 9–14.

——, (1959): "Determination of characteristic values," Quart. Appl. Math. **17**, 237–244.

LUZIN, N. N. (1931): "O metode Akademika A. N. Krylova sostavlenija vekovogo uravnenija," Izv. Akad. Nauk SSSR Otd. Mat. Estest., 903–958.

——, (1932): "O nekotoryh svoistvah peremeščajuščego množitelja v metode Akademika A. N. Krylova," Izv. Akad. Nauk SSSR, 595–638, 735–762, 1065–1102.

MacDUFFEE, CYRUS COLTON (1943): *Vectors and Matrices*. Mathematical Association of America, xi + 192 pp.

——, (1946): *The Theory of Matrices*. Chelsea Publishing Company, New York, v + 110 pp.

MACON, N., and SPITZBART, A. (1958): "Inverses of Vandermonde matrices," Amer. Math. Monthly **65**, 95–100.

MADIĆ, PETAR (1956): "Sur une méthode de résolution des systèmes d'équations algébriques linéaires," C.R. Acad. Sci. Paris **242**, 439–441.

MAGNIER, A. (1948): "Sur le calcul des matrices," C.R. Acad. Sci. Paris **226**, 464–465.

MANSION, P. (1888): "Rapport sur le mémoire de M. l'abbé B.-I. Clasen," Ann. Soc. Sci. Bruxelles **12**, (2), 50–59; Mathesis **9** (Suppl. 2), 32–40.

MARCUS, M. D. (1955): "A remark on a norm inequality for square matrices," Proc. Amer. Math. Soc. **6**, 117–119.

——, (1956): "An eigenvalue inequality for the product of normal matrices," Amer. Math. Monthly **63**, 173–174.

——, (1958): "On a determinantal inequality," Amer. Math. Monthly **65**, 266–268.

——, (1960): "Basic theorems in matrix theory," Nat. Bur. Standards Appl. Math. Ser. **57**, iv + 27 pp.

——, MINC, HENRY K., and MOYLS, BENJAMIN (1961): "Some results on nonnegative matrices," J. Res. Nat. Bur. Standards **65B**, 205–209.

MARÍK, JAN, and PTÁK, VLASTIMIL (1960): "Norms, spectra and combinatorial properties of matrices," Czechoslovak Math. J. **2**, 181–196.

MAJANC, L. S. (1945): "Metod utočnenija korneĭ vekovyh uravneniĭ vysokyh stepeneĭ i čislennogo analiza ih zavisimosti ot parametrov sootvetstvujuščih matric," Dokl. Akad. Nauk SSSR **50**, 121–124.

MEHMKE, R. (1892): "K" sposobu Zeĭdelja, služaščemu dlja rešenija sistemy lineinyh" uravneniĭ s" ves'ma bol'šim" čislom neizvestnyh" posredstvom posledovatel'nyh" približeniĭ. Izvlečenie iz pis'ma professoru Memke k" professoru Nekrasovy (German)," Mat. Sb. **16**, 342–345.

——, (1930a): "Praktische Lösung der Grundaufgaben über Determinanten, Matrizen, und lineare Transformationen," Math. Ann. **103**, 300–318.

——, (1930b): "Über die zweckmässigste Art, lineare Gleichungen durch Elimination aufzulösen," Z. Angew. Math. Mech. **10**, 508–514.

——, and NEKRASOV, P. A. (1892): "Rešenie lineinyh sistem uravneniĭ posredstvom posledovatel'nyh približenii," Mat. Sb. **16**, 437–459.

MENDELSOHN, N. S. (1955): *Some Elementary Properties of Ill-Conditioned Matrices and Linear Equations.* CARDE Technical Memorandum No. 120/55.

——, (1956): "Some properties of approximate inverses of matrices," Trans. Roy. Soc. Canada Section III **50**, 53–59.

MIKUSINSKI, J. (1957): "Sur quelques inégalités pour les déterminants," Bull. Acad. Polon. Sci. Sér. Sci. Math. Astr. Phys. (3) **5**, 699–700.

——, (1959): "Sur les extrema des déterminants," Ann. Polon. Math. **6**, 135–143.

MINKOWSKI, H. (1900): "Zur Theorie der Einheiten in den algebraischen Zahlkörpern," Nachr. Königlichen Ges. Wiss. Göttingen Math.-Phys. Kl., 90–93, Gesammelte Abh. **1**, 316–319.

MIRSKY L. (1955a): *An Introduction to Linear Algebra.* Clarendon Press, Oxford, xi + 433 pp.

——, (1955b): "An inequality for positive-definite matrices," Amer. Math. Monthly **62**, 428–430.

——, (1956a): "A note on normal matrices," Amer. Math. Monthly **63**, 479.

——, (1956b): "The spread of a matrix," Mathematika **3**, 127–130.

——, (1956c): "The norms of adjugate and inverse matrices," Arch. Math. **7**, 276–277.

——, (1957a): "On a generalization of Hadamard's determinantal inequality due to Szasz," Arch. Math. **8**, 274–275.

——, (1957b): "Inequalities for normal and Hermitian matrices," Duke Math. J. **24**, 591–600.

——, (1958a): "On the minimization of matrix norms," Amer. Math. Monthly **65**, 106–107.

——, (1958b): "Matrices with prescribed characteristic roots and diagonal elements," J. London Math. Soc. **33**, 14–21.

——, (1958c): "Maximum principles in matrix theory," Proc. Glasgow Math. Assoc. **4**, 34–37.

——, (1960): "Symmetric gauge functions and unitarily invariant norms," Quart. J. Math. (2) **11**, 50–59.

——, and RADO, R. (1957): "A note on matrix polynomials," Quart. J. Math. Oxford Ser. (2) **8**, 128–132.

VON MISES, R., and POLLACZEK-GEIRINGER, HILDA (1929): "Praktische Verfahren der Gleichungsauflösung," Z. Angew. Math. Mech. **9**, 58–77, 152–164.

MOORE, E. H. (1935): *General Analysis.* Part I. American Philosophical Society, Philadelphia. vii + 231 pp.

MORGENSTERN, DIETRICH (1956): "Eine Verschärfung der Ostrowski'schen Determinantenabschätzung," Math. Z. **66**, 143–146.

MORRIS, JOSEPH (1935): *A Successive Approximation Process for Solving*

Simultaneous Linear Equations. Aero. Res. Council, Report No. 1711.

——, (1946): "An escalator process for the solution of linear simultaneous equations," Philos. Mag. (7) **37,** 106–120.

——, (1947): *The Escalator Method in Engineering Vibration Problems.* John Wiley and Sons, Inc., New York, xv + 270 pp.

——, and HEAD, J. W. (1942): "Lagrangian frequency equations. An 'escalator' method for numerical solution (Appendix by G. Temple)," Aircraft Engrg. **14,** 312–316.

MORRISON, DAVID D. (1960): "Remarks on the unitary triangularization of a nonsymmetric matrix," J. Assoc. Comput. Mach. **7,** 185–186.

MOTT, J. L., and SCHNEIDER, HANS (1959): "Matrix algebras and groups relatively bounded in norms," Arch. Math. **10,** 1–6.

MOYLES, B. N., and MARCUS, M. D. (1955): "Field convexity of a square matrix," Proc. Amer. Math. Soc. **6,** 981–983.

MUIR, THOMAS (1906, 1911, 1920, 1923): *The Theory of Determinants in the Historical Order of Development.* Macmillan and Company, Ltd., London, Vol. 1, xi + 491 pp.; Vol. 2, xvi + 475 pp,; Vol. 3, xxvi + 503 pp.; Vol. 4, xxvi + 508 pp.

——, (1933): *Contributions to the History of Determinants. 1900–1920.* Blackie and Son, Ltd., London, xxiii + 408 pp.

——, and METZLER, W. H. (1933): *A Treatise on the Theory of Determinants.* Longmans, Green and Company, New York and London, 606 pp.

MÜNTZ, C. H. (1913a): "Solution direct de l'équation séculaire et de quelques problèmes analogues transcendents," C.R. Acad. Sci. Paris **156,** 43–46.

——, (1913b): "Sur la solution des équations séculaires et des équations intégrales," C.R. Acad. Sci. Paris **156,** 860–862.

——, (1917): "Zur expliziten Bestimmung der Hauptachsen quadratischer Formen und der Eigenfunktionen symmetrischer Kerne," Nachr. Akad. Wiss. Göttingen 136–140.

MURNAGHAN, F. D. (1932): "On the field of values of a square matrix," Proc. Nat. Acad. Sci. U.S.A. **18,** 246–248.

NANSON, E. J. (1901) "A determinantal inequality," Messenger of Math. **31,** 48–50.

National Bureau of Standards (1958): "Further contributions to the solution of simultaneous linear equations and the determination of eigenvalues," Nat. Bur. Standards Appl. Math. Ser. **49.**

National Physical Laboratory (1961): *Notes on Applied Science. No. 16. Modern Computing Methods* (Second edition). Her Majesty's Stationery Office, London, vi + 170 pp.

NEKRASOV", P. A. (1884): "Opredelenie neizvestnyh" po sposobu naimen'ših" kvadratov" pri ves'ma bol'šem" čisle neizvestnyh"," Mat. Sb. **12,** 189–204.

——, (1892): "K" voprosy o rešenii lineĭnoĭ sistemy uravneniĭ," Mat. Sb. **16,** 1–18.

VON NEUMANN, JOHN (1937): "Some matrix inequalities and metrization of matrix-space," Izv. Naučno-issledovatel'skogo Inst. Mat. Meh. Tomsk. Gos. Univ. 1, 286–299.

——, (1942): "Approximate properties of matrices of high finite order," Portugal. Math. 3, 1–62.

——, and GOLDSTINE, H. H. (1947): "Numerical inverting of matrices of high order," Bull. Amer. Math. Soc. 53, 1021–1099.

NEVILLE, E. H. (1948): "Ill-conditioned sets of linear equations," Philos. Mag. (7) 39, 35–48.

NEWING, R. A. (1937): "On the variation calculation of eigenvalues," Philos. Mag. 24, 114–127.

NEWMAN, MORRIS (1960): "Kantorovich's inequality," J. Res. Nat. Bur. Standards 64B, 33–34.

——, and TODD, JOHN (1958): "The evaluation of matrix inversion programs," J. Soc. Indust. Appl. Math. 6, 466–476.

NIKOLAEVA, M. V. (1949): "O relaksacionnom metode Sausella," Trudy Mat. Inst. Steklov 28, 160–182.

NUDEL'MAN, A. A., AND ŠVARCMAN, P. A. (1958): "O spektre proizvedenija unitarnyh matric," Uspehi Mat. Nauk 13(6), 111–117.

OEDER, ROBERT (1951): "Problem E 949," Amer. Math. Monthly 58, 37, 565.

OLDENBURGER, RUFUS (1940): "Infinite powers of matrices and characteristic roots," Duke Math. J. 6, 357–361.

OLKIN, INGRAM (1959): "Inequalities for the norms of compound matrices," Arch. Math. 10, 16–17.

OPPENHEIM, A. (1930): "Inequalities connected with definite Hermitian forms," J. London Math. Soc. 5, 114–119.

——, (1954): "Inequalities connected with definite Hermitian forms," Amer. Math. Monthly 61, 463–466.

ORTEGA, J. M. (1960): "On Sturm sequences for tridiagonal matrices," J. Assoc. Comput. Mach. 7, 260–263.

OSBORNE, ELMER E. (1958): "On acceleration and matrix deflation processes used with the power method," J. Soc. Indust. Appl. Math. 6, 279–287.

OSTROWSKI, ALEXANDER M. (1937a): "Sur la détermination des bornes inférieures pour une classe des déterminants," Bull. Sci. Math. (2) 61, 19–32.

——, (1937b): "Über die Determinanten mit überwiegender Hauptdiagonale, Comment. Math. Helv. 10, 69–96.

——, (1938a): "Sur quelques transformations de la série de Liouville-Newman," C.R. Acad. Sci. Paris 206, 1345–1347.

——, (1938b): "Sur l'approximation du déterminant de Fredholm par les déterminants des systèmes d'équations linéaires," Ark. Mat., Astronom. Fys. 26A, No. 14, 1–15.

——, (1950): "Sur la variation de la matrice inverse d'une matrice donnée C.R. Acad. Sci. Paris 231, 1019–1021.

——, (1951a): "Sur les conditions générales pour la régularité des matrices," Rend. Mat. e Appl. (5) **10**, 1–13.

——, (1951b): "Sur les matrices peu différentes d'une matrice triangulaire," C.R. Acad. Sci. Paris **233**, 1559–1560.

——, (1951c): "Ueber das Nichtverschwinden einer Klasse von Determinanten und die Lokalisierung der charakteristischen Wurzeln von Matrizen," Compositio Math. **9**, 209–226.

——, (1952a): "Note on bounds for determinants with dominant principal diagonal," Proc. Amer. Math. Soc. **3**, 26–30.

——, (1952b): "Bounds for the greatest latent root of a positive matrix," J. London Math. Soc. **27**, 253–256.

——, (1952c): *On the Convergence of Gauss' Alternating Procedure in the Method of Least Squares*. I. National Bureau of Standards Report 1857.

——, (1952d): *On the Convergence of Cyclic Linear Iterations for Symmetric and Nearly Symmetric matrices*. II. National Bureau of Standards Report 1759.

——, (1953): "On over and under relaxation in the theory of the cyclic single step iteration," Math. Tables Aids Comput. **7**, 152–159.

——, (1954a): "On the spectrum of a one-parametric family of matrices," J. Reine Angew. Math. **193**, 143–160.

——, (1954b): "On the linear iteration procedures for symmetric matrices," Rend. Mat. e Appl. **13**, 1–24.

——, (1954c): "On nearly triangular matrices," J. Res. Nat. Bur. Standards **52**, 319–345.

——, (1955a): "Note on bounds for some determinants," Duke Math. J. **22**, 95–102.

——, (1955b): "Sur les déterminants à diagonale dominante," Bull. Soc. Math. Belg. **1954**, 46–51.

——, (1955c): "Über Normen von Matrizen," Math. Z. **63**, 2–18.

——, (1956): "Determinanten mit überwiegender Hauptdiagonale und die absolute Konvergenz von linearen Iterationsprozessen," Comment. Math. Helv. **30**, 175–210.

——, (1957a): "Über die Stetigkeit von charakteristischen Wurzeln in Abhängigkeit von den Matrizenelementen," Jber. Deutsch. Math. Verein. **60**, Abt., 1, 40–42.

——, (1957b): "Über näherungsweise Auflösung von Systemen homogener linearer Gleichungen," Z. Angew. Math. Phys. **8**, 280–285.

——, (1958a): "On the bounds of a one-parametric family of matrices," J. Reine Angew. Math. **200**, 190–199.

——, (1958b): "On Gauss' speeding-up device in the theory of single step iteration," Math. Tables Aids Comput. **12**, 116–132.

——, (1958–1959): "On the convergence of the Rayleigh quotient iteration for the computation of characteristic roots and vectors. I, II, III, IV,"

Arch. Rational Mech. Anal. **1**, 233–241; **2**, 423–428; **3**, 325–340; **3**, 341–347.

——, (1959a): "On the convergence of the Rayleigh quotient iteration for the computation of the characteristic roots and vectors. V. (Usual Rayleigh quotient for non-Hermitian matrices and linear elementary divisors)," Arch. Rational Mech. Anal. **3**, 472–481.

——, (1959b): "On the convergence of the Rayleigh quotient iteration for the computation of characteristic roots and vectors. VI. (Usual Rayleigh quotient for nonlinear elementary divisors)," Arch. Rational Mech. Anal. **4**, 153–165.

——, (1959c): "On the convergence of Gauss' alternating procedure in the method of the least squares," Ann. Mat. Pura Appl. (IV) **48**, 229–236.

——, (1959d): "Über Eigenwerte von Produkten Hermitescher Matrizen," Abh. Math. Sem. Univ. Hamburg **23**, 60–68.

——, (1959e): "A quantitative formulation of Sylvester's law of inertia," Proc. Nat. Acad. Sci. U.S.A. **45**, 740–743.

——, (1959–1960): "Über Produkte Hermitescher Matrizen and Büschel Hermitescher Formen," Math. Z. **72**, 1–15.

——, (1960a): "On the eigenvector belonging to the maximal root of a non-negative matrix," Proc. Edinburgh Math. Soc. II **12**, 107–112.

——, (1960b): "Über geränderte Determinanten und bedingte Trägheitsindizes quadratischer Formen," Monatsh. Math. **64**, 51–63.

——, (1960c): *Solution of Equations and Systems of Equations.* Academic Press, New York and London, ix + 202 pp.

——, (1960d): "A quantitative formulation of Sylvester's law of inertia, II," Proc. Nat. Acad. Sci. U.S.A. **46**, 859–862.

——, (1960e): *A Regularity Condition for a Class of Partitioned Matrices.* MRC Technical Summary Report No. 132, Univ. of Wisc., Madison.

——, (1960f): *On Some Metrical Properties of Operator Matrices and Matrices Partitioned into Blocks.* MRC Technical Summary Report No. 138, Univ. of Wisc., Madison.

——, (1960g): "On the eigenvector belonging to the maximal root of a non-negative matrix," Proc. Edinburgh Math. Soc. (II) **12**, 107–112.

——, (1961a): *Note on a Theorem by Hans Schneider.* MRC Technical Summary Report No. 219, Univ. of Wisc., Madison.

——, (1961b): *On Lancaster's Decomposition of a Differential Matricial Operator.* MRC Technical Summary Report No. 223, Univ. of Wisc., Madison.

——, (1961c): "On some conditions for nonvanishing determinants," Proc. Amer. Math. Soc. **12**, 268–273.

——, (1961d): *On Some Inequalities in the Theory of Matrices.* MRC Technical Summary Report No. 217, Univ. of Wisc., Madison.

——, and SCHNEIDER, HANS (1960): "Bounds for the maximal characteristic root of a nonnegative irreducible matrix," Duke Math. J. **27**, 547–553.

——, and TAUSSKY, OLGA (1951): "On the variation of the determinant of a positive definite matrix," Neder. Akad. Wetensch. Proc. Ser. A **54**, 383–385.

PAIGE, L. J., and TAUSSKY, OLGA (Editors) (1953): "Simultaneous linear equations and the determination of eigenvalues," Nat. Bur. Standards Appl. Math. Ser. **29**, 126 pp.

PARKER, W. V. (1948a): "Characteristic roots and the field of values of a matrix," Duke Math. J. **15**, 439–442.

——, (1948b): "Sets of complex numbers associated with a matrix," Duke Math. J. **15**, 711–715.

——, (1950): "The matrix equation $AX = XB$," Duke Math. J. **17**, 43–51.

——, (1951): "Characteristic roots and field of values of a matrix," Bull. Amer. Math. Soc. **57**, 103–108.

——, (1955): "A note on a theorem of Roth," Proc. Amer. Math. Soc. **6**, 299–300.

PARODI, MAURICE (1949): "Sur les limites des modules des racines des équations algébriques," Bull. Sci. Math. **73**, 135–144.

——, (1950): "Quelques propriétés des matrices H," Ann. Soc. Sci. Bruxelles **64**, 22–25.

——, (1959): *La localisation des valeurs caractéristiques des matrices et ses applications.* Gauthier-Villars, Paris, xi + 172 pp.

PAUL, MANFRED (1957): *Zur Kenntnis des Weber-Verfahrens.* Tech. Hochsch. München, Diplom-Arbeit.

PENROSE, R. (1955): "A generalized inverse for matrices," Proc. Cambridge Philos. Soc. **51**, 406–413.

——, (1956): "On best approximate solutions of linear matrix equations," Proc. Cambridge Philos. Soc. **52**, 17–19.

PERES, Manuel (1952): "Sôbre a resolução dos sistemas de equações lineares simultâneas," Las Ciencias (Madrid) **17**, 443–449.

PERFECT, HAZEL (1951): "On matrices with positive elements," Quart. J. Math. (2) **2**, 286–290.

——, (1952): "On positive stochastic matrices with real characteristic roots," Proc. Cambridge Philos. Soc. **48**, 271–276.

——, (1955): "Methods of constructing certain stochastic matrices," Duke Math. J. **20**, 395–404.

——, (1956): "A lower bound for the diagonal elements of a nonnegative matrix," J. London Math. Soc. **31**, 491–493.

PERLIS, SAM (1952): *Theory of Matrices.* Addison-Wesley Publishing Co., Inc., Reading, Massachusetts, 237 pp.

PERRON, OSKAR (1908): "Zur Theorie der Matrizen," Math. Ann. **64**, 248–263.

——, (1951): *Algebra. I. Die Grundlagen. II. Theorie der algebraischen Gleichungen* (Dritte, verbesserte Auflage). Walter de Gruyter and Company, Berlin, viii + 301 pp., viii + 261 pp.

PETRIE, III, GEORGE W. (1953): "Matrix inversion and solution of simultaneous linear algebraic equations with the IBM 604 electronic calculating punch," Nat. Bur. Standards Appl. Math. Ser. **29**, 107–112.

PETRONE, LUIGI (1960): *Elementi di calcolo delle matrici.* Serie di statistica–Teoria e applicazioni, 14, Paolo Boringhieri, Torino, vi + 115 pp.

PHILLIPS, H. B. (1919): "Functions of matrices," Amer. J. Math. **41**, 266–278.

PICK, GEORGE (1922): "Über die Wurzeln der charakteristischen Gleichung von Schwingungsproblemen," Z. Angew. Math. Mech. **2**, 353–357.

PICONE, MAURO (1958a): "Sulla teoria delle matrici nel corpo complesso," Boll. Un. Mat. Ital. (3) **13**, 1–6.

——, (1958b): "Sulle maggioranti i numeri caratteristici di una matrice quadrata," Boll. Un. Mat. Ital. (3) **13**, 335–340.

PIETRZYKOWSKI, TOMASZ (1960): *Projection Method.* Zaktadu Aparatów Matematycznych Polskiej Akad. Nauk Praca A8.

POHLHAUSEN, E. (1921): "Berechnung der Eigenschwingungen statischbestimmter Fachwerke," Z. Angew. Math. Mech. **1**, 28–42.

POKORNÁ, OLGA (1955): "Řešeni soustav lineárních algebraických rovnic," Stroje na Zpracování Informací **3**, 139–196.

POPE, DAVID A., and TOMPKINS, C. (1957): "Maximizing functions of rotations—experiments concerning speed of diagonalization of symmetric matrices using Jacobi's method," J. Assoc. Comput. Mach. **4**, 459–466.

POTTERS, M. L. (1955): *A Matrix Method for the Solution of a Linear Second Order Difference Equation in Two Variables.* Math. Centrum, Amsterdam.

PRICE, G. BALEY (1947): "Some identities in the theory of determinants," Amer. Math. Monthly **54**, 75–90.

——, (1951): "Bounds for determinants with dominant principal diagonal," Proc. Amer. Math. Soc. **2**, 497–502.

PROSKURJAKOV, I. V. (1957): *Sbornik zadač po lineinoĭ algebre.* Gosudarstv. Izdat. Tehn.-Teoret. Lit., Moscow-Leningrad, 368 pp.

PTÁK, VLASTIMIL (1956): "Eine Bemerkung zur Jordansche Normalform von Matrizen," Acta Sci. Math. Szeged. **17**, 190–194.

——, (1958): "Ob odnoĭ kombinatornoĭ teoreme i ee primenenii k neotricatel'ným matricam," Čeho. Mat. Ž. **8**, 487–495.

——, and SEDLÁČEK, JIŘI (1958): "Ob indekse imprimitivnosti neotricatel'nyh matric, Čeho. Mat. Ž. **8**, 496–501.

PURCELL, EVERETT W. (1953): "The vector method of solving simultaneous linear equations," J. Math. Phys. **32**, 180–183.

QUADE, V. O. (1946): "Auflösung linearer Gleichungen durch Matrizeniteration," Ber. Math. Tübingen **1946**, 57–59.

RADOS, GUSTAV (1897): "Theorie der adjungierten Substitutionen," Math. Ann. **48**, 417–424.

REICH, EDGAR (1949): "On the convergence of the classical iterative method of solving linear simultaneous equations," Ann. Math. Statist. **20**, 448–451.

RICHARDSON, L. F. (1910): "The approximate arithmetical solution by finite differences of physical problems involving differential equations with an application to the stresses in a masonry dam," Philos. Trans. Roy. Soc. London Ser. A **210**, 307–357.

——, (1950): "A purification method for computing the latent columns of numerical matrices and some integrals of differential equations," Philos. Trans. Roy. Soc. London Ser. A **242**, 439–491.

RICHTER, HANS (1950): "Über Matrixfunktionen," Math. Ann. **122**, 16–34.

——, (1954): "Bemerkung zur Norm der Inversen einer Matrix," Arch. Math. **5**, 447–448.

——, (1958): "Zur Abschätzung von Matrizennormen," Math. Nachr. **13**, 178–187.

RILEY, JAMES D. (1956): "Solving systems of linear equations with a positive definite, symmetric, but possibly ill-conditioned matrix," Math. Tables Aids Comput. **9**, 96–101.

RINEHART, ROBERT F. (1955): "The equivalence of definitions of a matric function," Amer. Math. Monthly **62**, 395–414.

——, (1956): "The derivative of a matric function," Proc. Amer. Math. Soc. **7**, 2–5.

——, (1960): "Skew matrices as square roots," Amer. Math. Monthly **67**, 157 161.

ROHRBACK, HANS (1931): "Bemerkungen zu einem Determinantensatz von Minkowski," Jber. Deutsch. Math. Verein. **40**, 49–53.

ROMA, MARIA SOFIA (1946): "Il metodo dell'ortogonalizzazione per la risoluzione numerica dei sistemi di equazioni lineari algebriche," Ricerca Sci. **16**, 309–312.

——, (1950): "Sulla risoluzione numerica dei sistemi di equazioni algebriche lineari col metodo della ortogonalizzazione," Ricerca Sci. **20**, 1288–1290.

ROMANOVSKY, V. (1933): "Un théorème sur les zéros des matrices non-négatives," Bull. Soc. Math. France **61**, 213–219.

——, (1936): "Recherches sur les chaînes de Markoff," Acta Math. **66**, 147–251.

ROSSER, BARKLEY J. (1953): "Rapidly converging iterative methods for solving linear equations," Nat. Bur. Standards Appl. Math. Ser. **29**, 59–64.

——, LANCZOS, C., HESTENES, M. R., and KARUSH, W. (1951): "The separation of close eigenvalues of a real symmetric matrix," J. Res. Nat. Bur. Standards **47**, 291–297.

ROTA, GIAN-CARLO, and STRANG, W. GILBERT (1960): "A note on the joint spectral radius," Nederl. Akad. Wetensch. Proc. Ser. A **43**, 379–381.

ROTH, J. PAUL (1959): "An application of algebraic topology: Kron's method of tearing," Quart. Appl. Math. **17**, 1–24.

——, and SCOTT, D. S. (1956): "A vector method for solving linear equations and inverting matrices," J. Math. Phys. **35**, 312–317.

ROTH, WILLIAM E. (1954): "On the characteristic polynomial of the product of two matrices," Proc. Amer. Math. Soc. 5, 1–3.

——, (1956): "On the characteristic polynomial of the product of several matrices," Proc. Amer. Math. Soc. 7, 578–582.

RUTHERFORD, D. E. (1945): "Some continuant determinants arising in physics and chemistry. I.," Proc. Roy. Soc. Edinburgh Sect. A 62, 229–236.

——, (1951): "Some continuant determinants arising in physics and chemistry. II," Proc. Roy. Soc. Edinburgh Sect. A 63, 232–241.

RUTISHAUSER, HEINZ (1953): "Beiträge zur Kenntnis des Biorthogonalisierungs-Algorithmus von C. Lanczos," Z. Angew. Math. Phys. 4, 35–56.

——, (1954b): "Anwendungen des QD-Algorithmus," Z. Angew. Math. Phys. 5, 496–508.

——, (1954a): "Der Quotienten-Differenzen-Algorithmus," Z. Angew. Math. Phys. 5, 233–251.

——, (1954c): "Ein infinitesimales Analogon zum Quotient-Differenzen-Algorithmus," Arch. Math. 5, 132–137.

——, (1955a): "Une méthode pour la détermination des valeurs propres d'une matrice," C.R. Acad. Sci. Paris 240, 34–36.

——, (1955b): "Bestimmung der Eigenwerte und Eigenvektoren einer Matrix mit Hilfe des Quotienten-Differenzen-Algorithmus," Z. Angew. Math. Phys. 6, 387–401.

——, (1956): "Eine Formel von Wronski und ihre Bedeutung für den Quotienten-Differenzen-Algorithmus," Z. Angew. Math. Phys. 7, 164–169.

——, (1957): Der Quotienten-Differenzen-Algorithmus. Birkhäuser Verlag, Basel/Stuttgart, 74 pp.

——, (1958a): "Solution of eigenvalue problems with the LR-transformation," Nat. Bur. Standards Appl. Math. Ser. 49, 47–81.

——, (1958b): "Zur Bestimmung der Eigenwerte schiefsymmetrischer Matrizen," Z. Angew. Math. Phys. 9, 586–590.

——, (1959a): "Zur Matrizeninversion nach Gauss-Jordan," Z. Angew. Math. Phys. 10, 281–291.

——, (1959b): "Deflation bei Bandmatrizen," Z. Angew Math. Phys. 10, 314–319.

——, (1960): "Über eine kubisch konvergente Variante der LR-Transformation," Z. Angew. Math. Mech. 40, 49–54.

——, (1961): "Ein quadratisch konvergentes Verfahren zur Eigenwertbestimmung bei unsymmetrischen Matrizen I," Z. Angew. Math. Phys. 12, 568–571.

——, and BAUER, F. L. (1955): "Détermination des vecteurs propres d'une matrice par une méthode itérative avec convergence quadratique," C.R. Acad. Sci. Paris 240, 1680–1681.

SABROFF, RICHARD R., and HIGGINS, T. J. (1958): "A critical study of

Kron's method of 'tearing' V," Matrix Tensor Quart. **8,** 106–112.

SAIBEL, EDWARD (1943): "A modified treatment of the iterative method," J. Franklin Inst. **235,** 163–166.

——, and BERGER, W. J. (1953): "On finding the characteristic equation of a square matrix," Math. Tables Aids Comput. **7,** 228–236.

SAMELSON, KLAUS (1959): "Faktorisierung von Polynomen durch funktionale Iteration," Bayer. Akad. Wiss. Math.-Nat. Kl. Abh. **95,** 26 pp.

SAMUELSON, P. A. (1942): "A method of determining explicitly the coefficients of the characteristic equation," Ann. Math. Statist. **13,** 424–429.

SASSENFELD, H. (1950): "Zur Iteration bei linearen Gleichungen," Z. Angew. Math. Phys. **30,** 280–281.

——, (1951): "Ein hinreichendes Konvergenzkriterium und eine Fehlerabschätzung für die Iteration in Einzelschritten bei linearen Gleichungen," Z. Angew. Math. Mech. **31,** 92–94.

SAVAGE, I. RICHARD, and LUKACS, E. (1954): "Tables of inverses of finite segments of the Hilbert matrix," Nat. Bur. Standards Appl. Math. Ser. **39,** 105–108.

SCHECHTER, SAMUEL (1958) (1959): "Relaxation methods for linear equations," Comm. Pure Appl. Math. **12,** 313–335.

——, (1959): "On the inversion of certain matrices, Math. Tables Aids Comput." **13,** 73–77.

——, (1960): "Quasi-tridiagonal matrices and type-insensitive difference equations," Quart. Appl. Math. **18,** 285–295.

SCHMID, ERICH W. (1958): "Ein Iterationsverfahren zur modellmässigen Zuordnung von Molekülschwingungsspektren mit einer Anwendung auf die ebenen Schwingungen von Naphthalin," Z. Elektrochem. **62,** 1005–1019.

SCHMIDT, ERHARD (1908): "Über die Auflösung linearer Gleichungen mit unendlich vielen Unbekannten," Rend. Circ. Mat. Palermo **25,** 53–77.

SCHMIDT, R. J. (1941): "On the numerical solution of linear simultaneous equations by an iterative method," Philos. Mag. (7) **32,** 369–383.

SCHNEIDER, HANS (1952): "Theorems on normal matrices," Quart. J. Math. Oxford (2) **3,** 241–249.

——, (1953): "An inequality for latent roots applied to determinants with dominant principal diagonal," J. London Math. Soc. **28,** 8–20.

——, (1956): "A matrix problem concerning projections," Proc. Edinburgh Math. Soc. (2) **10,** 129–130.

——, (1958): "Note on the fundamental theorem on irreducible nonnegative matrices," Proc. Edinburgh Math. Soc. (2) **11,** 127–130.

——, and STRANG, W. G. (1962): "Comparison theorems for supremum norms," Numer. Math. **4,** 15–20.

SCHOPF, A. H. (1960): "On the Kantorovich inequality," Numer. Math. **2,** 344–346.

SCHREIER, OTTO, and SPERNER, EMANUEL (1951): *Introduction to Modern Algebra and Matrix Theory*, Translated by Martin Davis and Melvin Hausner. Chelsea Publishing Company, New York, viii + 378 pp.

SCHRÖDER, JOHANN (1953): "Eine Bemerkung zur Konvergenz der Iterationsverfahren für lineare Gleichungssysteme," Arch. Math. **4**, 322–326.

———, (1956a): "Neue Fehlerabschätzung für verschiedene Iterationsverfahren," Z. Angew. Math. Mech. **36**, 168–181.

———, (1956b): "Anwendung funktionalanalytischer Methoden zur numerischen Behandlung von Gleichungen," Z. Angew. Math. Mech. **36**, 260–261.

SCHULZ, GÜNTHER (1933): "Iterative Berechnung der reziproken Matrix," Z. Angew. Math. Mech. **13**, 57–59.

SCHUR, I. (1906): "Zur Theorie der vertauschbaren Matrizen," J. Reine Angew. Math. **130**, 66–76.

———, (1909a): "Lineare homogene Integralgleichungen," Math. Ann. **67**, 306–339.

———, (1909b): "Über die characteristischen Wurzeln einer linearen Substitution mit einer Anwendung auf die Theorie der Integralgleichungen," Math. Ann. **66**, 488–510.

———, (1917) (1918): "Über Potenzreihen, die im Innern des Einheitskreises beschränkt sind," J. Reine Angew. Math. **147**, 205–232; **148**, 122–145.

SCHWARZ, HANS RUDOLF (1955): "Critère de stabilité pour des systèmes à coefficients constants," C.R. Acad. Sci. Paris **241**, 15–16.

———, (1956): "Ein Verfahren zur Stabilitätsfrage bei Matrizen-Eigenwertproblemen," Z. Angew. Math. Phys. **7**, 473–500.

SCHWEITZER, P. (1914): "Egy egyenlötlenség az arithmetikai középéttekröl," Math. és Phys. Lapok **23**, 257–261.

SCHWERDTFEGER, HANS (1950): *Introduction to Linear Algebra and the Theory of Matrices*. P. Noordhoff N. V., Groningen, 280 pp.

———, (1953): "Problems in the theory of matrices and its applications," Austral. J. Sci. **15**, 112–115.

———, (1960): "Direct proof of Lanczos' decomposition theorem," Am. Math. Monthly **67**, 855–860.

———, (1961): "On the adjugate of a matrix," Portugal. Math. **20**, 39–41.

SCOTT, R. F. (1879): "On some symmetrical forms of determinants," Messenger of Math. **8**, 131–138, 145–150.

SEIDEL, L. (1874): "Über ein Verfahren die Gleichungen, auf welche die Methode der kleinsten Quadrate führt, sowie lineare Gleichungen überhaupt, durch successive Annäherung aufzulösen," Bayer. Akad. Wiss. Math. Phys. Kl. Abh. **11**, 81–108.

SEMENDJAEV, K. A. (1943): "O nahoždenii sobstvenyh značenii i invariantnyh mnogoobrazii matric posredstvom iteracii," Prikl. Mat. Meh. **7**, 193–222.

SHANKS, DANIEL (1955): "On analogous theorems of Fredholm and Frame and on the inverse of a matrix," Quart. Appl. Math. **13**, 95–98.

SHEFFIELD, R. D. (1958): "A general theory for linear systems," Amer. Math. Monthly **65**, 109–111.

SHELDON, J. W. (1959): "On the spectral norms of several iterative processes," J. Assoc. Comput. Mach. **6**, 494–505.

——, (1960): *Iterative Methods for the Solution of Elliptic Partial Differential Equations. Mathematical Methods for Digital Computers.* John Wiley and Sons, Inc., New York, pp. 144–156.

SHERMAN, JACK (1953): "Computations relating to inverse matrices," Nat. Bur. Standards Appl. Math. Ser. **29**, 123–124.

——, and MORRISON, W. J. (1949): "Adjustment of an inverse matrix corresponding to changes in the elements of a given column or a given row of the original matrix," Ann. Math. Statist. **20**, 621.

——, and ——, (1950): "Adjustment of an inverse matrix corresponding to a change in one element of a given matrix," Ann. Math. Statist. **21**, 124.

SHORTLEY, GEORGE H. (1953): "Use of Tschebyscheff-polynomial operators in the numerical solution of boundary-value problems," J. Appl. Phys. **24**, 392–396.

——, and WELLER, R. (1938): "Numerical solution of Laplace's equations," J. Appl. Phys. **9**, 334–348.

SMITH, T. (1927): "The calculation of determinants and their minors," Philos. Mag. (7) **3**, 1007–1009.

ŠMUL'JAN, YU. L. (1955): "Zamečanie po povodu stat'i Ju. M. Gavrilova ''O shodimosti iteracionnyh proccessov''," Izv. Akad. Nauk Ser. Mat. **19**, 191.

SOURIAU, JEAN-MARIE (1948): "Une méthode pour la décomposition spectrale et l'inversion des matrices," C.R. Acad. Sci. Paris **227**, 1010–1011.

——, (1959): *Calcul linéaire, "Euclid" Introduction aux Études Scientifiques.* Presses Universitaires de France, Paris, 263 pp.

STEIN, MARVIN L. (1952): "Gradient methods in the solution of systems of linear equations," J. Res. Nat. Bur. Standards **48**, 407–413.

STEIN, P. (1951a): "A note on inequalities for the norm of a matrix," Amer. Math. Monthly **58**, 558–559.

——, (1951b): "The convergence of Seidel iterants of nearly symmetric matrices," Math. Tables Aids Comput. **5**, 237–240.

——, (1952a): "A note on bounds of multiple characteristic roots of a matrix," J. Res. Nat. Bur. Standards **48**, 59–60.

——, (1952b): "Some general theorems on iterants," J. Res. Nat. Bur. Standards **48**, 82–83.

——, and ROSENBERG, R. L. (1948): "On the solution of linear simultaneous equations by iteration," J. London Math. Soc. **23**, 111–118.

STENZEL, H. (1922): "Über die Darstellbarkeit einer Matrix als Produkt von zwei symmetrischen Matrizen, als Produkt von zwei alternierenden Matrizen und als Produkt von einer symmetrischen und einer alternierenden Matrix," Math. Z. **15**, 1–25.

STIEFEL, EDUARD L. (1952): "Über einige Methoden der Relaxationsrechnung," Z. Angew. Math. Phys. **3**, 1–33.

——, (1952/53): "Ausgleichung ohne Aufstellung der Gausschen Normalgleichungen," Wiss. Z. Techn. Hochsch. Dresden **2**, 441–442.

——, (1953a): "Some special methods of relaxation technique," Nat. Bur. Standards Appl. Math. Ser. **29**, 43–48.

——, (1953b): "Zur Interpolation von tabellierten Funktionen durch Exponentialsummen und zur Berechnung von Eigenwerten aus den Schwarzschen Konstanten," Z. Angew. Math. Mech. **33**, 260–262.

——, (1955): "Relaxationsmethoden bester Strategie zur Lösung linearer Gleichungssysteme," Comment. Math. Helv. **29**, 157–179.

——, (1957): "Recent developments in relaxation techniques," Proc. Internat. Congress Math., Amsterdam, 1954 **1**, 384–391.

——, (1958): "Kernel polynomials in linear algebra and their numerical applications," Nat. Bur. Standards Appl. Math. Ser. **49**, 1–22.

——, (1961): *Einführung in die numerische Mathematik, Leitfäden der angewandten Mathematik und Mechanik, 2.* B. G. Teubner Verlagsgesellschaft, Stuttgart, 234 pp.

STIELTJES, T. J. (1886): "Sur les racines de l'équation $X_n = 0$," Acta Math. **9**, 385–400; Oeuvres **2**, 73–88.

STOJAKOVIĆ, MIRKO (1958): "Sur l'inversion d'une classe de matrices. Les mathématiques de l'ingénieur," Mém. Publ. Soc. Sci. Arts Lett. Hainaut Vol. hors séries, pp. 188–192.

STRACHEY, C., and FRANCIS, J. G. F. (1961): "The reduction of a matrix to codiagonal form by eliminations," Comput. J. **4**, 168–176.

SWANSON, C. A. (1961): "An inequality for linear transformations with eigenvalues," Bull. Amer. Math. Soc. **67**, 607–608.

SYNGE, J. L. (1944): "A geometrical interpretation of the relaxation method," Quart. Appl. Math. **2**, 87–89.

SZASZ, OTTO (1917): "Über eine Verallgemeinerung des Hadamardschen Determinantensatzes," Monatsh. Math. Phys. **28**, 253–257.

SZEGÖ, GABOR (1921): "Über orthogonale Polynome, die zu einer gegebenen Kurve der komplexen Ebene gehören," Math. Z. **9**, 218–270.

——, (1939): *Orthogonal polynomials.* Amer. Math. Soc. Colloquim Publication, New York, Vol. 23, ix + 401 pp.

TARNOVE, IVIN (1958): "Determination of eigenvalues of matrices having polynomial elements," J. Soc. Indust. Appl. Math. **6**, 163–171.

TAUSSKY, OLGA (1948): "Bounds for characteristic roots of matrices," Duke Math. J. **15**, 1043–1044.

——, (1949a): "A recurring theorem on determinants," Amer. Math. Monthly **56**, 672–676.

——, (1949b): "A remark concerning the characteristic roots of the finite segments of the Hilbert matrix," Quart. J. Math., Oxford Ser. (2) **20**, 80–83.

——, (1950): "Notes on numerical analysis—2. Note on the condition of matrices," Math. Tables Aids Comput. **4**, 111–112.

——, (Editor) (1954): "Contributions to the solution of linear systems and the determination of eigenvalues," Nat. Bur. Standards Appl. Math. Ser. **39**.

——, (1957a): "A determinantal inequality of H. P. Robertson. I," J. Washington Acad. Sci. **47**, 263–264.

——, (1957b): "Commutativity in finite matrices," Amer. Math. Monthly **64**, 229–235.

——, (1958): "On a matrix theorem of A. T. Craig and H. Hotelling," Nederl. Akad. Wetensch. Proc. Ser. A **61**, 139–141.

——, (1961): "A generalization of a theorem of Lyapunov," J. Soc. Indust. Appl. Math. **9**, 640–643.

——, and ZASSENHAUS, HANS (1959): "On the similarity transformation between a matrix and its transpose," Pacific J. Math. **9**, 893–896.

TAYLOR, ANGUS E. (1958): "The norm of a real linear transformation in Minkowski space," Enseignement Math. (2) **4**, 101–107.

TEMPLE, GEORGE (1929): "The computation of characteristic numbers and characteristic functions," Proc. London Math. Soc. (2) **29**, 257–280.

——, (1939): "The general theory of relaxation method applied to linear systems," Proc. Roy. Soc. London Ser. A **169**, 476–500.

——, (1952): "The accuracy of Rayleigh's method of calculating the natural frequencies of vibrating systems," Proc. Roy. Soc. London Ser. A **211**, 204–224.

TODD, JOHN (1949a): "The condition of a certain matrix," Proc. Cambridge Philos. Soc. **46**, 116–118.

——, (1949b): "The condition of certain matrices. I," Quart. J. Mech. Appl. Math. **2**, 469–472.

——, (1950): "Notes on modern numerical analysis—I. Solution of differential equations by recurrence relations," Math. Tables Aids Comput. **4**, 39–44.

——, (1953): "Experiments on the inversion of a 16×16 matrix," Nat. Bur. Standards Appl. Math. Ser. **29**, 113–115.

——, (1954a): "L. F. Richardson (1881–1953)," Math. Tables Aids Comput. **8**, 242–245.

——, (1954b): "The condition of certain matrices," Arch. Math. **5**, 249–257.

——, (1954c): "The condition of the finite segments of the Hilbert matrix," Nat. Bur. Standards Appl. Math. Ser. **39**, 109–116.

——, (1956): "A direct approach to the problem of stability in the numerical solution of partial differential equations," Comm. Pure Appl. Math. **9**, 597–612.

——, (1958): "The condition of certain matrices, III," J. Res. Nat. Bur. Standards **60**, 1–7.

——, (1961): "Computational problems concerning the Hilbert matrix," J. Res. Nat. Bur. Standards **65B**, 19–22.

TOEPLITZ, O. (1918): "Das algebraische Analogon zu einem Satze von Fejér," Math. Z. **2**, 187–197.

TREFFTZ, E. (1933): "Über Fehlerschätzung bei Berechnung von Eigenwerten," Math. Ann. **108**, 595–604.

TURING, A. M. (1948): "Rounding-off errors in matrix processes," Quart. J. Mech. Appl. Math. **1**, 287–308.

TURNBULL, H. W. (1929): *The Theory of Determinants, Matrices, and Invariants.* Blackie and Son, Ltd., London and Glasgow, xvi + 338 pp.

——, and AITKEN, A. C. (1930): *An Introduction to the Theory of Canonical Matrices.* Blackie and Son, Ltd., London and Glasgow, xiii + 192 pp.

UHLIG, J. (1960): "Ein Iterationsverfahren für ein inverses Eigenwertproblem endlicher Matrizen," Z. Angew. Math. Mech. **40**, 123–125.

UNGER, HEINZ (1950): "Nichtlineare Behandlung von Eigenwertaufgaben," Z. Angew. Math. Mech. **30**, 281–282.

——, (1951): "Orthogonalisierung (Unitarisierung) von Matrizen nach E. Schmidt und ihre praktische Durchführung," Z. Angew. Math. Mech. **31**, 53–54.

——, (1952a): "Zur Auflösung umfangreicher linearer Gleichungssysteme," Z. Angew. Math. Mech. **32**, 1–9.

——, (1952b): "Über direkte Verfahren bei Matrizeneigenwertproblemen," Wiss. Z. Tech. Hochsch. Dresden **2**, 449–456.

——, (1953): "Zur Praxis der Biorthonormierung von Eigen- und Hauptvektoren," Z. Angew. Math. Mech. **33**, 319–331.

VAN NORTON, ROGER (1960): *The Solution of Linear Equations by the Gauss-Seidel Method, Mathematical Methods for Digital Computers.* John Wiley and Sons, Inc., New York, 56–61 pp.

VARGA, RICHARD S. (1954): "Eigenvalues of circulant matrices," Pacific J. Math. **4**, 151–160.

——, (1957a): "A comparison of the successive overrelaxation method and semi-iterative methods using Chebyshev polynomials," J. Soc. Indust. Appl. Math. **5**, 39–46.

——, (1957b): *On a Lemma of Stieltjes on Matrices.* Westinghouse, WAPD-T-566.

——, (1959a): "Orderings of the successive overrelaxation scheme," Pacific J. Math. **9**, 925–939.

——, (1959b): "p-Cyclic matrices: A generalization of the Young-Frankel successive overrelaxation scheme," Pacific J. Math. **9**, 617–628.

——, (1960): *Factorization and Normalized Iterative Methods, Boundary Problems in Differential Equations.* University of Wisconsin Press, Madison, Wisconsin, 121–142 pp.

——, (1961): "On higher order stable implicit methods for solving parabolic partial differential equations," J. Math. Phys. **40**, 220–231.

——, (1962): *Matrix Iterative Analysis.* Prentice-Hall, Inc., Englewood Cliffs, New Jersey, xiii and 322 pp.

DE VEUBEKE, B. FRAEYS (1956): "Matrices de projection et techniques d'iteration," Ann. Soc. Sci. Bruxelles **70,** 37–61.

VINOGRADE, BERNARD (1950): "Note on the escalator method," Proc. Amer. Math. Soc. **1,** 162–164.

VOETTER, HEINZ (1952): "Über die numerische Behandlung der Eigenwerte von Säkulargleichungen," Z. Angew. Math. Phys. **3,** 314–316.

VOGEL, ALFRED (1950): "Zur Bestimmung der Eigenwerte einer Matrix durch Iteration," Z. Angew. Math. Mech. **30,** 174–182.

VOLTA, EZIO (1949): "Un nuovo metodo per la risoluzione rapida di sistemi di equazioni lineari," Atti Accad. Naz. Lincei, Rend. Cl. Sci. Fis., Mat. Nat. (8) **7,** 203–207.

VOROB'EV, YU. V. (1958): *Metod momentov v prikladnoi matematike.* Gosudarstv. Izdat. Fiz.-Mat. Lit., Moscow, 186 p.

WACHSPRESS, E. L., and HABETLER, G. J. (1960): "An alternating-direction-implicit iteration technique," J. Soc. Indust. Appl. Math. **8,** 403–423.

WALKER, A. G., and WESTON, J. D. (1949): "Inclusion theorems for the eigenvalues of a normal matrix," J. London Math. Soc. **24,** 28–31.

WASHIZU, KJUICHIRO (1953): "Geometrical representations of bounds of eigenvalues," J. Japan Soc. Appl. Mech. **5,** 29–32.

——, (1955): "On the bounds of eigenvalues," Quart. J. Mech. Appl. Math. **8,** 311–325.

WAYLAND, HAROLD (1945): "Expansion of determinantal equations into polynomial form," Quart. Appl. Math. **2,** 277–306.

WEBER, R. (1949): "Sur les méthodes de calcul employées pour la recherche des valeurs et vecteurs propres d'une matrice," Rech. Aéro. **1949,** No. 10, 57–60.

WEDDERBURN, J. H. M. (1925): "The absolute value of the product of two matrices," Bull. Amer. Math. Soc. **31,** 304–308.

——, (1934): *Lectures on Matrices.* American Mathematical Society, New York, vii + 200 p.

WEGNER, UDO (1951): "Bemerkungen zu den Iterationsverfahren für lineare Gleichungssysteme," Z. Angew. Math. Mech. **31,** 243–244.

——, (1953a): "Bemerkungen zur Matrizentheorie," Z. Angew. Math. Mech. **33,** 262–264.

——, (1953b): "Contributi alla teoria dei procedimenti iterativi per la risoluzione numerica dei sistemi di equazioni lineari algebriche," Atti Accad. Naz. Lincei Mem. Cl. Sci. Fis. Mat. Nat. (8) **4,** 1–48.

WEINBERGER, H. F. (1958): "Remarks on the preceding paper of Lax," Comm. Pure Appl. Math. **11,** 195–196.

WEINSTEIN, D. H. (1934): "Modified Ritz method," Proc. Nat. Acad. Sci. U.S.A. **20,** 529–532.

WEINZWEIG, A. I. (1960–61): "The Kron method of tearing and the dual method of identification," Quart. Appl. Math. **18**, 183–190.

WEISSINGER, JOHANNES (1951): "Über das Iterationsverfahren," Z. Angew. Math. Mech. **31**, 245–246.

——, (1952): "Zur Theorie und Anwendung des Iterationsverfahrens," Math. Nachr. **8**, 193–212.

——, (1953): "Verallgemeinerungen des Seidelschen Iterationsverfahrens," Z. Angew. Math. Mech. **33**, 155–163.

WELLSTEIN, JULIUS (1930): "Über symmetrische, alternierende und orthogonale Normalformen von Matrizen," J. Reine Angew. Math. **163**, 166–182.

WEYL, HERMANN (1912): "Das asymptotische Verteilungsgesetz der Eigenwerte linearer partieller Differentialgleichungen," Math. Ann. **71**, 441–479.

——, (1949): "Inequalities between the two kinds of eigenvalues of a linear transformation," Proc. Nat. Acad. Sci. U.S.A. **35**, 408–411.

WHITE, PAUL A. (1958): "The computation of eigenvalues and eigenvectors of a matrix," J. Soc. Indust. Appl. Math. **6**, 393–437.

WHITTAKER, E. T. (1917): "On the latent roots of compound matrices and Brill's determinants," Proc. Edinburgh Math. Soc. **35**, 2–9.

——, (1918): "On determinants whose elements are determinants," Proc. Edinburgh Math. Soc. **36**, 107–115.

WIEGMANN, N. A. (1948): "Normal products of matrices," Duke Math. J. **15**, 633–638.

WIELANDT, HELMUT (1944a): *Beiträge zur mathematischen Behandlung komplexer Eigenwertprobleme V: Bestimmung höherer Eigenwerte durch gebrochene Iteration.* Ber. B44/T. 37 der Aerodynamischen Versuchsanstalt Göttingen (1944), MS.

——, (1944b): "Das Iterationsverfahren bei nicht selbstadjungierten linearen Eigenwertaufgaben," Math. Z. **50**, 93–143.

——, (1948): "Ein Einschliessungssatz für charakteristische Wurzeln normaler Matrizen," Arch. Math. **1**, 348–352.

——, (1949a): "Die Einschliessung von Eigenwerten normaler Matrizen," Math. Ann. **121**, 234–241.

——, (1949b): "Zur Abgrenzung der selbstadjungierten Eigenwertaufgaben. I. Räume endlicher Dimensionen," Math. Nachr. **2**, 328–329.

——, (1950a): "Unzerlegbare, nicht negative Matrizen," Math. Z. **52**, 642–648.

——, (1950b): "Lineare Scharen von Matrizen mit reellen Eigenwerten," Math. Z. **53**, 219–225.

——, (1953): "Inclusion theorems for eigenvalues," Nat. Bur. Standards Appl. Math. Ser. **29**, 75–78.

——, (1954): "Einschliessung von Eigenwerten Hermitescher Matrizen nach dem Abschnittsverfahren," Arch. Math. **5**, 108–114.

——, (1955a): "An extremum property of sums of eigenvalues," Proc. Amer. Math. Soc. **6**, 106–110.

——, (1955b): "On eigenvalues of sums of normal matrices," Pacific J. Math. **5**, 633–638.

WILF, HERBERT S. (1959): "Matrix inversion by the annihilation of a rank," J. Soc. Indust. Appl. Math. **7**, 149–151.

——, (1960a): "Almost diagonal matrices," Amer. Math. Monthly **67**, 431–434.

——, (1960b): *Matrix Inversion by the Method of Rank Annihilation, Mathematical Methods for Digital Computers.* John Wiley and Sons, Inc., New York, 73–77 pp.

WILKINSON, J. H. (1954): "The calculation of the latent roots and vectors of matrices on the Pilot Model of the A.C.E.," Proc. Cambridge Philos. Soc. **50**, 536–566.

——, (1955): "The use of iterative methods for finding the latent roots and vectors of matrices," Math. Tables Aids Comput. **9**, 184–191.

——, (1958a): "The calculation of the eigenvectors of codiagonal matrices," Comput. J. **1**, 90–96.

——, (1958b): "The calculation of eigenvectors by the method of Lanczos," Comput. J. **1**, 148–152.

——, (1959): "Stability of the reduction of a matrix to almost triangular and triangular forms by elementary similarity transformations," J. Assoc. Comput. Mach. **6**, 336–359.

——, (1960a): "Householder's method for the solution of the algebraic eigenproblem," Comput. J. **3**, 23–27.

——, (1960b): "Error analysis of floating-point computation," Numer. Math. **2**, 319–340.

——, (1960c): *Rounding errors in algebraic processes. Information processing.* UNESCO, Paris; R. Oldenbourg, Munich; Butterworths, London, pp. 44–53.

——, (1961): "Error analysis of direct methods of matrix inversion," J. Assoc. Comput. Mach. **8**, 281–330.

——, (1961–62): "Rigorous error bounds for computed eigensystems," Comput. J. **4**, 230–241.

WILLIAMSON, JOHN (1931): "The latent roots of a matrix of special type," Bull. Amer. Math. Soc. **37**, 585–590.

WILLIAMSON, J. H. (1954): The characteristic polynomial of AB and BA, Edinburgh Math. Notes No. **39**, 13.

WITTMEYER, H. (1936a): "Einfluss der Änderung einer Matrix auf die Lösung des zugehörigen Gleichungssystems, sowie auf die characteristischen Zahlen und die Eigenvektoren," Z. Angew. Math. Mech. **16**, 287–300.

——, (1936b); "Über die Lösung von linearen Gleichungssysteme durch Iteration," Z. Angew. Math. Mech. **16**, 301–310.

——, (1955): "Berechnung einzelner Eigenwerte eines algebraischen linearen Eigenwertproblems durch 'Störiteration'," Z. Angew. Math. Mech. **35**, 442–452.

WOODBURY, MAX (1950): *Inverting Modified Matrices.* Memorandum Report 42, Statistical Research Group, Princeton.

WORCH, G. (1932): "Über die zweckmässigste Art, lineare Gleichungen durch Elimination aufzulösen," Z. Angew. Math. Mech. **12**, 175–181.

WREN, F. L. (1937): "Neo-Sylvester contractions and the solution of systems of linear equations," Bull. Amer. Math. Soc. **43**, 823–834.

YOUNG, DAVID M. (1954a): "Iterative methods for solving partial difference equations of elliptic type," Trans. Amer. Math. Soc. **76**, 92–111.

——, (1954b): "On Richardson's method for solving linear systems with positive definite matrices," J. Math. Phys. **32**, 243–255.

——, (1956): *On the Solution of Linear Systems by Iteration.* Proc. Sixth Symposium Appl. Math., pp. 283–298.

ZÜRMÜHL, RUDOLF (1944): "Das Eliminationsverfahren von Gauss zur Auflösung linearer Gleichungssysteme," Ber. Inst. Praktische Math., Techn. Hochsch. Darmstadt Mitteilung Nr. 774, 11–14.

——, (1949): "Zur numerischen Auflösung linearer Gleichungssysteme nach dem Matrizenverfahren von Banachiewicz," Z. Angew. Math. Mech. **29**, 76–84.

——, (1961): *Matrizen und ihre technischen Anwendungen* (Dritte, neubearbeitete Auflage). Springer-Verlag, Berlin, Göttingen, Heidelberg, 459 pp.

Index

A CATALOGUE OF SELECTED DOVER BOOKS
IN ALL FIELDS OF INTEREST

A CATALOGUE OF SELECTED DOVER BOOKS IN ALL FIELDS OF INTEREST

CELESTIAL OBJECTS FOR COMMON TELESCOPES, T. W. Webb. The most used book in amateur astronomy: inestimable aid for locating and identifying nearly 4,000 celestial objects. Edited, updated by Margaret W. Mayall. 77 illustrations. Total of 645pp. 5⅜ x 8½.
20917-2, 20918-0 Pa., Two-vol. set $9.00

HISTORICAL STUDIES IN THE LANGUAGE OF CHEMISTRY, M. P. Crosland. The important part language has played in the development of chemistry from the symbolism of alchemy to the adoption of systematic nomenclature in 1892. ". . . wholeheartedly recommended,"—Science. 15 illustrations. 416pp. of text. 5⅝ x 8¼. 63702-6 Pa. $6.00

BURNHAM'S CELESTIAL HANDBOOK, Robert Burnham, Jr. Thorough, readable guide to the stars beyond our solar system. Exhaustive treatment, fully illustrated. Breakdown is alphabetical by constellation: Andromeda to Cetus in Vol. 1; Chamaeleon to Orion in Vol. 2; and Pavo to Vulpecula in Vol. 3. Hundreds of illustrations. Total of about 2000pp. 6⅛ x 9¼.
23567-X, 23568-8, 23673-0 Pa., Three-vol. set $27.85

THEORY OF WING SECTIONS: INCLUDING A SUMMARY OF AIRFOIL DATA, Ira H. Abbott and A. E. von Doenhoff. Concise compilation of subatomic aerodynamic characteristics of modern NASA wing sections, plus description of theory. 350pp. of tables. 693pp. 5⅜ x 8½.
60586-8 Pa. $8.50

DE RE METALLICA, Georgius Agricola. Translated by Herbert C. Hoover and Lou H. Hoover. The famous Hoover translation of greatest treatise on technological chemistry, engineering, geology, mining of early modern times (1556). All 289 original woodcuts. 638pp. 6¾ x 11.
60006-8 Clothbd. $17.95

THE ORIGIN OF CONTINENTS AND OCEANS, Alfred Wegener. One of the most influential, most controversial books in science, the classic statement for continental drift. Full 1966 translation of Wegener's final (1929) version. 64 illustrations. 246pp. 5⅜ x 8½. 61708-4 Pa. $4.50

THE PRINCIPLES OF PSYCHOLOGY, William James. Famous long course complete, unabridged. Stream of thought, time perception, memory, experimental methods; great work decades ahead of its time. Still valid, useful; read in many classes. 94 figures. Total of 1391pp. 5⅜ x 8½.
20381-6, 20382-4 Pa., Two-vol. set $13.00

THE PHILOSOPHY OF HISTORY, Georg W. Hegel. Great classic of Western thought develops concept that history is not chance but a rational process, the evolution of freedom. 457pp. 5⅜ x 8½. 20112-0 Pa. $4.50

LANGUAGE, TRUTH AND LOGIC, Alfred J. Ayer. Famous, clear introduction to Vienna, Cambridge schools of Logical Positivism. Role of philosophy, elimination of metaphysics, nature of analysis, etc. 160pp. 5⅜ x 8½. (Available in U.S. only) 20010-8 Pa. $2.00

A PREFACE TO LOGIC, Morris R. Cohen. Great City College teacher in renowned, easily followed exposition of formal logic, probability, values, logic and world order and similar topics; no previous background needed. 209pp. 5⅜ x 8½. 23517-3 Pa. $3.50

REASON AND NATURE, Morris R. Cohen. Brilliant analysis of reason and its multitudinous ramifications by charismatic teacher. Interdisciplinary, synthesizing work widely praised when it first appeared in 1931. Second (1953) edition. Indexes. 496pp. 5⅜ x 8½. 23633-1 Pa. $6.50

AN ESSAY CONCERNING HUMAN UNDERSTANDING, John Locke. The only complete edition of enormously important classic, with authoritative editorial material by A. C. Fraser. Total of 1176pp. 5⅜ x 8½.
20530-4, 20531-2 Pa., Two-vol. set $16.00

HANDBOOK OF MATHEMATICAL FUNCTIONS WITH FORMULAS, GRAPHS, AND MATHEMATICAL TABLES, edited by Milton Abramowitz and Irene A. Stegun. Vast compendium: 29 sets of tables, some to as high as 20 places. 1,046pp. 8 x 10½. 61272-4 Pa. $14.95

MATHEMATICS FOR THE PHYSICAL SCIENCES, Herbert S. Wilf. Highly acclaimed work offers clear presentations of vector spaces and matrices, orthogonal functions, roots of polynomial equations, conformal mapping, calculus of variations, etc. Knowledge of theory of functions of real and complex variables is assumed. Exercises and solutions. Index. 284pp. 5⅝ x 8¼. 63635-6 Pa. $5.00

THE PRINCIPLE OF RELATIVITY, Albert Einstein et al. Eleven most important original papers on special and general theories. Seven by Einstein, two by Lorentz, one each by Minkowski and Weyl. All translated, unabridged. 216pp. 5⅜ x 8½. 60081-5 Pa. $3.50

THERMODYNAMICS, Enrico Fermi. A classic of modern science. Clear, organized treatment of systems, first and second laws, entropy, thermodynamic potentials, gaseous reactions, dilute solutions, entropy constant. No math beyond calculus required. Problems. 160pp. 5⅜ x 8½.
60361-X Pa. $3.00

ELEMENTARY MECHANICS OF FLUIDS, Hunter Rouse. Classic undergraduate text widely considered to be far better than many later books. Ranges from fluid velocity and acceleration to role of compressibility in fluid motion. Numerous examples, questions, problems. 224 illustrations. 376pp. 5⅝ x 8¼. 63699-2 Pa. $5.00

THE AMERICAN SENATOR, Anthony Trollope. Little known, long unavailable Trollope novel on a grand scale. Here are humorous comment on American vs. English culture, and stunning portrayal of a heroine/villainess. Superb evocation of Victorian village life. 561pp. 5⅜ x 8½.
23801-6 Pa. $6.00

WAS IT MURDER? James Hilton. The author of *Lost Horizon* and *Goodbye, Mr. Chips* wrote one detective novel (under a pen-name) which was quickly forgotten and virtually lost, even at the height of Hilton's fame. This edition brings it back—a finely crafted public school puzzle resplendent with Hilton's stylish atmosphere. A thoroughly English thriller by the creator of Shangri-la. 252pp. 5⅜ x 8. (Available in U.S. only)
23774-5 Pa. $3.00

CENTRAL PARK: A PHOTOGRAPHIC GUIDE, Victor Laredo and Henry Hope Reed. 121 superb photographs show dramatic views of Central Park: Bethesda Fountain, Cleopatra's Needle, Sheep Meadow, the Blockhouse, plus people engaged in many park activities: ice skating, bike riding, etc. Captions by former Curator of Central Park, Henry Hope Reed, provide historical view, changes, etc. Also photos of N.Y. landmarks on park's periphery. 96pp. 8½ x 11. 23750-8 Pa. $4.50

NANTUCKET IN THE NINETEENTH CENTURY, Clay Lancaster. 180 rare photographs, stereographs, maps, drawings and floor plans recreate unique American island society. Authentic scenes of shipwreck, lighthouses, streets, homes are arranged in geographic sequence to provide walking-tour guide to old Nantucket existing today. Introduction, captions. 160pp. 8⅞ x 11¾. 23747-8 Pa. $6.95

STONE AND MAN: A PHOTOGRAPHIC EXPLORATION, Andreas Feininger. 106 photographs by *Life* photographer Feininger portray man's deep passion for stone through the ages. Stonehenge-like megaliths, fortified towns, sculpted marble and crumbling tenements show textures, beauties, fascination. 128pp. 9¼ x 10¾. 23756-7 Pa. $5.95

CIRCLES, A MATHEMATICAL VIEW, D. Pedoe. Fundamental aspects of college geometry, non-Euclidean geometry, and other branches of mathematics: representing circle by point. Poincare model, isoperimetric property, etc. Stimulating recreational reading. 66 figures. 96pp. 5⅝ x 8¼.
63698-4 Pa. $2.75

THE DISCOVERY OF NEPTUNE, Morton Grosser. Dramatic scientific history of the investigations leading up to the actual discovery of the eighth planet of our solar system. Lucid, well-researched book by well-known historian of science. 172pp. 5⅜ x 8½. 23726-5 Pa. $3.50

THE DEVIL'S DICTIONARY. Ambrose Bierce. Barbed, bitter, brilliant witticisms in the form of a dictionary. Best, most ferocious satire America has produced. 145pp. 5⅜ x 8½. 20487-1 Pa. $2.25

HISTORY OF BACTERIOLOGY, William Bulloch. The only comprehensive history of bacteriology from the beginnings through the 19th century. Special emphasis is given to biography-Leeuwenhoek, etc. Brief accounts of 350 bacteriologists form a separate section. No clearer, fuller study, suitable to scientists and general readers, has yet been written. 52 illustrations. 448pp. 5⅝ x 8¼. 23761-3 Pa. $6.50

THE COMPLETE NONSENSE OF EDWARD LEAR, Edward Lear. All nonsense limericks, zany alphabets, Owl and Pussycat, songs, nonsense botany, etc., illustrated by Lear. Total of 321pp. 5⅜ x 8½. (Available in U.S. only) 20167-8 Pa. $3.95

INGENIOUS MATHEMATICAL PROBLEMS AND METHODS, Louis A. Graham. Sophisticated material from Graham *Dial*, applied and pure; stresses solution methods. Logic, number theory, networks, inversions, etc. 237pp. 5⅜ x 8½. 20545-2 Pa. $4.50

BEST MATHEMATICAL PUZZLES OF SAM LOYD, edited by Martin Gardner. Bizarre, original, whimsical puzzles by America's greatest puzzler. From fabulously rare *Cyclopedia*, including famous 14-15 puzzles, the Horse of a Different Color, 115 more. Elementary math. 150 illustrations. 167pp. 5⅜ x 8½. 20498-7 Pa. $2.75

THE BASIS OF COMBINATION IN CHESS, J. du Mont. Easy-to-follow, instructive book on elements of combination play, with chapters on each piece and every powerful combination team—two knights, bishop and knight, rook and bishop, etc. 250 diagrams. 218pp. 5⅜ x 8½. (Available in U.S. only) 23644-7 Pa. $3.50

MODERN CHESS STRATEGY, Ludek Pachman. The use of the queen, the active king, exchanges, pawn play, the center, weak squares, etc. Section on rook alone worth price of the book. Stress on the moderns. Often considered the most important book on strategy. 314pp. 5⅜ x 8½. 20290-9 Pa. $4.50

LASKER'S MANUAL OF CHESS, Dr. Emanuel Lasker. Great world champion offers very thorough coverage of all aspects of chess. Combinations, position play, openings, end game, aesthetics of chess, philosophy of struggle, much more. Filled with analyzed games. 390pp. 5⅜ x 8½. 20640-8 Pa. $5.00

500 MASTER GAMES OF CHESS, S. Tartakower, J. du Mont. Vast collection of great chess games from 1798-1938, with much material nowhere else readily available. Fully annotated, arranged by opening for easier study. 664pp. 5⅜ x 8½. 23208-5 Pa. $7.50

A GUIDE TO CHESS ENDINGS, Dr. Max Euwe, David Hooper. One of the finest modern works on chess endings. Thorough analysis of the most frequently encountered endings by former world champion. 331 examples, each with diagram. 248pp. 5⅜ x 8½. 23332-4 Pa. $3.75

THE CURVES OF LIFE, Theodore A. Cook. Examination of shells, leaves, horns, human body, art, etc., in *"the* classic reference on how the golden ratio applies to spirals and helices in nature . . . . "—Martin Gardner. 426 illustrations. Total of 512pp. 5⅜ x 8½.   23701-X Pa. $5.95

AN ILLUSTRATED FLORA OF THE NORTHERN UNITED STATES AND CANADA, Nathaniel L. Britton, Addison Brown. Encyclopedic work covers 4666 species, ferns on up. Everything. Full botanical information, illustration for each. This earlier edition is preferred by many to more recent revisions. 1913 edition. Over 4000 illustrations, total of 2087pp. 6⅛ x 9¼.   22642-5, 22643-3, 22644-1 Pa., Three-vol. set $25.50

MANUAL OF THE GRASSES OF THE UNITED STATES, A. S. Hitchcock, U.S. Dept. of Agriculture. The basic study of American grasses, both indigenous and escapes, cultivated and wild. Over 1400 species. Full descriptions, information. Over 1100 maps, illustrations. Total of 1051pp. 5⅜ x 8½.   22717-0, 22718-9 Pa., Two-vol. set $15.00

THE CACTACEAE,, Nathaniel L. Britton, John N. Rose. Exhaustive, definitive. Every cactus in the world. Full botanical descriptions. Thorough statement of nomenclatures, habitat, detailed finding keys. The one book needed by every cactus enthusiast. Over 1275 illustrations. Total of 1080pp. 8 x 10¼.   21191-6, 21192-4 Clothbd., Two-vol. set $35.00

AMERICAN MEDICINAL PLANTS, Charles F. Millspaugh. Full descriptions, 180 plants covered: history; physical description; methods of preparation with all chemical constituents extracted; all claimed curative or adverse effects. 180 full-page plates. Classification table. 804pp. 6½ x 9¼.
23034-1 Pa. $12.95

A MODERN HERBAL, Margaret Grieve. Much the fullest, most exact, most useful compilation of herbal material. Gigantic alphabetical encyclopedia, from aconite to zedoary, gives botanical information, medical properties, folklore, economic uses, and much else. Indispensable to serious reader. 161 illustrations. 888pp. 6½ x 9¼. (Available in U.S. only)
22798-7, 22799-5 Pa., Two-vol. set $13.00

THE HERBAL or GENERAL HISTORY OF PLANTS, John Gerard. The 1633 edition revised and enlarged by Thomas Johnson. Containing almost 2850 plant descriptions and 2705 superb illustrations, Gerard's *Herbal* is a monumental work, the book all modern English herbals are derived from, the one herbal every serious enthusiast should have in its entirety. Original editions are worth perhaps $750. 1678pp. 8½ x 12¼.
23147-X Clothbd. $50.00

MANUAL OF THE TREES OF NORTH AMERICA, Charles S. Sargent. The basic survey of every native tree and tree-like shrub, 717 species in all. Extremely full descriptions, information on habitat, growth, locales, economics, etc. Necessary to every serious tree lover. Over 100 finding keys. 783 illustrations. Total of 986pp. 5⅜ x 8½.
20277-1, 20278-X Pa., Two-vol. set $11.00

AMERICAN BIRD ENGRAVINGS, Alexander Wilson et al. All 76 plates. from Wilson's *American Ornithology* (1808-14), most important ornithological work before Audubon, plus 27 plates from the supplement (1825-33) by Charles Bonaparte. Over 250 birds portrayed. 8 plates also reproduced in full color. 111pp. 9⅜ x 12½.   23195-X Pa. $6.00

CRUICKSHANK'S PHOTOGRAPHS OF BIRDS OF AMERICA, Allan D. Cruickshank. Great ornithologist, photographer presents 177 closeups, groupings, panoramas, flightings, etc., of about 150 different birds. Expanded *Wings in the Wilderness*. Introduction by Helen G. Cruickshank. 191pp. 8¼ x 11.   23497-5 Pa. $6.00

AMERICAN WILDLIFE AND PLANTS, A. C. Martin, et al. Describes food habits of more than 1000 species of mammals, birds, fish. Special treatment of important food plants. Over 000 illustrations. 500pp. 5⅜ x 8½.
                                           20793-5 Pa. $4.95

THE PEOPLE CALLED SHAKERS, Edward D. Andrews. Lifetime of research, definitive study of Shakers: origins, beliefs, practices, dances, social organization, furniture and crafts, impact on 19th-century USA, present heritage. Indispensable to student of American history, collector. 33 illustrations. 351pp. 5⅜ x 8½.   21081-2 Pa. $4.50

OLD NEW YORK IN EARLY PHOTOGRAPHS, Mary Black. New York City as it was in 1853-1901, through 196 wonderful photographs from N.-Y. Historical Society. Great Blizzard, Lincoln's funeral procession, great buildings. 228pp 9 x 12.   22907-6 Pa. $8.95

MR. LINCOLN'S CAMERA MAN: MATHEW BRADY, Roy Meredith. Over 300 Brady photos reproduced directly from original negatives, photos. Jackson, Webster, Grant, Lee, Carnegie, Barnum; Lincoln; Battle Smoke, Death of Rebel Sniper, Atlanta Just After Capture. Lively commentary. 368pp. 8⅜ x 11¼.   23021-X Pa. $8.95

TRAVELS OF WILLIAM BARTRAM, William Bartram. From 1773-8, Bartram explored Northern Florida, Georgia, Carolinas, and reported on wild life, plants, Indians, early settlers. Basic account for period, entertaining reading. Edited by Mark Van Doren. 13 illustrations. 141pp. 5⅜ x 8½.   20013-2 Pa. $5.00

THE GENTLEMAN AND CABINET MAKER'S DIRECTOR, Thomas Chippendale. Full reprint, 1762 style book, most influential of all time; chairs, tables, sofas, mirrors, cabinets, etc. 200 plates, plus 24 photographs of surviving pieces. 249pp. 9⅞ x 12¾.   21601-2 Pa. $7.95

AMERICAN CARRIAGES, SLEIGHS, SULKIES AND CARTS, edited by Don H. Berkebile. 168 Victorian illustrations from catalogues, trade journals, fully captioned. Useful for artists. Author is Assoc. Curator, Div. of Transportation of Smithsonian Institution. 168pp. 8½ x 9½.
                                           23328-6 Pa. $5.00

YUCATAN BEFORE AND AFTER THE CONQUEST, Diego de Landa. First English translation of basic book in Maya studies, the only significant account of Yucatan written in the early post-Conquest era. Translated by distinguished Maya scholar William Gates. Appendices, introduction, 4 maps and over 120 illustrations added by translator. 162pp. 5⅜ x 8½.
23622-6 Pa. $3.00

THE MALAY ARCHIPELAGO, Alfred R. Wallace. Spirited travel account by one of founders of modern biology. Touches on zoology, botany, ethnography, geography, and geology. 62 illustrations, maps. 515pp. 5⅜ x 8½.
20187-2 Pa. $6.95

THE DISCOVERY OF THE TOMB OF TUTANKHAMEN, Howard Carter, A. C. Mace. Accompany Carter in the thrill of discovery, as ruined passage suddenly reveals unique, untouched, fabulously rich tomb. Fascinating account, with 106 illustrations. New introduction by J. M. White. Total of 382pp. 5⅜ x 8½. (Available in U.S. only)   23500-9 Pa. $4.00

THE WORLD'S GREATEST SPEECHES, edited by Lewis Copeland and Lawrence W. Lamm. Vast collection of 278 speeches from Greeks up to present. Powerful and effective models; unique look at history. Revised to 1970. Indices. 842pp. 5⅜ x 8½.   20468-5 Pa. $8.95

THE 100 GREATEST ADVERTISEMENTS, Julian Watkins. The priceless ingredient; His master's voice; 99 44/100% pure; over 100 others. How they were written, their impact, etc. Remarkable record. 130 illustrations. 233pp. 7⅞ x 10 3/5.   20540-1 Pa. $5.95

CRUICKSHANK PRINTS FOR HAND COLORING, George Cruickshank. 18 illustrations, one side of a page, on fine-quality paper suitable for watercolors. Caricatures of people in society (c. 1820) full of trenchant wit. Very large format. 32pp. 11 x 16.   23684-6 Pa. $5.00

THIRTY-TWO COLOR POSTCARDS OF TWENTIETH-CENTURY AMERICAN ART, Whitney Museum of American Art. Reproduced in full color in postcard form are 31 art works and one shot of the museum. Calder, Hopper, Rauschenberg, others. Detachable. 16pp. 8¼ x 11.
23629-3 Pa. $3.00

MUSIC OF THE SPHERES: THE MATERIAL UNIVERSE FROM ATOM TO QUASAR SIMPLY EXPLAINED, Guy Murchie. Planets, stars, geology, atoms, radiation, relativity, quantum theory, light, antimatter, similar topics. 319 figures. 664pp. 5⅜ x 8½.
21809-0, 21810-4 Pa., Two-vol. set $11.00

EINSTEIN'S THEORY OF RELATIVITY, Max Born. Finest semi-technical account; covers Einstein, Lorentz, Minkowski, and others, with much detail, much explanation of ideas and math not readily available elsewhere on this level. For student, non-specialist. 376pp. 5⅜ x 8½.
60769-0 Pa. $4.50

THE COMPLETE WOODCUTS OF ALBRECHT DURER, edited by Dr. W. Kurth. 346 in all: "Old Testament," "St. Jerome," "Passion," "Life of Virgin," Apocalypse," many others. Introduction by Campbell Dodgson. 285pp. 8½ x 12¼. 21097-9 Pa. $7.50

DRAWINGS OF ALBRECHT DURER, edited by Heinrich Wolfflin. 81 plates show development from youth to full style. Many favorites; many new. Introduction by Alfred Werner. 96pp. 8⅛ x 11. 22352-3 Pa. $5.00

THE HUMAN FIGURE, Albrecht Dürer. Experiments in various techniques—stereometric, progressive proportional, and others. Also life studies that rank among finest ever done. Complete reprinting of Dresden Sketchbook. 170 plates. 355pp. 8⅜ x 11¼. 21042-1 Pa. $7.95

OF THE JUST SHAPING OF LETTERS, Albrecht Dürer. Renaissance artist explains design of Roman majuscules by geometry, also Gothic lower and capitals. Grolier Club edition. 43pp. 7⅞ x 10¾ 21306-4 Pa. $3.00

TEN BOOKS ON ARCHITECTURE, Vitruvius. The most important book ever written on architecture. Early Roman aesthetics, technology, classical orders, site selection, all other aspects. Stands behind everything since. Morgan translation. 331pp. 5⅜ x 8½. 20645-9 Pa. $4.50

THE FOUR BOOKS OF ARCHITECTURE, Andrea Palladio. 16th-century classic responsible for Palladian movement and style. Covers classical architectural remains, Renaissance revivals, classical orders, etc. 1738 Ware English edition. Introduction by A. Placzek. 216 plates. 110pp. of text. 9½ x 12¾. 21308-0 Pa. $10.00

HORIZONS, Norman Bel Geddes. Great industrialist stage designer, "father of streamlining," on application of aesthetics to transportation, amusement, architecture, etc. 1932 prophetic account; function, theory, specific projects. 222 illustrations. 312pp. 7⅞ x 10¾. 23514-9 Pa. $6.95

FRANK LLOYD WRIGHT'S FALLINGWATER, Donald Hoffmann. Full, illustrated story of conception and building of Wright's masterwork at Bear Run, Pa. 100 photographs of site, construction, and details of completed structure. 112pp. 9¼ x 10. 23671-4 Pa. $5.50

THE ELEMENTS OF DRAWING, John Ruskin. Timeless classic by great Viltorian; starts with basic ideas, works through more difficult. Many practical exercises. 48 illustrations. Introduction by Lawrence Campbell. 228pp. 5⅜ x 8½. 22730-8 Pa. $3.75

GIST OF ART, John Sloan. Greatest modern American teacher, Art Students League, offers innumerable hints, instructions, guided comments to help you in painting. Not a formal course. 46 illustrations. Introduction by Helen Sloan. 200pp. 5⅜ x 8½. 23435-5 Pa. $4.00

THE ANATOMY OF THE HORSE, George Stubbs. Often considered the great masterpiece of animal anatomy. Full reproduction of 1766 edition, plus prospectus; original text and modernized text. 36 plates. Introduction by Eleanor Garvey. 121pp. 11 x 14¾.   23402-9 Pa. $6.00

BRIDGMAN'S LIFE DRAWING, George B. Bridgman. More than 500 illustrative drawings and text teach you to abstract the body into its major masses, use light and shade, proportion; as well as specific areas of anatomy, of which Bridgman is master. 192pp. 6½ x 9¼. (Available in U.S. only)
22710-3 Pa. $3.50

ART NOUVEAU DESIGNS IN COLOR, Alphonse Mucha, Maurice Verneuil, Georges Auriol. Full-color reproduction of *Combinaisons ornementales* (c. 1900) by Art Nouveau masters. Floral, animal, geometric, interlacings, swashes—borders, spots—all incredibly beautiful. 60 plates, hundreds of designs. 9⅜ x 8-1/16.   22885-1 Pa. $4.00

FULL-COLOR FLORAL DESIGNS IN THE ART NOUVEAU STYLE, E. A. Seguy. 166 motifs, on 40 plates, from *Les fleurs et leurs applications decoratives* (1902): borders, circular designs, repeats, allovers, "spots." All in authentic Art Nouveau colors. 48pp. 9⅜ x 12¼.
23439-8 Pa. $5.00

A DIDEROT PICTORIAL ENCYCLOPEDIA OF TRADES AND INDUSTRY, edited by Charles C. Gillispie. 485 most interesting plates from the great French Encyclopedia of the 18th century show hundreds of working figures, artifacts, process, land and cityscapes; glassmaking, papermaking, metal extraction, construction, weaving, making furniture, clothing, wigs, dozens of other activities. Plates fully explained. 920pp. 9 x 12.
22284-5, 22285-3 Clothbd., Two-vol. set $40.00

HANDBOOK OF EARLY ADVERTISING ART, Clarence P. Hornung. Largest collection of copyright-free early and antique advertising art ever compiled. Over 6,000 illustrations, from Franklin's time to the 1890's for special effects, novelty. Valuable source, almost inexhaustible.
Pictorial Volume. Agriculture, the zodiac, animals, autos, birds, Christmas, fire engines, flowers, trees, musical instruments, ships, games and sports, much more. Arranged by subject matter and use. 237 plates. 288pp. 9 x 12.
20122-8 Clothbd. $14..50

Typographical Volume. Roman and Gothic faces ranging from 10 poin⁀ to 300 point, "Barnum," German and Old English faces, script, logotypes, scrolls and flourishes, 1115 ornamental initials, 67 complete alphabets, more. 310 plates. 320pp. 9 x 12.   20123-6 Clothbd. $15.00

CALLIGRAPHY (CALLIGRAPHIA LATINA), J. G. Schwandner. High point of 18th-century ornamental calligraphy. Very ornate initials, scrolls, borders, cherubs, birds, lettered examples. 172pp. 9 x 13.
20475-8 Pa. $7.00

ART FORMS IN NATURE, Ernst Haeckel. Multitude of strangely beautiful natural forms: Radiolaria, Foraminifera, jellyfishes, fungi, turtles, bats, etc. All 100 plates of the 19th-century evolutionist's *Kunstformen der Natur* (1904). 100pp. 9⅜ x 12¼. 22987-4 Pa. $5.00

CHILDREN: A PICTORIAL ARCHIVE FROM NINETEENTH-CENTURY SOURCES, edited by Carol Belanger Grafton. 242 rare, copyright-free wood engravings for artists and designers. Widest such selection available. All illustrations in line. 119pp. 8⅜ x 11¼.
23694-3 Pa. $4.00

WOMEN: A PICTORIAL ARCHIVE FROM NINETEENTH-CENTURY SOURCES, edited by Jim Harter. 391 copyright-free wood engravings for artists and designers selected from rare periodicals. Most extensive such collection available. All illustrations in line. 128pp. 9 x 12.
23703-6 Pa. $4.50

ARABIC ART IN COLOR, Prisse d'Avennes. From the greatest ornamentalists of all time—50 plates in color, rarely seen outside the Near East, rich in suggestion and stimulus. Includes 4 plates on covers. 46pp. 9⅜ x 12¼. 23658-7 Pa. $6.00

AUTHENTIC ALGERIAN CARPET DESIGNS AND MOTIFS, edited by June Beveridge. Algerian carpets are world famous. Dozens of geometrical motifs are charted on grids, color-coded, for weavers, needleworkers, craftsmen, designers. 53 illustrations plus 4 in color. 48pp. 8¼ x 11. (Available in U.S. only) 23650-1 Pa. $1.75

DICTIONARY OF AMERICAN PORTRAITS, edited by Hayward and Blanche Cirker. 4000 important Americans, earliest times to 1905, mostly in clear line. Politicians, writers, soldiers, scientists, inventors, industrialists, Indians, Blacks, women, outlaws, etc. Identificatory information. 756pp. 9¼ x 12¾. 21823-6 Clothbd. $40.00

HOW THE OTHER HALF LIVES, Jacob A. Riis. Journalistic record of filth, degradation, upward drive in New York immigrant slums, shops, around 1900. New edition includes 100 original Riis photos, monuments of early photography. 233pp. 10 x 7⅞. 22012-5 Pa. $7.00

NEW YORK IN THE THIRTIES, Berenice Abbott. Noted photographer's fascinating study of city shows new buildings that have become famous and old sights that have disappeared forever. Insightful commentary. 97 photographs. 97pp. 11⅜ x 10. 22967-X Pa. $5.00

MEN AT WORK, Lewis W. Hine. Famous photographic studies of construction workers, railroad men, factory workers and coal miners. New supplement of 18 photos on Empire State building construction. New introduction by Jonathan L. Doherty. Total of 69 photos. 63pp. 8 x 10¾.
23475-4 Pa. $3.00

UNCLE SILAS, J. Sheridan LeFanu. Victorian Gothic mystery novel, considered by many best of period, even better than Collins or Dickens. Wonderful psychological terror. Introduction by Frederick Shroyer. 436pp. 5⅜ x 8½. 21715-9 Pa. $6.00

JURGEN, James Branch Cabell. The great erotic fantasy of the 1920's that delighted thousands, shocked thousands more. Full final text, Lane edition with 13 plates by Frank Pape. 346pp. 5⅜ x 8½. 23507-6 Pa. $4.50

THE CLAVERINGS, Anthony Trollope. Major novel, chronicling aspects of British Victorian society, personalities. Reprint of Cornhill serialization, 16 plates by M. Edwards; first reprint of full text. Introduction by Norman Donaldson. 412pp. 5⅜ x 8½. 23464-9 Pa. $5.00

KEPT IN THE DARK, Anthony Trollope. Unusual short novel about Victorian morality and abnormal psychology by the great English author. Probably the first American publication. Frontispiece by Sir John Millais. 92pp. 6½ x 9¼. 23609-9 Pa. $2.50

RALPH THE HEIR, Anthony Trollope. Forgotten tale of illegitimacy, inheritance. Master novel of Trollope's later years. Victorian country estates, clubs, Parliament, fox hunting, world of fully realized characters. Reprint of 1871 edition. 12 illustrations by F. A. Faser. 434pp. of text. 5⅜ x 8½. 23642-0 Pa. $5.00

YEKL and THE IMPORTED BRIDEGROOM AND OTHER STORIES OF THE NEW YORK GHETTO, Abraham Cahan. Film *Hester Street* based on *Yekl* (1896). Novel, other stories among first about Jewish immigrants of N.Y.'s East Side. Highly praised by W. D. Howells—Cahan "a new star of realism." New introduction by Bernard G. Richards. 240pp. 5⅜ x 8½. 22427-9 Pa. $3.50

THE HIGH PLACE, James Branch Cabell. Great fantasy writer's enchanting comedy of disenchantment set in 18th-century France. Considered by some critics to be even better than his famous *Jurgen*. 10 illustrations and numerous vignettes by noted fantasy artist Frank C. Pape. 320pp. 5⅜ x 8½. 23670-6 Pa. $4.00

ALICE'S ADVENTURES UNDER GROUND, Lewis Carroll. Facsimile of ms. Carroll gave Alice Liddell in 1864. Different in many ways from final Alice. Handlettered, illustrated by Carroll. Introduction by Martin Gardner. 128pp. 5⅜ x 8½. 21482-6 Pa. $2.50

FAVORITE ANDREW LANG FAIRY TALE BOOKS IN MANY COLORS, Andrew Lang. The four Lang favorites in a boxed set—the complete *Red, Green, Yellow* and *Blue* Fairy Books. 164 stories; 439 illustrations by Lancelot Speed, Henry Ford and G. P. Jacob Hood. Total of about 1500pp. 5⅜ x 8½. 23407-X Boxed set, Pa. $15.95

GEOMETRY, RELATIVITY AND THE FOURTH DIMENSION, Rudolf Rucker. Exposition of fourth dimension, means of visualization, concepts of relativity as Flatland characters continue adventures. Popular, easily followed yet accurate, profound. 141 illustrations. 133pp. 5⅜ x 8½.

23400-2 Pa. $2.75

THE ORIGIN OF LIFE, A. I. Oparin. Modern classic in biochemistry, the first rigorous examination of possible evolution of life from nitrocarbon compounds. Non-technical, easily followed. Total of 295pp. 5⅜ x 8½.

60213-3 Pa. $4.00

PLANETS, STARS AND GALAXIES, A. E. Fanning. Comprehensive introductory survey: the sun, solar system, stars, galaxies, universe, cosmology; quasars, radio stars, etc. 24pp. of photographs. 189pp. 5⅜ x 8½. (Available in U.S. only)

21680-2 Pa. $3.75

THE THIRTEEN BOOKS OF EUCLID'S ELEMENTS, translated with introduction and commentary by Sir Thomas L. Heath. Definitive edition. Textual and linguistic notes, mathematical analysis, 2500 years of critical commentary. Do not confuse with abridged school editions. Total of 1414pp. 5⅜ x 8½.   60088-2, 60089-0, 60090-4 Pa., Three-vol. set $18.50